DATE DUE

Waste Management and Research Center
Library
One E. Hazelwood Drive
Champaign, IL 61820
(217) 333-8957

DEMCO

Y0-BDY-278

Environmental Engineering

Series Editors: U. Förstner, R. J. Murphy, W. H. Rulkens

Springer

Berlin
Heidelberg
New York
Barcelona
Hong Kong
London
Milan
Paris
Singapore
Tokyo

Thomas T. Shen

Industrial Pollution Prevention

2nd completely revised and enlarged edition

With 26 Figures and 14 Tables

 Springer

Series Editors

Prof. Dr. U. Förstner   Arbeitsbereich Umweltschutztechnik
Technische Universität Hamburg-Harburg
Eißendorfer Straße 40
D-21073 Hamburg, Germany

Prof. Robert J. Murphy   Dept. of Civil Engineering and Mechanics
College of Engineering
University of South Florida
4202 East Fowler Avenue, ENG 118
Tampa, FL 33620–5350, USA

Prof. Dr. ir. W. H. Rulkens   Wageningen Agricultural University
Dept. of Environmental Technology
Bomenweg 2, P.O. Box 8129
NL-6700 EV Wageningen, The Netherlands

Author

Professor Thomas T. Shen
146 Fernbank Avenue
Delmar, NY 12054
USA

ISBN 3-540-65208-6 Springer-Verlag Berlin Heidelberg New York

Cataloging-in-Publication Data applied for

Die Deutsche Bibliothek – CIP-Einheitsaufnahme
Shen, Thomas T.:
Industrial pollution prevention: with 14 tables / Thomas T. Shen. - 2., completely rev. and enl. ed. - Berlin ;
Heidelberg ; New York ; Barcelona ; Hong Kong ; London ; Milan ; Paris ; Singapore ; Tokyo : Springer, 1999
  ISBN 3-540-65208-6

Typesetting: MEDIO, Berlin
Coverdesign: Struve & Partner, Heidelberg
SPIN: 10683533   61/3020 - 5 4 3 2 1 0 – Printed on acid-free paper.

Foreword

Sustainable economic development is necessary to improve the standard of living and quality of life in the developing countries. It is also required to maintain or enhance their attributes for those lands already commercially developed. Past experience indicates that one of the most important elements in the economic growth is the development of industry. It should be also recognized that developing industry, if pursued according to the traditional means, entails the additional inefficient consumption of limited national resource and generation of large amount of residue that called industrial pollution.

In general, industry has three generations of pollution problems. The first-generation pollution problem is from the manufacturing facilities. The second-generation pollution problem is related to the use of the products after leaving the site of manufacturing. The final disposal of the used or unused products represents the third-generation pollution problem for industry.

The traditional way of controlling pollution by industry is building costly waste treatment facilities added-on to the end of manufacturing processes. Industry also has to commit continuous funding to maintain and operate these facilities for their entire life span. The waste treatment facilities were supposed to solve the manufacturing related pollution problem for industry, i.e., the first-generation problem. In fact, these facilities, in according to the Law of Conservation, do not make the pollution disappeared. They only transform, accumulate and generate residues of different forms which create separate control problems of their own. This becomes a Catch 22 situation. It take resources to remove pollution; pollution removal generates residue. It take more resources to dispose of these residues. Disposal of these residues also produces pollution. It can be visualized that this approach will not be able to provide a final solution to the industry's first-generation pollution problem and leaves the second and third generation pollution problem untouched.

It becomes very clear that, solving the industrial pollution problem more efficiently and effectively, a new approach is needed. The concept of pollution prevention was born. The primary purpose of this book is to provide industry with the needed information and methods to plan and implement pollution prevention programs, and with successful examples. The book can also be used as a course manual or references for education and training purposes.

The very first coordinated company-wide industrial pollution prevention program was initiated in 1975 in the 3M Company, a multi-national enterprise with operations in over 50 countries. The basic concept of this program, i.e., Pollution Prevention Pays Program (3P Program, in short), was that eliminating or reducing the pollution at the source will eliminate or reduce the clean-up costs and it will also provide some solutions to the second- and third-generation pollution problems. Since most pollutants are actually valuable raw materials passing through the manufacturing processes, preventing the generation of pollution will also conserve raw materials and make the manufacturing processes more efficiently and less costly.

The 3P program in 3M Company consisted of two parts. Part I is Process Environmental Assessment which is to solve the first-generation pollution problems. The Product Environmental Assessment (Part II) is to looking for solutions to the second- and third-generation pollution problems.

In the first year, the 3P program achieved significant environmental gain and produces a saving of some 20 million dollars. At the invitation of the United Nations Economic Commission (UNEC) for Europe, the results of 3P program were presented to the Conference of Non-waste Technology and Production in Paris in 1976. The presentation was very well received. The UNEC requested if a book could be prepared about this 3P new approach so that many industries in other countries could be benefited. Due to the many other commitments, I declined to the request but committed to prepare a 30-page booklet "Low- or Non-Pollution Technology Through Pollution Prevention". It was published and widely distributed by the United Nations Environment Program (UNEP) at its Annual Meeting in Nairobi, Kenya. Meanwhile Dr. Michael Rayston of the Center For Education in International Management in Geneva volunteered to prepare a book with the title of "Pollution Prevention Pays" and I was honored to write the Foreword. Since then, many countries and many international organizations and industries established their own instruments to promote and implement the preventive concept in their environmental programs, including the United Nations Environmental Program (UNEP) and Organization of Economic Cooperation and Development (OECD).

In the United States, the US Environmental Protection Agency (USEPA) and US Department of Commerce also indicated their interest in the pollution prevention approach. In 1977, they jointly sponsored four regional conferences in Chicago, Boston, Dallas and San Francisco to encourage other industry eliminating pollution at the sources. However, U.S. Congress did not recognize the merit of the preventive approach until the hazardous wastes became an issue. Because the distribution of hazardous wastes was so widely scattered and the control technology was lacking, the Congress called for preventing and minimizing the generation hazardous waste when the Resources Conservation and Recovery Act (RCRA) was amended in 1984. In 1989, the USEPA established an Office of Pollution Prevention and issued a national policy statement on pollution prevention. Finally, the Congress adopted the Pollution Prevention Act in 1990. USEPA also funded the American Institute For Pollution Prevention

(AIPP) to promote the preventive approach and I was honored to be the first chairman.

It took 20 years before the benefit of pollution prevention is fully recognized. The Clinton Administration declared that pollution prevention is the "Corner Stone" of its environmental program. Chapter 15 of this book has compiled a number of successful pollution prevention examples. In spite of these positive results, a large number of industries are not taking the advantage of this preventive approach as yet. This is especially true for the small- and medium-sized companies, and even more true for many of new and old industrial facilities in most developing countries. It is my sincere hope that the book will provide additional encouragement to those who may want to start their own pollution prevention programs.

The very first worldwide environmental concern was initiated at the World Conference on Human and the Environment sponsored in the United Nations in Stockholm in 1972. The most visible result of the Conference was the creation of the United Nations Environment Program (UNEP). As UNEP promoting environmental enhancement program around the world, it encountered tremendous resistance during the 1970's when many countries suffered much economic difficulties. It was then realized that when governments are busy fighting the unemployment and other economic and social issues, environment becomes an unimportant matter. In responses to this significant revelation, the United Nations General Assembly created a World Commission on Environment and Development in 1984. After 3-year deliberation, this Commission issued its final report "Our Common Future" which called for "Sustainable Development". This concept was further confirmed by the United Nations Conference on Environment and Development which was held in Rio de Janeiro in 1992. The Conference concluded that "Sustainable Development" is the realistic way to meet the needs of the present without compromising the ability of future generation to meet their own needs. Of course, maintaining a livable environment is a very important element in the "Sustainable Development".

To follow-up the recommendation of "Sustainable Development", the United Nations established an UN Commission on Sustainable Development. The Clinton Administration also formed a President's Commission on Sustainable Development. It is clear that "Sustainable Development" is the new challenge for the years to come and industry has a major role to play as described in Chapter 12 of this book.

It is my believe that the most important contribution of industry to the "Sustainable Development" is the developing and marketing environmentally compatible products with minimum use of non-renewable resources. Industry can achieve this by intelligently applying the pollution prevention concept with product life cycle analysis and assessment approach in the product design, and considering environmental problems not constrains but opportunities, and producing no or minimum environmental impact as the design objective. In other words, pollution prevention remains as a very valuable tool and serves as a foundation on which new programs and measures can be developed and implemented to meet this new challenge of sustainable development. More detailed discussions are presented in Chapters 5, 12, and 13 which involves some of the programs and measusures already in practice.

It is essential that other sectors of society, in addition to industry, must also cooperate to accomplish this goal of "Sustainable Development". This includes government and the general public as discussed in Chapters 10, 13 and 14 of this book. The government must build resource and environmental concerns into its existing and new policies, regulations, programs, covering not only in the environmental field but also in the areas related to economic development, land use, city planning, transportation, agriculture, mining and energy development. The public must demand and support appropriate governmental actions and private initiatives to encourage environmentally friendly programs. The public should also open to the idea of modifying life style and consuming habits to reduce their own impact to the environment and the associated non-renewable resource.

Experience taught us that the success or failure of any program, in most cases, depends on the getting understanding and proper use of the needed knowledge. This also is true for pollution prevention programs required knowledge which can be secured by formal education or through self-learning. In this information age, many sources of information can be easily accessed through various electronic devices. More discussions are presented in Chapters 14 and 16.

Experience also revealed that it does not always require highly sophisticated technology nor large sum of expenditure for preventing the pollution at its source. When pollution prevention is properly implemented, industry will not only be able to meet their environmental requirements efficiently but also work toward the goal of the new challenge "Sustainable Development".

The advantages of pollution prevention are many as described in this book. Dr. Shen took tremendous effort of collecting, reviewing, revising and compiling a large amount of new information and data for this new edition. This makes the book more update, complete and efficient for readers of various professions in planning and implementing pollution prevention activities which are building blocks toward the new challenge of sustainable development. Because of the immense challenge and tremendous opportunity presented, the second edition of Dr. Shen's book is especially timely and welcome.

Dr. Joseph T. Ling, Ph.D.
Vice President (Retired)
Environmental Affairs, 3M Company

Preface to the Second Edition

Since "Industrial Pollution Prevention" was published in 1995, I have been very gratified by its use in many universities, governments and industries worldwide. The book was translated into Chinese by the Industrial Development Bureau, Ministry of Economic Affairs in Taiwan and into Korean by the Sigma Press in Korea. I have been using the book chapters to lecture in universities, at seminars and workshops in the United States and Asia-Pacific countries. Thus, I have received valuable feedback and suggestions for improving the book contents from readers of various professions throughout the United States and abroad.

For over 40 years, I have observed and been involved with environmental protection activities. The luck of the genes has equipped me to observe and learn. I have found that the problem of environmental pollution (air, water, or soil) can never be solved, but it can be minimized if we give special attention to the fact that pollution problems are caused not only by pollutants and wastes, but also by environmental unfriendly products and services. I have experienced the quality of our environment decreases worldwide, despite the various factual and scientific warnings. The multimedia pollution prevention solutions have been ignored by leaders in government and in industry. It is useless to state a problem or potential barriers without also offering suggestions for pollution prevention solutions.

The first publication of this text was in 1995, this second edition exists because of the rapid change of environmental management strategies, the market demand of need, and the encouragement of my publisher (Springer-Verlag International). In view of new knowledge, environmental legislation, and higher expectations of the people, many professionals, public officials and individuals concerned with protection of the environment and natural resources, are finding themselves unprepared to deal effectively with the current and emerging pollution problems resulted from toxic chemicals and hazardous wastes. Hopefully, this second edition will enhance appeal, clarity and close some of the gaps and questions which characterized the incompleteness of the subject at the time of the first edition.

This second edition reduces six chapters into sections and adds six new chapters to encompass and modify some of the existing materials with current materials of pollution prevention sciences, technologies, legislation and management practices. The new chapters are: Chapter 5 Total Environment Quality Management; Chapter 9 U.S. P2 Laws, Regulations, Programs; Chapter 10 State, City, and Local P2 Programs; Chapter 14 Education and Research; Chapter 15 P2 in the U.S. Defense Department; and Chapter 16 Sources of P2 Information. The book defines pollution prevention in broad terms to include waste minimization, source reduction, design for environment, and cleaner production. It provides more updated materials and information with the hope that this book will not only serve to make more explicit to the established professional, but also help stimulate the student toward career opportunities in this vital area of industrial pollution prevention. It will assist readers in understanding the principles and practices as well as the logic, benefits, and barriers of pollution prevention.

I would like to especially thank Dr. Joseph T. Ling, Retired Vice President of 3M Company and Dr. Granville Sewell (deceased) of Columbia University for their early encouragement and inspiration. Dr. Joseph Laznow's contribution of Chapter 15 is most grateful. He describes the successful case studies of integrating pollution prevention at the Department of Defense military installations. I also thank Dr. David Kline of USEPA, Dr. Bill Batt of Economic Consultant, Dr. Rao Kolluru of CH2M Hill, Dr. Vic Walker and Dr. Ernest Siew of State University of New York for their review and comment of the new chapters; Mr. Randall Law for his time and effort to solve my various computer problems; and Dr. Grace E. Shen for her editorial review. The indispensable ingredient in the publishing of this second edition book is my wife Cynthia. For more than 40 years, she has supported my efforts in every conceivable way – beginning from my graduate studies until my semi-retirement. She keeps me alive and healthy.

Special thanks also go to Springer-Verlag, Heidelberg, Germany, in particular to Dr. Hubertus Riedesel, and Ms. Erdmuthe Raufelder for providing a fruitful relationship with me.

Thomas T. Shen
Albany, New York

Contents

1
Introduction

The overall purpose of this second edition of Industrial Pollution Prevention (P2) is to meet the rapid changing pollution prevention strategies and market demand. It also seems the logic time to update and revise all chapters with new information as well as rearrange the book contents by condensing six chapters edition into sections and embracing those sections into the new chapters. The second remains with 16 chapters consisting of six new chapters: total environmental quality management; laws, regulations, programs and strategies; state, city and local P2 programs; education and research; P2 in the U.S. Defense Department; and sources of P2 information. The contents of the 10 existing chapters are updated with new information as well as major and minor revisions.

The second edition calls the reader's attention to the concept and practices of pollution management rather than waste management only; to the understanding of pollution problems caused not only by pollutants and wastes but also by environmental unfriendly products and services. It provides more updated materials and information for professionals in regulatory agencies, industries and academic institutes with the hope that this book will assist readers in understanding the principles and practices as well as the logic, benefits, and barriers of pollution prevention.

The book may not be prescriptive or comprehensive. It describes the need of a broader approach for environmental policies, strategies, and programs with emphasis on toxic chemicals and hazardous waste pollution problems in terms of both wastes and products. It explains how to identify and assess opportunities for preventing pollution by environmental audit and feasibility analyses. It illustrates the basic steps involved in developing and implementing an adequate pollution prevention plan. It remains the reader's responsibility to go beyond these basics to develop a pollution prevention program for any specific circumstance. It is a general approach for governments, academia and industries in any business and geographical area. The appendices at the end of the book provide forms, worksheets, checklists, and reading suggestions.

1.1
The environmental challenge

The deterioration of environmental quality has existed as a serious problem under the ever-increasing impacts of exponentially increasing population and of industrializing society. Environmental contamination of air, water, soil, and food has become a threat to the continued existence of many plant and animal communities of the ecosystem. This may ultimately threaten the very survival of the human race. It is obvious that civilization will continue to require increasing amounts of fuel, transportation, industrial chemicals, fertilizers, pesticides, and countless other products; and that it will continue to produce waste products of all kinds. What is urgently needed is a total system approach to modern civilization through which the pooled talents of scientists and engineers, in cooperation with social scientists and the medical profession, can be focused on the development of order and equilibrium in the presently disparate segments of the human environment. Most of the skills and tools that are needed are already in existence. We surely hope a technology that has created such manifold environmental problems is also capable of solving them. This is necessary if we can develop and apply new preventive technologies to benefit man both now and in future generations without transferring an environmental problem from one media (air, water, or land) to another. Individual ingenuity and investigation will not be silent by such practicality, but will be challenged and stimulated to always consider new developments and alternatives.

Today's global economy has reinforced the geographic separation among resource extraction, production, and consumption. Hence, those who reap the economic benefits of using natural resources often do not bear the environmental costs. The United Nations Conference on Environment and Development in Rio de Janeiro in June 1992 focused on these issues. This new awareness led to an international agenda for sustainable development and various non-binding agreements, but we must further prevent environmental degradation (Carnegie, 1992). We need a balance between technological innovation and environmental enhancement, as well as a balance between economic development and environmental preservation. Economic growth supplies the financial and technological resources necessary for environmental enhancement; while its opposite, the struggle for bare survival, places strains on environmental preservation. Sustainable development requires that we recognize the inter-relationship between economic and environmental goals. We can implement sustainable development and design for sustainability through pollution prevention which includes broadly and aims similarly the terms such as waste minimization, source reduction, cleaner production, and design for environment. (Shen, 1992; Ling, 1998).

Over the past 35 years, we have come to recognize the seriousness of the environmental problems worldwide. Although we have tried to prevent and control environmental problems, we are only beginning to understand their complexities. In the past, environmental management strategy focused on pollution control – waste removal, treatment and disposal techniques. Pollution control has improved environmental quality to a certain extent, but in general, it not only fails to eliminate a pollutant but also often transfers it from one medium to another. Waste treatment processes have produced a large amount of sludge and residue that need to be treated again prior to disposal so that they will not create secondary pollution.

There are three generations of industrial pollution problems which must be solved if our environmental quality is to be protected. The first generation of industrial pollution problems is the release of waste (gaseous, liquid and solid) within the plant facility. The second generation pollution problems deal with industrial product use that includes product transport, storage, and distribution. The disposal of used and unused products constitutes the third generation of industrial pollution problems. Government environmental management strategies largely emphasize the control of waste releasing from the manufacturing facilities with little concern toward managing industrial product transport, storage, distribution and consumption which have proved to be as important as waste management. The increase of toxic chemicals and hazardous wastes in our society and our environment has become man's most urgent environmental pollution problem now and in the future. In order to solve this urgent problem, we will have to take immediate action on a worldwide scale of a magnitude never before undertaken by mankind. These three generations of industrial pollution problems must be properly managed.

As the environmental concern deepens, we have to move further up the production chain: first, end-of-pipe solution to primary pollutants; later, internal process modifications to reduce waste and eventually redesign the products to allow maximal recycling of processing materials and minimization of waste production after the products are used. At some point, we may have to redesign the whole system that is providing the service in question if the environmental consequences of the present system become unbearable. The transport system has been cited as an area in search of new options. In this domain, we are entering a complex chain of new thinking in which the importance of the technical dimension decreases and the social, cultural and political dimensions grow in importance as we go along.

Most of the current research and development activity is devoted to efforts toward reducing and reusing industrial waste streams. The rising cost of waste handling makes a decrease of industrial waste an economic necessity for most companies. Certain useful products, while leaving the manufacturing plant for distribution through transport and storage as well as for consumption and after-use disposal, can cause pollution. Examples of such products are explosives, toxic chemicals, pesticides, herbicides, food and drugs, plastics, disposal items, automobiles, motor-cycles, trains, airplanes, and others. Unfriendly environmental services such as engineering design and construction, management practices, education and training can cause long-term, poor administration, and mental pollution. In industrial production, virtually the only remaining option is to decrease the formation of waste streams at the source by internal process modifications. This may involve a change of raw materials as well as redesigning the product itself. The potential of such new engineering for improving the environment is considerable, as demonstrated by several successful examples. Virtually the only limitation is that posed by our lack of imagination (Shen, 1996).

Few would disagree that economic development and environmental protection will top the national agenda in the future. Since industries release the largest amount of highly toxic waste, we must focus on industrial pollution through pollution prevention programs and projects. Such programs and projects need to be adequately funded and closely linked with the policy-making process. Our objective is to establish a society-oriented approach towards sustainable development and to live in a stable, fruitful, peaceful, and healthy existence on this planet. To achieve this objective, we need to create a society prepared to prevent pollution – a society that does not waste our natural resources, and a society that exists in harmony with nature.

1.2
Source of pollution

We must recognize that all sectors of our society generate waste: industry, government, agriculture, mining, energy, transportation, construction, and consumers. Wastes contain pollutants which are discarded process materials or chemicals. Pollution is caused by these pollutants, releasing into the environment beyond the assimilation capacity of the environment. Among those sources, industry is the primary target of all waste generators because of its quantity and toxicity. Industrial activities such as production, distribution, transport, storage, consumption of goods and services have known to be the most critical sources of environmental pollution problems, especially the chemical industry.

Traditionally, we have been controlling the releases of pollutants and neglecting the production of environmental unfriendly products and services that also caused environmental pollution. For example, DDT, CFCs, asbestos, leaded gasoline, certain kinds of plastics, medicines, cosmetics, fertilizers, pesticides, and herbicides, as well as discarded by-products and used-products are known to cause environmental and health problems. As to environmental unfriendly service, it refers to those activities such as government policies, regulations, and implementation plans; consulting services in product and process design, equipment manufacturing and supply; and education and training that result in adverse impacts of the environmental quality. In other words, we are currently preventing and minimizing the release of wastes and pollutants from all sources; however, we are only beginning to improve the quality of products and services through the concept of designing for the environment and sustainability.

In the U.S., dozens of national data sources provide a partial picture of waste generation and management and pollutant emissions. These data are collected by public agencies such as the Environmental Protection Agency, the Department of Interior, the Department of Energy, and by private-industry groups such the Chemical Manufacturers Association and the American Petroleum Institute.

Each of these data sources focuses on a particular aspect of wastes and emissions; none provides a global view.

1.3
Industrial pollution problems

The increase of toxic chemicals and hazardous wastes in our environment has become man's most urgent environmental pollution problem both now and in the future. In order to solve this urgent problem, we will have to take immediate action on a worldwide scale of a magnitude never before undertaken by mankind. The three generations of industrial pollution problems must be properly managed through preventive modern technologies and proactive management skills, focusing on not only industrial wastes and pollutants, but also industrial products and services.

1.3.1
Industrial growth

The scale of industrial production worldwide seems set for inexorable growth. Ecosystems in all parts of Europe, in particular central Europe, have declined severely since 1990. Destruction of habitats and eutrophication are seen as the most important causes (RIVM, 1992). Developing countries clearly aim to achieve the levels of material prosperity enjoyed in the developed countries, and they intend to do it by industrialization. Since their goal represents market growth to companies in the developed countries, and is directly in line with current democratic and economic rhetoric, it seems politically inevitable. Leaving aside environmental concerns, simple equity argues that it is also morally unavoidable. We are witnessing the evolution of a fully industrialized world, with global industrial production, global markets, global telecommunication, global transportation, and global prosperity. This prospect brings the realization that current patterns of industrial production will not be adequate to sustain environmentally safe growth on such a scale and are therefore all but obsolete (Tibbs, 1991).

As information about pollution costs accumulates, it has become increasingly difficult to argue for completely unregulated industrial growth in poor countries. Table 1-1 presents estimates of relative air pollution severity for 45 major world metropolitan areas. Extremely high air pollution levels are currently prevalent in China, India and Indonesia. Laboratory and epidemiological studies suggest that air pollutant exposure at such levels can obstruct breathing and increase the incidence of coughs, colds, asthma, bronchitis, and emphysema, and sometimes even heart and lung disease. The presence of many known toxins, carcinogens, and mutagens in existing emissions has also been well documented (WRI, 1990).

Table 1-1 Air pollution index: major world metropolitan areas

	Country	Metropolitan area	Total index
1	China	Shenyang	100
2	China	Beijing	82
3	Indonesia	Jakarta	80
4	China	Kian	71
5	India	Calcutta	66
6	India	Delhi	59
7	China	Guangzhou	56
8	China	Shanghai	47
9	Italy	Milan	47
10	Philippines	Manila	45
11	India	Bombay	37
12	Korea, Rep	Seoul	34
13	Thailand	Bangkok	30
14	Span	Madrid	27
15	Yugoslavia	Zagreb	25
16	Hong Kong	Hong Kong	22
17	France	Gourdan	19
18	Chile	Santiago	16
19	Finland	Helsinki	12
20	Brazil	Rio de Janeiro	11
21	Brazil	Sao Paulo	10
22	Belgium	Brussels	10
23	United States	Birmingham	9
24	Portugal	Lisbon	9
25	Malaysia	Kuala Lumpur	9
26	German, Fed Rep	Frankfurt	8
27	Israel	Tel Aviv	8
28	Canada	Montreal	8
29	Poland	Wroclaw	7
30	United Kingdom	Glasgow	7
31	Poland	Warsaw	6
32	United Kingdom	London	6
33	United States	Chattanooga	5
34	Greece	Athens	5
35	Australia	Sydney	5
36	United States	New York	5
37	Canada	Hamilton	4
38	United States	St. Louis	3

Table 1-1 (continued)

	Country	Metropolitan area	Total index
39	United States	Chicago	2
40	Netherlands	Amsterdam	2
41	Ireland	Dublin	2
42	Canada	Toronto	2
43	New Zealand	Christchurch	1
44	German, Fed Rep	Munich	1
45	Japan	Tokyo	1

Source: WRI (1990)

Society at large incurs costs when wastes are discarded into the environment, although the market frequently does not reflect those costs or the value of re-using wastes. Industries should seek the most efficient and effective way to achieve environmental protection. Market prices should include both the costs of producing goods and services as well as the social costs of depleting resources and harming the environment. Improved methods of accounting for such costs should address this oversight.

Developing countries are encountering toxic emission problems far worse than the developed countries. A good example is the use of leaded gasoline for automobiles in developing countries. It contributes to serious lead contamination of air in metropolitan areas which may cause severe and permanent brain damage; even death. Another good example is provided by industrial chemical production in the industrialization process. One major subsector, synthetic organic chemicals, is essentially a new industry, having entered volume production only since World War II. The associated hazards are suggested by the high ranking of synthetic organic chemicals and many other chemical subsectors among U.S. sources of toxic, carcinogenic, and mutagenic pollution as shown in Tables 1-2 and 1-3 (WB, 1992). Table 1-4 presents the U.S. production of synthetic organic chemical products, 1969–1990 (CEQ, 1992).

Table 1-2 Toxic estimates: U.S. Inventory, 1987

Industry	Lbs/$'000 of value added	Lbs/$'000 of gross output
Industrial chemicals	99.71	52.42
Primary metals	56.80	21.48
Paper products	56.46	25.87
Petroleum refining	41.43	5.79
Textiles	13.45	5.47
Leather, fur products	12.18	5.82
Rubber, plastic products	6.26	3.20

Table 1-2 (continued)

Industry	Lbs/$'000 of value added	Lbs/$'000 of gross output
Metal products	4.06	2.06
Pottery, glass products	3.54	1.92
Electrical machinery	3.10	1.72
Furniture	2.95	1.60
Transport equipment	2.45	0.99
Food products	2.35	0.87
Other manufacturing	2.08	1.13
Wood products	1.26	0.52
Professional, scientific eqt.	1.14	0.75
Non-electrical machinery	0.83	0.45
Tobacco products	0.73	0.50
Printing, publishing	0.71	0.47
Wearing apparel	0.14	0.07

Sources: USEPA, Toxics Release Inventory, 1987
USDOC, Census of Manufactures, 1987
USIDO, Industry and Development, Global Report, 1990/91

Table 1-3 Top ten industries producing hazardous wastes in the United States, 1987

SIC	Category	Volume (million tons)
2869	Industrial organic chemicals	60–80
2800	General chemical manufacturing	40–50
2911	Petroleum refining	20–30
2892	Explosive	10–15
2821	Plaxtic materials/resins	6–10
2879	Agricultural chemicals	5–8
2865	Cyclic crudes, intermediates	5–8
2816	Inorganic pigments	3–5–5
2812	Alkalis, chlorine	2.5–4.5

Sources: USEPA, Toxics Release Inventory, 1987
USDOC, Census of Manufactures, 1987
USIDO, Industry and Development, Global Report, 1990/91

Table 1-4 U.S. production of synthetic organic chemical products, 1969–1990

Product	Year											
	1969	1971	1973	1975	1977	1979	1981	1983	1985	1987	1989	1990
	billion liters											
Crude coal tar	na	na	na	na	2.24	2.23	1.76	1.07	na	0.72	0.59	0.60
	million kilograms											
Dyes	109	111	129	93	120	121	104	111	101	116	140	117
Organic pigm'ts	28	26	31	23	31	40	34	35	37	43	50	52
Medic'l chem.	91	101	106	94	109	142	111	106	102	118	129	144
Flavor/perfume	54	44	53	46	68	88	75	79	69	57	64	60
Rubber	137	147	182	127	182	179	127	133	118	173	176	179
Plasticizers	635	680	862	635	816	953	862	771	771	907	976	891
Pesticides	499	499	590	726	635	635	649	463	562	472	572	557
	billion kilograms											
Petroleum	32	37	41	35	57	55	50	50	47	54	50	52
Cyclic intermed.	13	14	16	14	8	22	21	20	21	25	25	24
Plastic & resin	8	10	14	11	16	19	18	20	23	27	26	30
Elastomers	2	2	3	2	3	3	2	2	2	2	2	2
Surface ag'ts	2	2	2	2	2	2	2	2	2	3	3	4
Miscellaneous	34	36	45	39	48	55	53	52	53	44	48	50

Source: U.S. International Trade Commission. Synthetic Organic Chemicals, United States Production and Sales, (Washington, DC: ITC, annual).
Notes: Petroleum = Primary products from petroleum and natural gas for chemical conversion. Cyclic intermed. = Cyclic intermediates which are synthetic organic chemicals derived principally from petroleum and natural gas and from coal-tar crudes produced by destructive distillation (pyrolysis) of coal. Most are used in the manufacture of more advanced synthetic organic chemicals and finished products, such as dyes, medicinal chemicals, elastomers (synthetic rubber), pesticides, and plastics.

Rapid hazardous pollutant buildups in recently industrialized countries may provide even more persuasive evidence. Developing countries are encountering complex toxic emission problems far earlier in the industrialization process than the developed countries. Industrial chemical production, for example, currently grows at about 7.5% per year in developing countries, versus 2.5% in developed countries. Growth is particularly rapid in Asia, where the rapid increase of organic chemicals consumption has stimulated ambitious plans for local capacity expansion (WB, 1992).

During 1980–1988, domestic use of ethylene, propylene, and benzene grew 16.3%, 12.3%, and 13.4% respectively in Korea; 14.1%, 14.0%, and 14.2% in India; 9.8%, 12.2%, and 11.7% in China; and 10.5%, 7.5%, and 5.6% in Indonesia. The growth has been higher since 1988. In the absence of significant pollution regulation, the emissions intensity of the new production in Asia is likely to exceed that in the U.S. (UNIDO, 1990).

Industrial pollution problems not only deplete resources and damage the environment, but also hurt the economy. The U.S. EPA estimated that U.S. spending for pollution abatement and control by both public and private sectors was about $100 billion in 1991. Tables 1-5, 1-6, and 1-7 present U.S. pollution abatement expenditures by type of pollution and sectors in 1972–1990. The U.S. government has begun to implement a variety of policies to ensure that economic actors recognize the full cost of using natural resources and, accordingly, better manage these resources.

Table 1-5 U.S. pollution abatement expenditures, by type of pollution, 1981–1990

Year	Air		Water		Solid waste	
	Current	Constant	Current	Constant	Current	Constant
	billion $					
1981	26.3	28.5	20.8	25.2	10.2	13.4
1982	26.1	27.2	21.1	24.2	9.9	12.1
1983	27.6	28.5	22.5	24.7	10.3	12.0
1984	30.4	31.0	24.7	26.2	11.8	13.2
1985	30.1	32.0	26.7	27.4	12.7	13.7
1986	32.1	33.3	28.2	28.7	14.3	14.8
1987	30.7	30.7	30.6	30.6	15.9	15.9
1988	32.5	31.9	30.7	30.1	18.6	17.9
1989	30.9	29.2	33.4	31.6	21.7	19.8
1990	29.5	27.3	37.3	34.4	24.1	20.9

Source: Rutledge GL and ML Leonard, Pollution abatement and control expenditures, 1972–1990, Survey of Current Business, Table 9, pp. 35–38, (Washington, DC: DOC, BEA, June 1992). Notes: Constant=1987 $. Expenditures cover most, but not all, pollution abatement and control activities, which are defined as those resulting from rules, policies and conventions, and formal regulations restricting the release of pollutants into common-property media such asthe air and water. Solid waste management includes the collection and disposal of solid waste and the alteration of production processes that generate less solid waste. Data for 1981–89 are revised estimates. Data for 1990 are new estimates. Estimates do not include interest costs.

Table 1-6 U.S. pollution abatement and control expenditures, 1972–1990

Year	Pollution abatement		Regulation & monitoring		Research & development		Total	
	Current	Constant	Current	Constant	Current	Constant	Current	Constant
					billion $			
1972	15.9	43.1	0.4	1.0	0.8	2.3	17.0	46.3
1973	18.0	46.3	0.5	1.2	0.9	2.3	19.4	49.8
1974	22.0	48.7	0.6	1.3	1.0	2.3	23.6	52.3
1975	26.7	53.7	0.7	1.3	1.1	2.3	28.4	57.3
1976	29.9	56.6	0.7	1.4	1.3	2.5	31.9	60.5
1977	32.8	57.8	0.8	1.5	1.5	2.7	35.1	62.0
1978	36.9	60.3	0.9	1.8	1.6	2.8	39.5	64.9
1979	42.8	62.4	1.1	1.6	1.8	2.7	45.6	66.8
1980	48.4	63.0	1.3	1.9	1.8	2.4	51.5	67.3
1981	53.4	62.6	1.4	1.8	1.7	2.2	56.5	66.5
1982	53.4	59.5	1.4	1.7	1.8	2.1	56.6	63.2
1983	56.3	60.9	1.4	1.6	2.3	2.6	60.0	65.1
1984	62.8	66.0	1.4	1.5	2.3	2.5	66.4	70.0
1985	67.3	67.3	1.3	1.4	2.4	2.5	70.9	72.7
1986	70.1	70.1	1.5	1.6	2.6	2.6	74.2	76.4
1987	72.5	72.5	1.5	1.5	2.6	2.6	76.7	76.7
1988	76.6	76.6	1.7	1.6	2.8	2.7	81.1	79.1
1989	80.6	80.6	1.8	1.7	3.0	2.7	85.4	80.1
1990	85.1	85.1	1.8	1.6	3.1	2.7	90.0	81.8

Source: Rutledge GL and ML Leonard, Pollution abatement and control expenditures, 1972–1990, Survey of Current Business, Table 9, pp. 35–38, (Washington, DC: DOC, BEA, June 1992). Notes: Constant=1987 $. Expenditures are for goods and services that U.S. residents use to produce cleaner air and water and to manage solid waste. Pollution abatement directly reduces emissions by preventing the generation of pollutants, by recycling the pollutants, orby treating the pollutants prior to discharge. Regulation and monitoring are government activities that stimulate and guide action to reduce pollutant emissions. Research and development by business and government not only support abatement but also help increase the efficiency of regulation and monitoring. Data for 1972–1989 are revised estimates. Data for 1990 are new estimates. Estimates do not include interest costs. Totals may not agree with detail because of independent rounding.

Table 1-7 U.S. pollution abatement and control expenditures by sectors, 1972–1990

Year	Pollution abatement					Regulation & monitoring			Research & development	
	Private		Government			Fed.	State	Prv't	Fed.	State
	Pers'l	Bus.	Fed.	State	Other					
1972	3.45	30.53	0.40	3.67	5.03	0.48	0.48	1.44	0.55	0.27
1973	4.54	32.14	0.54	3.69	5.34	0.63	0.56	1.49	0.68	0.17
1974	4.95	32.63	0.69	3.62	6.81	0.74	0.6	1.42	0.78	0.09
1975	6.17	33.11	0.94	3.74	9.72	0.74	0.61	1.28	0.93	0.10
1976	6.73	34.88	0.95	3.71	10.33	0.74	0.67	1.39	1.02	0.09
1977	7.15	37.1	0.92	3.73	8.90	0.74	0.78	1.56	1.04	0.10
1978	7.41	38.15	0.81	3.89	10.03	0.99	0.8	1.69	0.99	0.10
1979	7.16	40.22	0.84	3.93	10.27	0.91	0.72	1.76	0.87	0.12
1980	7.31	40.15	0.68	4.05	10.82	1.11	0.76	1.50	0.82	0.10
1981	8.49	40.47	0.63	3.98	8.98	1.04	0.77	1.33	0.81	0.04
1982	8.52	37.79	0.65	4.07	8.43	0.98	0.73	1.32	0.70	0.04
1983	10.02	38.04	0.91	4.20	7.70	0.92	0.69	1.82	0.71	0.04
1984	10.97	41.18	1.05	4.38	8.42	0.81	0.69	1.81	0.63	0.04
1985	11.78	42.41	1.30	4.65	8.67	0.61	0.75	1.86		0.02
1986	12.69	43.80	1.40	4.99	9.29	0.74	0.85	1.93	0.67	0.03
1987	10.88	44.50	1.24	5.36	10.54	0.70	0.82	1.99	0.63	0.03
1988	11.83	45.96	1.34	5.88	9.78	0.81	0.83	1.99	0.64	0.03
1989	10.15	47.76	1.27	6.50	10.04	0.78	0.88	1.99	0.72	0.03
1990	8.67	49.69	1.23	6.99	10.89	0.77	0.87	2.01	0.68	0.04

Source: Rutledge GL and ML Leonard, Pollution abatement and control expenditures, 1972–1990, Survey of Current Business, Table 9, pp. 35–38, (Washington, DC: DOC, BEA, June 1992). Notes: Expenditures are attributed to the sector that performs the air or water pollution abatement or solid waste collection and disposal. State also includes expenditures by local authorities but excludes agricultural production except feedlot operations. Pers'l = Personal consumption (all of which is used to purchase and operate motor vehicle emission abatement devices). Bus. = Business. Fed. = Federal government. Prv't = Private (=business spending). Other government spending refers to government enterprise fixed capital for publicly owned electric utilities and public sewer systems. Data do not include interest costs. Totals may not agree with detail on preceding tables because of independent rounding.

1.3.2
Knowledge of pollution prevention

Millions of people are concerned about pollution and waste, garbage, or trash, but there are over 5 billion people living on the planet. Many people see severe environmental problems, but still many others do not. Even those who see pollution problems may not know and understand all the impacts on: what happens to waste and pollution, and what the impacts are on public health and the environment. Waste seems natural both biologically and socially. If waste is inevitable, then all forms of waste have to go somewhere. Most people will say "not in my backyard."

Pollution exists because the environment can absorb only a limited amount of pollutants and wastes. Some hazardous or toxic by-products and chemicals are termed hazardous pollutants because they have toxic characteristics and the environment cannot assimilate them. The others are known as non-hazardous or degradable pollutants, most of which have a critical threshold beyond which they cannot be assimilated. Both hazardous and non-hazardous pollutants have caused environmental pollution in many areas worldwide. Most professionals in under-developing countries are lacking the knowledge of how to deal with hazardous wastes and pollutants. Environmental professionals in developed countries have been debating the cost effectiveness of innovative hazardous waste management approaches. We clearly need cost-effective government regulation of pollution, but controversy continues over which environmental management strategies should be implemented.

Many wastes or pollutants will have multiple adverse effects. There may also be synergistic interactions and cumulative exposures from different wastes and pollutants in the environment. A toxic chemical may be in drinking water, food, household and workplace air, and consumer products. Rarely do government agencies and others take into account cumulative exposures of people and the environment to harmful chemicals. Thousands of toxic pollutants are not regulated at all. Further, for many harmful chemicals and types of radiation, there no safe level of exposure; this means that even very small exposures can eventually result in serious and perhaps fatal health effects. Today's rapidly changing technologies, industrial products and practices may generate wastes that, if improperly managed, could threaten public health and the environment. Industrial pollution is created when factories release harmful process by-products and wastes into the environment beyond the assimilation of the environment. Many wastes, when mixed, can produce a hazard through heat generation, fire, explosion or release of toxic substances. These wastes are generally considered incompatible. Waste generators should describe and characterize their wastes accurately, by including information as to the type and the nature of the wastes, chemical compositions, hazardous properties, and special handling instructions.

Current environmental pollution problems – air pollution, water pollution, and soil contamination – have resulted also from past decision-makers who did not consider impacts on the natural environment. These decisions "borrowed" from the environment and the debt has now come due in the form of expensive clean-ups and other remedial action. The formation of toxic or hazardous waste constitutes a problem of its own. We are dealing with hazardous wastes that present a real danger to human health or the environment. We should have dealt with the underlying causes of the problem, not just the symptoms. Since we must have natural assets to support our economy and a healthy human population, decision makers need to have a sustainable development concept and to integrate environmental and economic considerations based on a total cost system with sound pollution cost information, including pollution control cost and pollution damage cost.

Regulatory agencies have an obligation to designated competent authorities to be responsible for the supervision and administration of operations for the disposal of toxic and dangerous waste. Unfortunately, these authorities, responsible in given areas, do not have the knowledge to plan, organize and supervise such operations. They are responsible for issuing permits for the storage, treatment, and/or deposit of toxic and dangerous waste, and for controlling such undertakings and those responsible for the transport of the waste. However, many countries still lack the legal framework for dealing properly with hazardous wastes. Although the regulations differ from country to country, there are certain basic similarities. In the regulations, the scope of application is indicated by giving general information about the waste or by listing separately the wastes considered to be hazardous, using criteria such as origin of the waste, toxic substances present, and their properties. Up to now, no generally accepted list of toxic substances has been agreed upon by the working group of United Nations, although such work is in preparation. Many countries rely on the principle that the generator of the waste is responsible.

Some people still believe that a waste management or pollution control action by end-of-pipe strategies can prevent pollution, such as a wastewater treatment plant, flue gas cleaning system, land disposal or incineration. This is because such equipment or system does limit the release of harmful pollutants compared to uncontrolled waste discharge into the environment. Like using medication to avoid feeling pain from a headache, infection, or surgical incision, waste management and pollution control methods attempt, although imperfectly, to minimize the effect or pain of releasing pollutants into the environment. True, some releases and effects are curtailed. But the original toxic or destructive environmental wastes remain hazardous or are transformed into different hazardous substances to some degree.

1.3.3
Human behavior

There are still many people who do not like to learn new ideas for pollution prevention programs that can help national environmental goals and coincide with industry's interest. Some people resist change without even thinking. The human behavior for change in pollution prevention needs new knowledge and education, with creative and probing personalities. It is no longer acceptable to rest on environmental experiences gained decades ago in wastewater treatment or air pollution control systems. The future factory will consider the environmental implications of a new product or service. It will design pollution management strategies into its research experiments. It will insist on pollution prevention in the acquisition of new components or systems from others so that waste generation in the life cycle of its product will be minimized. The future factory will constantly use preventive techniques to reduce wastes and seek solutions. It is critical that people must know what to do. If behavior changes are required, the responsibility for change rests on thoughtful management (Shen, 1997).

Practicing prevention on a daily basis requires considerable personal commitment and discipline. Individuals have begun to take more responsibility for their own health. It became practical and cost effective to change habits in order to stay healthy. People pay more attention to diet, exercise, and the avoidance of addictive drugs and alcohol. Preventive health care requires personal commitment and discipline, along with options in the marketplace and supportive government policies and programs. A preventive approach for environmental pollution will require even more commitment and discipline because the benefits are likely to seem more distant and uncertain to most people.

1.4
Changing environmental management concept

Dr. Michael Royston of The Environmental Management Center for Education in International Management in Geneva promoted the concept of an integrated approach to waste management and pollution control. He suggested the six dimensions of the environment in his book, namely its technological, economic, physical, cultural, social and political aspects (Royston, 1979). It is true that pollution problems can be seen as a technical problem, and then become an economic problem, and ultimately end up as a political problem. Dr. Royston's theory of the six dimensions of the environment are briefly stated below:

(1)Technological environment

It lies in the application of non-waste technology to pollution problems. Financial assistance and economic incentive must be provided for research designed to produce non-waste technology.

(2)Economic environment

The prerequisites of a successful strategy are the internalization of all environmental damages caused by any party in the economics of a particular operation, the provision of economic incentives to encourage the cleanup of the environment and to create the economic benefits which result from a cleanup operation. A system of waste releasing charges was suggested.

(3)Physical environment

The basic principle is to work with the eco-system. It is important to know the assimilation capacity of a locality and to use this knowledge in the setting of local standards which need to be set in conjunction with those who know most about the local environment.

(4)Cultural environment

The most important and critical aspects of the cultural environment is education. The most effective way for a community to combat pollution is by means of a communal activity in which citizens, government and enterprise all work together toward a common effort. Without an understanding of the nature of the problem, of the constraints under which different parties are working, and of what solutions are possible, such an approach cannot function.

(5)Social environment

Every culture around the world has an anti-pollution ethic. In most countries, dirt and pollution have universally been condemned by all sectors of the society. Even today, most people believe that the only way for economic growth is, in fact, to waste more and more, to build-in obsolescence, to keep the wheels of industry turning by pouring raw materials and energy into a system which provides for short-lived, intrinsically energy- and resource-consuming products. This unhealthy belief must be changed for resources conservation and manufacturing more environmentally sound products for consumption.

In the social sphere, some industrial development activities result in the transfer of wastes and residuals from an industry to its community. The benefits of the industry is only matched by the costs of its community. In particular there should be a positive approach to the sharing of the costs and benefits that are associated with development and pollution. For those communities need to receive benefits proportionate to the costs which they bear. In any industrial development, the developer might be putting in money, capital and technology, but the community is also putting in capital – the air, water and land that will be modified by that particular development.

(6)Political environment

The resolution of pollution problems is a political matter. The basic rule of participatory planning is suggested. We seem to be simultaneously moving into a new political era all around the world. The way in which we reach decisions and plan our activities is changing. This is partly due to the social changes that are taking place and partly to the level of education. Societies are beginning to realize that technological education does not necessarily confer wisdom, and the technological issues are not necessarily the most important issues society has to deal with. Government is no longer a top-down process with all concerned individuals and groups making their decisions. People are concerned about the future, they want to know about the future of the world, and that of their children, and they want to have a say in that future (Royston, 1979).

We may also think of the two ways for classifying human environment: (1) according to dimension, or the order of magnitude of the environment's state in time and space (i.e., micro-, meso-, and macro-); and (2) according to type of environmental system (i.e., physical, biological, social and behavioral). These, in turn, are associated with their respective spheres. According to the environmental dimension, the human environment may be considered as local, regional, national and global. According to the type of environmental system, the human environment may be divided into physical, biological, social and behavioral spheres. Each environmental sphere can be specified with respect to time and/or space dimensions, as:

- Physical (geosphere) – air, water, and land;
- Biological (biosphere) -animal, plant and protist;
- Social (sociosphere) – political, economic, and cultural; and
- Behavioral (behaviorsphere) – percepual and conceptual ability in man and animals.

For the past 40 years, environmental management concept has been undergoing a dramatic transformation. We have been looking for the best and most cost-effective methods to manage pollutants in the waste (gaseous, liquid, and solid) from entering the environment. Our waste management concept has been applying rules and regulations to control of waste materials through chemical, physical, biological, thermal, mechanical, or electrical treatment and disposal method after they were generated. For example, in the early 1970's, the U.S. government adopted a series of pollution control laws aimed at media-specific problems (air pollution, water pollution, waste disposal) and a regulatory framework of specific goals, deadlines, and technological controls. Today, while the regulatory framework of waste management has accomplished significant pollution reduction, such "command-and-control" measures are often viewed as inadequate and needlessly expensive for each increment of improvement.

Since the late 1980s, the concept of pollution management has grown. Pollution management takes a broad preventive approach, looking at three generations of industrial pollution problems. This concept prevents not only waste generation from the conventional manufacturing facilities, but also pollution from product distribution, storage, transport and use as well as from products after useful life for disposal. In other words, waste management focuses on pollutants and wastes, while pollution management deals with not only pollutants and wastes, but also products and services. Therefore, we need to apply the pollution prevention concept to manage both wastes and products before they create pollution problems.

1.5
Environmental management strategy

Environmental management strategies are gradually transforming as more and more professionals accept the pollution prevention concept. The difference between waste management strategy and pollution management strategy is that the former emphasizes controlling pollutants in the wastes and minimizing waste generation at the sources; while pollution management strategy is not just to prevent and control pollutants and wastes, but also the search for better answers for manufacturing and consuming environmental friendly products and services in a broader sense. Sometimes, the pursuit of pollution prevention involves subtle reveals of institutional trends, reveals which in small ways inspire considerable improvements. Therefore, environmental management strategy needs to upgrade beyond waste management strategies to a broad approach of pollution management strategies (Shen, 1997).

We must recognize that waste management strategies deal mostly with physical and biological spheres; while pollution management strategies deal with all types of spheres in the environmental system. Throughout our life-span, we are constantly adjusting our environmental for our comfort, health, and survival. Accordingly, any meaningful discussion of the environment must consider the interaction and interdependence of man and his environment. Thus, the total environmental concept and pollution management strategy with preventive approaches will provide the most cost-effective tools for achieving sustainable development goals (USEPA, 1997).

Industrial pollution prevention involves waste minimization, source reduction, cleaner production, design for environment and preventive service. The concept of industrial ecology, a relatively new terminology, considers environmental management more than operations at a single manufacturing facility. However, in practice, both industrial pollution prevention and industrial ecology use less raw materials, energy, and natural resources; substitute nontoxic chemicals for the hazardous or toxic materials currently used in the process; and redesign a manufacturing line to take advantage of newer and cleaner technologies and process equipment. It is important to realize that pollution prevention management is a very broad term and applies beyond industrial sectors to a variety of economic sectors and institutional settings. Many organizations and institutions can apply pollution prevention concept which not only reduces the generation of waste materials, but also produces the environmental friendly products and services. During past few years, considerable progress and success have been made in attaining pollution prevention in various sectors of our society.

References

Carnegie Corp. (1992) Environmental Research and Development: Strengthening the Federal Infrastructure. A Report of The Carnegie Commission on Science, Technology, and Government, December 1992

CEQ (1992) The 23rd Annual Report of the Council on Environmental Quality. The U.S. Government Printing Office, ISBN 0-16-041612-4. Washington, D.C.

Hinrichsen D (1987) A Reader's Guide of Our Common Future. The International Institute for Environment and Development (IIED) publication with support of the Norwegian Government.

Ling JT (1998) Industrial Waste Management. A speech published in the Vital Speeches of the Day, Vol.LXIV, No. 9, pp. 284-288, February 18, 1998.

RIVM (1991) National Environmental Outlook 2, 1990-2010. National Institute of Public Health and Environmental Protection. Bilthoven, The Netherlands, May, 1992. ISBN 90-6960-02205.

Royston MG (1979). Pollution Prevention Pays. Pergamon Press, Elsford, NY, pp. 137-153.

Shen TT (1992) "Motivating Decision Makers Toward Sustainable Development", presented at the Science and Technology Conference in St. Louis, MO., June 13-15.

Shen TT (1966) "Critical Environmental Issues Worldwide and Pollution Prevention Strategies", presented at the Academia Sinica in Taipei, Taiwan, December 10.

Shen TT (1997) "Five Categories to Avoid Industrial Pollution Prevention Barriers". Proceedings of Asian-Pacific Industrial Waste Minimization Conference in Taipei, Taiwan, December 12-18.

Shen TT (1997) "Environmental Management in China", presented at the A&WMA 90th Annual Meeting and Exhibition in Toronto, Canada, June 8-13.

Tibbs BC (1991) Industrial Ecology: An Environmental Agenda for Industry. World Earth Review, P.5, Sausalito, Ca., Winter 1991.

UNIDO (1990) Industry and Development: Global Report, 1990/1991.

WRI (1990) World Resources, 1990-1991. Published by the World Resources Institute in collaboration with UNDP and UNDP, New York.

USEPA (1992) Facility Pollution Prevention Guide. Office of R&D, Washington, D.C., EPA/600/R-92/088, May 1992.

USEPA (1997) Pollution Prevention 1997 – A National Progress Report, EPA-742-R-97-000, June 1997.

World Bank (1992) The Economics of Industrial Pollution Control: An International Perspective. Publication of the World Bank, Industrial Series Paper No. 60.

2
Industrial pollution prevention

In the rush toward industrialization, it became expedient for people to produce more products and to consume more. They forgot that this is an orderly universe and an orderly world which works according to scientific, i.e., natural laws which are inviolable. Now the extent of environmental destruction has forced us to recognize that everything in the environment has an effect on everything else. With that recognition, the first feeble efforts at a multidisciplinary approach to problems has begun. We have yet to recognize that the design of a viable socio-economic order can not be accomplished by sociologists who know nothing of economics, economists who know nothing of sociology, and lawyers who know only their allegiance to the laws of the status quo. Few of them recognize that the socio-economic order and the environment are completely dependent on each other.

Pollution prevention in the industrial sector is hardly a new concept. Industrial operations traditionally have adopted a variety of waste reduction techniques to lower costs of production to increase profits. However, only in recent years, economical incentives and the corresponding emphasis on prevention as a management priority have grown more rapidly. In the United States, before major environmental legislation, there was little economic incentive for industries to properly manage their wastes. Disposing wastes into the environment were cheap, if not free, because the social costs from the pollution were not assessed at the source. Industrialists pursued waste reduction as long as it was profitable. Continued use of the environment as a given, led to pollution awareness in the 1960s and end-of-pipe controls. Investment costs for such control technologies were partially offset by depreciation tax credits, industrial revenue bonds, and grants to municipalities for wastewater treatment facilities (Koenigsberger, 1986).

Dr. Joseph Ling, Vice President (retired) of 3M Company once said: "Pollution controls solve no problem. They only alter the problem, shifting it from one form to another, contrary to this immutable law of nature: the form of matter may be changed, but matter does not disappear… It is apparent that conventional controls, at some point, create more pollution than they remove and consume resources out of proportion to the benefits derived…What emerges is an environmental paradox. It takes resources to remove pollution; pollution removal generates residue; it makes more resources to dispose of this residue and disposal of residue also produces pollution." Pollution prevention priorities are source reduction in a hierarchy of options addressing pollutants and wastes. Following source reduction, in descending order of preference, are recycling and reuse, treatment, and disposal. All of these options are valid, and current economical and environmental data even support a combination of them. Nonetheless, source reduction remains the preferred approach for the following reasons:

1. Individual pollutant controls in stack gas and wastewater treatment processes do not always address cross-media impacts, for example, these treatment processes generate sludge that needs to be treated again and create a waste or sludge disposal problem. Once pollutants enter the environment, they will cycle throughout the air, water and soil. Unless destroyed, pollutants will continue to transfer from one medium to another as shown in Fig. 2-1 (Shen and Sewell, 1986).

2. Both non-point pollution sources (e.g., municipal and industrial waste treatment and disposal facilities, and urban storm water runoffs) and small, dispersed point sources (e.g., hospitals, research laboratories, academic institutes and commercial establishments) contribute significantly to society's harmful pollution. Such sources are difficult to regulate through traditional large-source emission standards (Shen et al., 1993).

3. Because pollution prevention saves energy and resources, in most cases, it is more cost-effective in the long-run than direct regulation. At a time when economic competitiveness is a national priority and total pollution control costs to businesses and public agencies have grown to billions of dollars annually, we need an economically sound approach like pollution prevention.

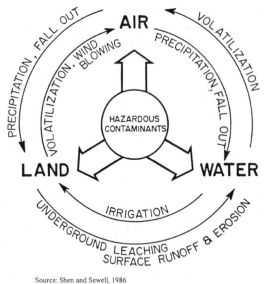

Fig. 2-1 Hazardous constituent cycle

Source: Shen and Sewell, 1986

Sustainable development has been universally accepted as our common environmental goal. To implement sustainable development, it requires promotion and application of pollution prevention through source reduction and clean technologies.

According to the findings of the Industrial Pollution Prevention Project (IP3), the four most important general motivators for pollution prevention in industry are: economics, technical and financial assistance, open communication, and flexibility (especially regulatory flexibility). The IP3 found that the key trigger for pollution prevention is a stringent regulation or enforcement action (EPA, 1995). The desire to avoid being subject to regulations provided the most critical impetus for pollution prevention, only motivating source reduction initiatives but also ensuring their success in the marketplace. Environmental pressures from regulations and from consumers and professional advocacy campaigns created opportunities for industry and business to gain competitive advantage in domestic and international markets.

2.1
What is pollution?

Pollution changes the physical, chemical and biological characteristics of our air, land, and water. Pollution harms human life, the life of other species, and also degrades living conditions and cultural assets, while wasting or deteriorating raw material resources (NRC, 1966). Pollution exists because the environment can absorb only a limited amount pollutants and wastes. Some hazardous wastes or toxic by-products and chemicals are termed hazardous pollutants because they have toxic characteristics and the environment cannot assimilate them. In other words, pollution is the undesirable change in the physical, chemical or biological characteristics of air, water or land that may or will harmfully effect human life or that of other desirable species, our industrial processes, living conditions or cultural assets, or that may or will waste or deteriorate raw material resources. Pollution causes risks to environmental quality, human health, and damage to natural resources.

For producers in the economic system, "waste is a nonproductive stream of material or energy for which the cost of recovery, collection, and transport... to another use is greater than the value as an input." (Bower, 1990). For society, materials and energy become wastes or residues when the cost of putting them to use, exceeds the cost of discarding them into the environment. Society incurs the costs when these residuals are discarded into the environment, although the market frequently does not reflect the costs or the value of reusing residuals. Improved methods of accounting for such costs have begun to address this oversight.

Since some waste is inevitable in the use of materials and energy, manufacturers should apply the concept and practices of design for the environment by choosing materials and technologies so as to minimize the quantity of residuals which harm life-sustaining elements in the natural environment. Design for the environment is a process which aims at minimizing or prevention environmental damage as a design objective in the first place. This signifies not only waste and pollution created during development and manufacture of a product but, also use and dispose of the used-product by the consumer. In other words, it considers the environmental effect of a product through the entire life-cycle of that product. The word "product" can be an airplane, automobile, bridge, building, eco-industrial park, land-use project, city-planning, etc. Indeed, the long-range goal should be to discard only the quantity and quality of residuals that can be sustained by the environment's natural self-cleaning capacity. Useful products can also become wastes over time as they deteriorate or are discarded, causing pollution problems(Shen, 1997a).

2.2
What is Pollution prevention?

Professionals in various organizations define pollution prevention based on their own understanding and applications, resulting with somewhat different interpretations. Basically, it means prevent or reduce the sources of pollution before problems occur. Many organizations in the United States define pollution prevention in terms of the Pollution Prevention Act hierarchy or Resource Conservation and Recovery Act waste management hierarchy, though there is much debate over whether to include recycling in the definition. Some define pollution prevention more narrowly as source reduction or toxics use reduction. And, others view pollution prevention more conceptually as any process that involves continuous improvement and movement up the environmental management hierarchy.

The term "pollution prevention" started in 1976, when Dr. Joseph Ling of 3M Company talked about a new program of "Pollution Prevention Pays" or 3P program during the first United Nations Economic Commission for Europe Seminar on "Principles and Creation of Non-waste Technology" held in Paris. The 3P program has been based on technological and management advances that (a) reduced environmental releases and (b) resulted in production costs lower than those associated with the previous more polluting method. It has taken the world a long time, to work into the pollution prevention mode. While the United Nations meeting set out principles of pollution prevention in 1976, it was not until 1988 that the US Environmental Protection Agency (USEPA) established its Pollution Prevention Office, and it was not until USEPA upgraded this office to a status where pollution prevention was substantively interjected into environmental policy decision-making (Purcell, 1990).

In the United States, the term "pollution prevention" legally means "source reduction." That is any practice which: (1) reduces the amount of any hazardous substance, pollutant, or contaminant entering any waste stream or otherwise released into the environment prior to recycling, treatment, or disposal; (2) reduces the hazards to public health and the environment associated with the release of such substances, pollutants, or contaminants. (3) increases efficiency in the use of raw materials, energy, water, or other resources, or (4) protects natural resources by conservation. Pollution prevention also includes equipment or technology modifications; process or procedure modifications; reformulation or redesign of products; substitution of raw materials for cleaner production; improvements of managerial services in housekeeping, maintenance, training, or inventory control.

In practice, pollution prevention approaches can be applied not only to industrial sectors, but also to all pollution-generating activities, including such as in energy production and consumption, transportation, agriculture, construction, land use, city planning, government activities and consumer behavior. In the energy sector, pollution prevention can reduce environmental damages from extraction, processing, transport, and combustion of fuels. Pollution prevention approaches include: (1) increasing efficiency in energy use; (2) substituting fossil fuels by renewable energies; and (3) design changes that reduce the demand for energy. Pollution prevention involves broadly all sectors of the societal activities and deals with not only wastes (gaseous, liquid, and solid), but also products and services. As mentioned previously, products refer to environmental unfriendly products such as pesticides, CFC, leaded gasoline, plastics, disposal diapers, and many toxic chemicals. Services refer to governmental, professional and non-professional services such as management, education, design, operation and maintenance activities that avoid or reduce pollution.

Pollution prevention includes practices that reduce or eliminate the creation of pollutants through increased efficiency in the use of raw materials, energy, water or other resources, or protection of natural resources by conservation. It calls for the judicious use of resources through source reduction, energy efficiency, re-use of input materials during production, and reduced water consumption. Two general methods of source reduction can be used in a pollution prevention program: product changes and process changes which essentially aim for cleaner production. They reduce the volume and toxicity of wastes and of end products during their life-cycle and at disposal. Fig. 2-2 provides some examples of source reduction.

Pollution prevention involves waste minimization, source reduction, design for the environment, and clean technology. Waste minimization refers to the reduction or elimination of the waste generation(the total volume or toxicity) at the source, usually within a process. Source reduction is defined as any practice which reduces the amount of any hazardous substance, pollutant, or contaminant entering any waste stream or otherwise released into the environment prior to recycling, treatment, or disposal. It includes practice which reduces the hazards to public health and the environment associated with the release of such substances, pollutants, or contaminants. The term includes: equipment or technology modifications, process or procedure modifications, re-formulation or redesign of products, substitution of raw materials, and improvements in housekeeping, maintenance, training and inventory control.

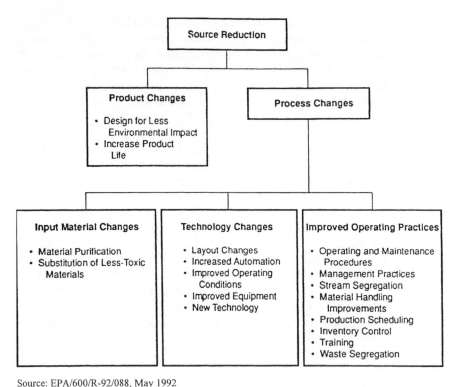

Source: EPA/600/R-92/088, May 1992

Fig. 2-2 Source reduction methods

Design for the environment is a process which aims at minimizing or preventing environmental damage as a design objective in the first place. Clean technology aims at products design and manufacturing processes that use less raw materials, energy, and water; generates less or no wastes (gas, liquid, and solid); and recycle waste as useful materials in a closed system. It may alter existing manufacturing processes to reduce generation of wastes. Clean technology includes the use of materials, processes, or practices that reduce or eliminate the creation of pollutants or wastes.

As the pollution prevention concept has evolved, successful case studies reveal key characteristics of pollution prevention, which contrast sharply with this concept of pollution control, the dominant environmental policy for the past two decades worldwide. Since a number of pollution control measures apply only after wastes have been generated, they cannot be called pollution prevention. Pollution prevention does not include any practice which alters the physical, chemical, or biological characteristics or the volume of a hazardous substance, pollutant, or contaminant through a process or activity which itself is not integral to and necessary for the production of a product or the providing of service. Legally, pollution prevention does not include treating waste, recycling off-site, concentrating hazardous or toxic constituents to reduce volume, diluting constituents to reduce hazards or toxicity, and transferring hazardous or toxic constituents from one environmental medium to another. Recycling, energy recovery, treatment, and disposal are not included within the definition of pollution prevention. Some practices commonly described as "in-process recycling" may qualify as pollution prevention. Recycling that is conducted in an environmentally sound manner shares many of the advantages of prevention – it can reduce the need for treatment or disposal, and conserve energy and resources.

2.3
What is industrial pollution prevention?

Industrial pollution prevention is only one sector of environmental pollution prevention. Other sectors of pollution prevention includes agriculture, commerce, mining, construction, transport, energy, and consumers as discussed previously. Nevertheless, industrial section is the most important sector of all other sectors, simply because of the ever-increasing quantity and toxicity of many industrial wastes and products that will be discussed in Chapter 3.

Industrial pollution prevention analyzes product-life-cycle impacts to the environment and considers all aspects of product design, manufacturing, consumption, and then through recycling, reuse, or disposal of discarded products. The benefits of industrial pollution prevention will be highlighted in the next section.

2.4
Benefits of pollution prevention

In the late 1980s, U.S. legislation has increasingly provided economic incentives to industries to prevent the generation of wastes. First, costs of pollution control, cleanup, and liability rose. Second, costs of resource inputs, energy and raw materials also rose, further encouraging their efficient use. Finally, the public increasingly pressured industries to decrease pollution, evidenced by new requirements under Section 313 of the Superfund Amendments, requiring industries to report annual releases of toxic chemicals (USEPA, 1989). Studies have shown that these factors have all worked to elevate the priority of pollution prevention in the industrial sector.

Pollution prevention helps national environmental goals and coincides with industry's interests. Businesses will have strong economic incentives to reduce the toxicity and sheer volume of the waste they generate. Reducing wastes provides upstream benefits because it reduces ecological damage from raw material extraction and pollutants releasing during the production process and during waste recycling, treatment, and disposal operations. A company with an effective, ongoing pollution prevention plan, may well be the lowest-cost producer, and as a result, have a significant competitive edge. Furthermore, costs per unit produced will drop as pollution prevention measures reduce liability risks and operating costs, pollution prevention measures will enhance the company's public image, public health, and overall environmental benefits (USEPA, 1992).

2.4.1
Reduced risk of liability

Industry can decrease the risk of both civil and criminal liabilities by reducing the volume and the potential toxicity of the gaseous, liquid, and solid wastes released into the environment. Industry should look at both hazardous and non-hazardous wastes. Since toxicity definitions and regulations change, reducing the total volume of wastes, including even non-hazardous wastes, is a sound long-term management policy for the following reasons (USEPA, 1992):

(1) Environmental regulations require that industrial plants document waste releases that must comply with emission/effluent standards. Companies that produce excessive waste risk face heavy fines, while their managers may be subject to fines and imprisonment for mismanaged potential pollutants.

(2) Civil liability increases for generating hazardous waste and other potential pollutants, because waste handling affects public health and property values in the communities surrounding production and disposal sites. Even if current waste regulations do not cover certain materials, they may in the future present a risk for civil litigation.

(3) Workers' compensation costs and risks directly relate to the volume of waste materials produced; not only hazardous waste, but also non-hazardous waste.

2.4.2
Cost savings

An effective pollution prevention program can yield cost savings that will more than offset program development and implementation costs. Cost reductions may involve immediate savings that appear directly on the balance sheet, or they may involve anticipated savings in terms of avoiding potential future costs. Cost savings are particularly noticeable when the costs result from the treatment, storage, or services that produce the waste, such as:

(1) Materials costs can be reduced by adopting production and packaging procedures that consume fewer resources, thereby creating less waste. As wastes

are reduced, the percentage of raw materials converted to finished products increases, with a proportional decrease in materials costs.

(2) Waste management and disposal costs are an obvious and readily measured potential savings to be realized from pollution prevention. Environmental regulations mandate special in-plant handling procedures and specific treatment and disposal methods for toxic wastes. The costs of complying with regulatory requirements and reporting on waste disposition are direct costs to businesses. Higher taxes for public services like landfill management, represent indirect costs to businesses. These costs will continue to increase at higher rates, but pollution prevention will reduce waste management costs.

(3) A pollution prevention assessment can find out where to reduce production costs. When a multi-disciplinary group examines production processes, it may uncover unrealized opportunities to increase efficiency. Production scheduling, material handling, inventory control, and equipment maintenance can be optimized to reduce all types of waste and production costs.

(4) Pollution prevention will lower energy costs in various production lines. A thorough assessment of how various operations interact will also reduce the energy used to operate the overall factory.

(5) Plant cleanup costs may result from complying with future regulations; or selling production facilities, off-site waste storage, or disposal sites. Pollution prevention will minimize these future costs because less waste would be generated.

USEPA conducted various case studies to answer the question, "Why do so few firms appear to find it cost-effective to adopt a prevention strategy if pollution prevention pays?" The case studies highlighted two pollution prevention projects among many others: a white water and fiber reuse project at a coated fine paper mill, and a conversion from solvent/heavy metal-free coating at a paper coating mill. Researchers compared a typical "company analysis" which contains costs typically accounted for by the firm, with a total cost assessment of the same project, in which a full accounting was made of less tangible, longer term, and indirect costs and savings.

In both studies, the total cost assessment approach showed markedly different results, in terms of estimating net present value of the projects, the internal rate of return on investment, and the simple payback for the capital expenditure. For each of feasibility analyses, the total cost assessment approach makes the pollution prevention project a far better investment than conventional direct feasibility analysis would indicate. In conclusion, the studies suggest that in some cases the expense of preparing the total cost assessment itself may be prohibitive, total cost assessment would still serve as a valuable tool for translating discretionary judgments into concrete dollar values during the capital budgeting process.

The realization of the benefits of pollution prevention must be based on a recognition that regulatory programs and activities could introduce to the environmental incentives to encourage voluntary action, while providing for improvements in environmental protection, including: (1) long term commitment; (2) comprehensive environmental accounting; (3) environmental capability of regulated parties varies widely; (4) government constraints limit pollution prevention; and (5) regulatory stimuli are needed to promote innovation and diffusion (USEPA, 1997).

2.4.3
Improved company image

As society increasingly emphasizes environmental quality, a company's policies and practices for controlling waste will increasingly influence employees and community attitudes.

(1) Employees will feel positive toward their company when management provides a safe work environment and acts as a responsible member of the community. By participating in pollution prevention activities, employees can interact positively with each other and with management. Helping to implement and maintain a pollution prevention program will increase their sense of identity with company goals. This positive atmosphere will retain a competitive workforce and attract new highly-qualified employees.

(2) The community will endorse companies that operate and publicize a thorough pollution prevention program. Most communities actively resist the siting of new waste disposal facilities in their areas. In addition, they are growing more conscious of the monetary costs of treatment and disposal. When a company creates environmentally compatible products and avoids excessive consumption and discharge of material and energy resources, rather than concentrate solely on treatment and disposal, it will greatly enhance its image within the community and toward potential customers.

2.4.4
Public health and environmental benefits

It should be noted that much of our regulatory standards in specifying acceptable limits for specific pollutants rely on hypothetical extrapolations of laboratory data of health risks. The laboratory results when translated to public health with average values and specific conditions or factors, we are making hypothesis and general assumptions. The use of hypothetical arguments as the basis for restrictive regulatory action is acceptable only when represents a justified threat to public health. However, these hypotheses and limits must be readily modified if new evidence or data so demand.

Public health and environmental benefits of pollution prevention can be evaluated by toxic release inventory and material accounting survey. Such inventory and survey provide data to indicate primarily the progress of pollution prevention by calculating the reduction of material use for production processes and reduction of pollutants entering into the environment. Detailed discussion of toxic release inventory and material accounting survey will be presented in Chapter 9.

Among all the benefits, the economic benefits of pollution prevention have proven to be the most compelling argument for industry and business to undertake prevention projects. Cost savings from pollution prevention come not only from avoiding environmental costs like hazardous waste disposal fees, but also from avoiding costs that are often more challenging to count, like those resulting from injuries to workers and ensuing losses in productivity. In that sense, prevention is not only an environmental activity, but also a tool to promote worker health and safety (Shen, 1997a).

2.5
Potential pollution prevention barriers

This Section analyzes the current concerns and barriers of industrial pollution prevention activities with illustrative examples. We must recognize potential barriers to pollution prevention. The purpose is that barriers can be overcome if they are clearly understood and analyzed. Five broad categories of pollution prevention barriers are identified for discussion: mental, technical, financial, regulatory, and institutional. They are closely inter-related. The purpose of this section is to provide information and suggestions as how to avoid or reduce the identified barriers for efficient implementation of industrial pollution prevention programs.

The recognition of the hazards, liabilities, and costs associated with toxic chemical and hazardous waste is relatively new to the industrial community. Industrial operations traditionally have adopted a variety of waste reduction techniques to lower costs of production and to increase profits. However, only in recent years, economical incentives and the corresponding emphasis on prevention of toxic chemical production and hazardous waste generation as a management priority have grown more rapidly. Many companies initiate only portions of a pollution prevention program, undertaking some specific projects, but then finding it difficult to maintain momentum. Often this is because there is not full commitment to the program. Unless pollution prevention is strongly supported by decision-makers and integrated into the company's business practices, early successes with pollution prevention projects may not be enough to build and maintain a vigorous pollution prevention program.

For industry, the most important part of the Pollution Prevention Act has been to collect source reduction and recycling data. Facilities that file actual toxic chemical release forms must now file information on past and future source reduction and recycling. The USEPA's Pollution Prevention Strategy is designed to accomplish two main goals: (1) to provide guidance and focus for current and future efforts to incorporate pollution prevention principles and programs in existing USEPA regulatory and non-regulatory programs, and (2) to set forth a program that will achieve specific pollution prevention objectives within a set, reasonable time frame. The first goal has caused some pollution prevention barriers such as organizational structures, budget allocation, implementation processes, and full commitment to a complete environmental pollution program.

Although various government incentives for pollution prevention are available, sometimes implementation can be inhibited by barriers. Exact circumstances differ from facility to facility, some of the most common barriers identified include lack of technical information, lack of funds, and lack of management support. Fears of future regulations, shortage of skilled staff, and reluctance to change are other barriers that have also been observed. This section identifies potential barriers while implementing pollution prevention programs and suggests how to avoid the barriers (Shen, 1997b).

2.5.1
Mental barrier

The first and most important barrier of all to pollution prevention is mental resistance to change without even thinking. Many like the old ways of doing things and there is often an attitude of "If isn't broken, don't fix it." These mental attitudes and fear of new approaches must be overcome before effective pollution prevention can be implemented. For example, our environmental programs are too much emphasis on managing pollution from "industrial waste" and neglecting pollution from "industrial products and services." The reason is that people do not have the opportunity or desire to learn the three generations of industrial pollution problems which are critical to facilitate implementation of pollution prevention programs. The first generation pollution problem is from the manufacturing facilities; the second generation pollution problem is related to the use of the products after leaving the manufacturing facilities; and the final disposal of the used or unused products represents the third generation pollution problem for industry. Government and industry must consider preventing pollution problems from not only waste generation, but also industrial products and services as well.

There are still many people who do not like to learn new ideas or do not care for pollution prevention program that can help national environmental goals and coincide with industry's interests. They have no knowledge of pollution prevention that will have strong economic incentives for industry to reduce the toxicity and volume of the waste they generate. They do not know that a company with an effective, ongoing pollution prevention program, may be the lowest-cost producer, and as a result, have a significant competitive edge. Furthermore, they have no idea about pollution prevention measures will enhance the company's public image, public heath, liability, and overall environmental benefits.

Every industrial activity is linked to hundreds of other transactions and activities that can result environmental impacts. For example, a large manufacturing industry may have hundreds of suppliers. The industry may actually manufacture hundreds of different products for a wide variety of customers, with different needs and cultural characteristics. Accordingly, each customer may use the product very differently, which is a big consideration if the use and maintenance of the product may be a source of potential environmental impact. When the product is finally disposed, the product may end up in a landfill, an incinerator, on the road side, or in a river that supplies drinking water to a small community. Another example, the success of market-driven pollution prevention initiatives depends in large part on consumer awareness and knowledge of pollution prevention issues. To use the product and service market effectively as a pollution prevention tool, there must be some assurance that pollution prevention benefit claims made about processes and products are truthful and result in real environmental quality improvements. The recent rush to market green products has achieved some success because of environmentally conscious consumers who translate their environmental values and fears into purchase decisions.

The mental change required for pollution prevention needs new knowledge and education, with creative and probing personalities. It is no longer acceptable to rest on environmental experiences gained decades ago in wastewater treatment or air pollution control systems. The future factory will consider the environmental implications of a new product or service. It will design pollution prevention into its research experiments. It will insist on pollution prevention in the acquisition of new components or systems from others so that waste generation in the life cycle of its product will be minimized. The future factory will constantly use preventive techniques to reduce wastes and seek solutions. It is critical that people must know what to do. If mental changes are required, the responsibility for change rests on management and with adequate pollution prevention education and training.

2.5.2
Technical barrier

Current pollution prevention technology generally employs conventional engineering approaches. Even as priority shifts from waste treatment and pollution control to waste minimization and source reduction, engineers will still employ available technology at first to achieve their objectives. In the future, however, designs will become more innovative for the environment. Industrial preventive technologies can be generalized into five groups: Improved plant operations, in-process recycling, process modification, materials and product substitutions, and materials separations. Pollution prevention technologies will be further discussed in Chapter 4. Information will be needed on alternative and preventive procedures that should be considered, how to integrate them in the production process, and what side effects are possible. This is because businesses and industry often need further information and technical resources in order to overcome barriers to adopting pollution prevention practices.

Industrial operations traditionally have adopted a variety of waste treatment and waste reduction techniques to lower costs of production and to increase profits. However, only in recent years, economical incentives and the corresponding emphasis on prevention as a management priority have grown more rapidly. Product quality or customer acceptance concerns might cause resistance to change. Potential product quality degradation can be avoided by contacting customers and verifying customer needs, testing the new process or product, and increasing quality control during manufacture.

A fundamental change of industrial processes is never easy, especially where it also involves a major change in critical raw material input to the manufacturing process. All of this leads to the need for formal project and facility analyses, capture and identify costs, benefits, uncertainties, risks, schedules, and relationships to other company plans and programs, such as R&D, expansion, diversification, and marketing of new products. Without formal analyses, people pay incorrectly conclude that they have exhausted their pollution prevention opportunities or that the costs of implementing pollution prevention are too high, or they may even pursue projects which are either technically, economically, or environmentally ill advised.

Limited flexibility in the manufacturing process may pose another technical barrier. A proposed pollution prevention option may involve modifying the work flow or the product, or installing new equipment; implementation could require a production shutdown, with loss of production time. The new operation might not work as expected or might create a bottleneck that could slow production. In addition, the production facility might not have space for pollution prevention equipment. These technical barriers can be overcome by having design and production personnel take part in the planning process and by using tested technology or setting up pilot operations.

For the past several years, new preventive technologies are being developed to cope with changing waste characteristics, such as focusing on industrial toxics, energy efficiency, and system approach. Industry needs to establish or strengthen information network and to encourage employees to watch for information in the technical journals, newsletters, and computer Internet. Sources of pollution prevention information will be highlighted in Chapter 16. However, information resources could be a problem for small- or medium-sized industries and business. They may not have ready access to central source of information on pollution prevention techniques. Government must help provide technical assistance and specific pollution prevention information for small- and medium-sized industries and businesses.

2.5.3
Financial barrier

Limited financial resources for pollution prevention programs could be a barrier, even for options that will ultimately be profitable. Financial barriers include:

inaccurate market signals; lack of a full accounting system to illustrate pollution prevention benefits; incomplete cost/benefit analysis; inappropriately short time horizons; fear of market share loss and/or consumer pressure; limited access to necessary resources; and worker fear of job loss. New patterns of value systems, accounting system and effective communication with appropriate feedback and interaction may reduce those barriers. Information needs to be shared and employees need to know that sharing this information is crucial in implementing pollution prevention.

The lack of immediately obvious economic benefits is a common barrier even though such benefits exist. Saving and making money is an immutable motivation for all people. But because the true cost of waste have either been at the end of the pipe, instead of upstream where wastes are produced, and because many costs have been externalized to society as a whole, workers in industry may not see the economic benefits of pollution prevention. Industrial managers have learned that changing accounting procedures to allocate waste costs to profit centers where wastes are produced is critical. Also, providing incentives such as rewards for successful efforts is a powerful tool for workers who has reduced generation of wastes and cut costs for the company.

Cost-benefit analysis procedures for pollution prevention project should be defined. Many proposed pollution prevention options will have to analyze costs. For example, additional or replacement equipment may need to be purchased, staff training may be required, or alternative raw materials may cost more. Some of these additional costs can be justified readily because they clearly will be cost-effective and will have short payback times. However, many will not be so clear-cut and will need more sophisticated analysis. We need to help establish environment value and full accounting systems that can clearly illustrate economic benefits of pollution prevention.

2.5.4
Regulatory barrier

Pollution prevention is defined as the "use of processes, materials or products that avoid, reduce or control pollution, which may include recycling, treatment, process changes, control mechanisms, efficient use of resources and material substitution." However, the major regulatory functions of environmental programs are still administered under media-specific commend-and-control regulatory structure. The exiting end-of-pipe pollution control regulations have caused delay in implementation of pollution prevention programs. These regulations, consumed a great portion of environmental management resources, have reduced short-term pollution, but have not really eliminated pollutants. These regulations might transfer pollutants from one medium to another. Regulatory inflexibility and uncertainty also can be barriers to some pollution prevention opportunities. For example, changing to another feed material in a manufacturing plant may require changing the existing permits. In addition, it may be necessary to learn what regulations might apply to proposed alternative input materials.

Incentives and rewards for pollution prevention generally can be an advantage to industry. Case studies have shown that implementation of pollution prevention within the industry is successful when employees are involved and receive incentives for their efforts in pollution prevention programs. The particular incentive mechanisms selected can crucially influence how these issues are addressed. Escalating regulatory and economic consequences have forced companies to consider establishing pollution prevention programs to eliminate more stringent future regulations and the limited treatment options. Incentives may be classified as economic benefits, enhanced public image and relations, regulatory compliance, and reduction in liability.

The use of price regulations (such as raw materials, goods and services) and green tax to achieve environmental objectives can provide strong incentives for technological innovation and behavioral change, and offer good prospects for achieving environmental objectives in a cost-effective manner, but the price should better reflect their full environmental and social costs. In contrast to earlier years, the relations between government and industry are now less confrontational and more cooperative. Industry should work with government agencies and other concerned people to identify and reduce legislative and regulatory barriers. Industry should also make positive constructive recommendations for change which can improve environmental quality and enhance industrial competitiveness. Working with the appropriate regulatory bodies early in the planning process will help overcome many regulatory barriers. It would be advisable to amend the existing laws, regulations, and implementation processes that reflect sustainable development policy, multimedia pollution prevention strategy, incentives, partnership approach, and also product life-cycle concept, value system, and a full cost accounting.

2.5.5
Institutional barrier

General resistance to change and friction among elements with the organization may arise. These can result from many factors, such as (1) lack of top management support; (2) lack of awareness of corporate goals and objectives; (3) lack of clear communication of priorities; (4) organizational structures separating environmental decisions; and (5) individual or organizational resistance to change. Poor internal communication can be result from (1) lack of involvement of affected workers; (2) reward system not focusing on pollution prevention; (3) firms lacking the technical ability to apply preventive methods; (4) lack of information about sources of waste releases, alternative strategies and resources; and (5) frequent changes to output, product design, and other factors making implementation more difficult.

Management is concerned with production costs, efficiency, productivity, return on investment, and present and future liability. Workers are concerned about job security, pay, and workplace health and safety. The extent to which these issues are addressed in the pollution prevention program will affect the success of the pollution prevention program. To gain the support of staff at all levels very early in the pollution prevention effort is vital to avoid institutional barriers. When a new pollution prevention program is implemented, a feedback mechanism for monitoring effectiveness and making corrections is essential. Continuous communication will reduce resistance to change. Resistance can be positive or negative and should not necessarily be treated a barrier.

The development and implementation of a pollution prevention program can involve a significant change in a company's traditional business practices. Whereas regulatory compliance has usually been the responsibility of an environmental department or coordinator, pollution prevention must involve employees from all areas of the plant and levels of management, including hourly workers. A pollution prevention program can also become the primary strategy a company uses to achieve and maintain regulatory compliance. In fact, it may allow companies to get out of the regulatory system altogether.

Lack of clarity about how to share leadership on various aspects of multimedia pollution prevention approached creates some confusion. For example, it is widely acknowledged that it is difficult to share data across offices, but it is not clear whose role it is to take the lead in resolving this issue. Lack of clarity about one's leadership role on prevention can result in little risk-taking and a reluctance to elevate pollution prevention policy issues. This, in turn, can limit the organization's creativity in finding ways to define the functional relationships among offices as a matrix that promotes cross-media and pollution prevention outcomes.

In summary, major barriers to implement pollution prevention programs are identified as mental, technical, financial, regulatory and institutional barriers. They are closely inter-related and can be avoided or reduced by: strengthening education and outreach programs; providing information and communication network; amending regulations and implementation processes; applying environment value and full costing systems; emphasizing incentives for pollution prevention; improving organizational structure and management system to ensure interaction and cooperative spirit as well as gaining the support of decision-makers and staff at all levels in very early stage of pollution prevention effort. While there is consensus about many of the changes needed to reduce barriers and increase incentives for preventing pollution, some decision-makers believe that much remains to be learned such as the regulatory amendments, institutional changes, preventive technologies, and integration of pollution prevention into the current regulatory and environmental management systems. We must deal with not only pollutants and wastes, but also products and services that create pollution problems. Re-education with adequate modern information system is the key(Shen, 1997b).

References

Bower BT (1990) "Economic, engineering and policy options for waste reduction," in: National Research Council, Committee on Opportunities in Applied Environmental Research and Development of the Waste Reduction Workshop Report, published by National Research Council, p. 3

Koenigsberger MD (1986) "For Pollution Prevention and Industrial Efficiency", presented at Governor's Conference on Pollution Prevention Pays, Nashville, TN, March 1986

NRC (1966) Waste Management and Control. National Academy of Sciences, National Research Council, published by National Academy Press, p. 3

Purcell AH (1990) "Pollution Prevention in the 21st Century: Lessons From Geneva". Proceedings of International Conference on Pollution Prevention, June 10–13, 1993, Washington, DC, p. 529

Shen TT and GH Sewell (1986) "Control of Toxic Pollutants Cycling", Proceeding of the International Symposium on Environmental Pollution and Toxicology, published by Hong Kong Baptist College in Hong Kong, September 9–11, 1986

Shen TT, CE Schmidt and TR Card (1993) Assessment and Control of VOC Emissions From Waste Treatment and Disposal Facilities, Chapter 1. Van Norstrand Reinhold Publisher, New York

Shen TT (1997a) "Industrial Pollution Prevention", presented at the 1997 International Chinese Sustainable Development Conference in Los Angeles, California, July 4–5. Proceedings prepared by the ITRI of Taiwan, 1997.

Shen TT (1997b) "Five Categories to Avoid Industrial Waste Minimization and Pollution Prevention Barriers", presented at the Asian-Pacific Conference on Industrial Waste Minimization and Sustainable Development in Taipei International Convention center, December 14–18, Taipei, Taiwan.

USEPA (1991) The National Report on Toxics in the Community – National and Local Perspectives. Chapter 5: Pollution Prevention. EPA 560/4-91-014, September 1991

USEPA (1992) Facility Pollution Prevention Guide. Office of R&D, Washington, D.C., EPA/600/R-92/088, May 1992

USEPA (1995). Industrial Pollution Prevention Project (IP3): Summary Report, EPA-820-R-95-007, July 1995. Also see: Incentives and Disincentives, EPA-820-R-94-004, August 1994.

USEPA (1997). Pollution Prevention 1997 – A National Progress Report, EPA-742-R-97-000, June 1997.

3
Toxic chemicals and processes wastes

Environmental issues, particularly concerned about toxic chemicals and hazardous wastes from industrial processes, are receiving more attention in policy,
program and media dimensions in both developed and developing countries.
Pollution from toxic chemicals and hazardous wastes is more extensive and difficult to manage than originally believed. Balancing industrial development and
environmental protection encourages a manage-for-results approach. When
toxic chemicals and hazardous wastes contaminate the environment, all life is
exposed to the potential high risks. There is evidence that such risks may also
cause subtle harm which does not show up until it is too late to prevent human
disease or ecological disruption. To add to the problem, significant amounts of
toxic pollutants are cycling and recycling in the environment not only from industrial sources, but also from industrial waste treatment and disposal activities
(Shen, 1991).

Many businesses do not know that certain chemicals they use are toxic and that
the amounts of toxic wastes releasing into the environment at different stages in
their life-cycle from extraction or manufacture to use and disposal. Nor have they
explored the opportunities for cleaner production and products. Public inventories begin to stimulate reduction of pollution at its source by providing information to anyone who needs it to make a decision. The data provided by these inventories will help the groups of people involved in choosing the materials we use and
help measure to what extent businesses are moving toward processes and products that avoid putting toxic chemicals into the environment in the first place.
Since toxic chemicals are moving all over the world, it is essential that national
governments and non-government organizations should work to establish and
improve the existing toxic chemical inventories in their own countries. They
should also cooperate through international organizations to develop software for
inventories and to make technical assistance available so developing country governments and citizens can adapt and use the software. (WWF, 1994).

Industrial pollution problems mostly result from mismanagement caused by
a varying of reasons: carelessness, indifference, or ignorance, as well as lack of
measurement and monitoring methods to provide baseline or background data
to serve as a technical basis for engineering prevention and control, and as verification of adverse effects of health and the environment.

3.1
Industrial processes

The industrial wastes (gas, liquid, and solid) are generated from many different processes. The amount and toxicity of waste releases vary with specific industrial processes. Figure 3-1 shows a typical industrial process which produces wastes containing different types of pollutants, depending on the input materials and process designs. Thus, process information is critical to make accurate and reliable assessment of the potential for pollution prevention. Generally, essential process information includes process description with flow diagram, operating materials, water and energy inputs, flow measurements, equipment list and specifications, hazardous waste manifests, emission inventories, permits, material safety data sheets, operating logs, and production schedules.

Industrial wastes from various types of processes can be prevented, including use, manufacture, or store raw materials and products, Five types of pollutant releases in all industrial processes are: (1) Releases from routine production operation and energy source; (2) Releases from limited process upset; (3) Releases from fugitive emissions; (4) Releases from accidents and mishandling; and (5) Releases from storage facilities.

In the United States, the Chemical and Allied Products industry accounted for the largest share of total toxic chemical releases and transfers in industrial processes as reported in toxic release inventory (TRI). The Chemical industry also had the largest number of facilities reporting to TRI. The TRI is considered the

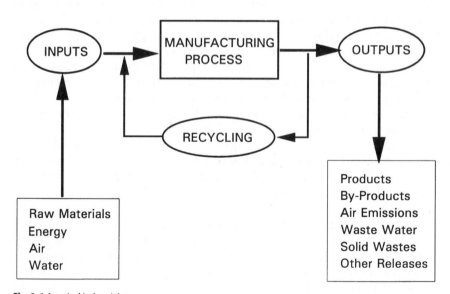

Fig. 3-1 A typical industrial process

most comprehensive national database on toxic chemicals released and transferred to all parts of the environment. It has affected public policy since its inception in 1986. Governmental officials have used TRI data to help set environmental priorities and to shape regulations. Perhaps the most important use of TRI data is in the development of pollution prevention initiatives. A number of state legislatures, motivated by TRI data, have passed pollution prevention laws. TRI data have also helped focus attention on facilities where pollution prevention efforts can be most effectively directed.

The chemical industry dominated TRI releases and transfers in the United States. It generated wastes more than 3.5 times greater than the second-ranked primary metals industry in 1989. Other reported TRI industrial processes, in order of their waste quantity, including plastics, transportation, paper, measure/photo, electrical, fabricated materials, stone/clay, machinery, petroleum, lumber, miscellaneous, textiles, printing, furniture, food, leather apparel, and tobacco (more discussion in Section 3.4). Many small- to medium-sized industrial processes controlled their toxic chemicals and wastes by installing treatment and disposal systems, but they began to prevent pollution in their facilities. These processes may be grouped in three categories:

1. General such as aqueous cleaning, solvent cleaning, disinfecting, sterilizing, venting, disposable, and off-spec materials;
2. Fermentation such as fermenter and process in formation; and
3. Chemical synthesis, natural product extraction, formulation, including solvent-based processes, and aqueous-based processes.

Waste information is critical to pollution prevention. This is especially true for small- and medium-sized facilities, including material handling and process operations. Waste from material handling consists of off-spec materials; obsolete raw materials; spills and leaks; paper, plastic, and metal disposals; pipeline and tank drainage; laboratory wastes; evaporative losses; and empty container cleaning. Waste from process operations include tank cleaning, container cleaning, blender cleaning, and process equipment cleaning.

For hazardous air pollutants (HAPs) from industrial processes, they take the form of vapor or particulate. Specific source categories are divided into nine general classifications listed in the USEPA's HAPs Control Technology Document (USEPA, 1991):

(1) Solvent usage operations include processes dependent on solvents, such as surface coating and dry cleaning operations.

(2) Metallurgical industries include processes associated with the manufacture of metals, such as primary aluminum production.

(3) Synthetic organic chemical manufacturing processes include operations associated with the manufacture of organic chemicals.

(4) Inorganic chemical manufacturing processes include operations associated with the manufacture of inorganic chemicals.

(5) Chemical products industries include industries using chemicals in the formulation of products.

(6) Mineral and wood products industries include operations such as asphalt batch plants and kraft pulp mills.

(7) Non-metallic mineral products industries involve the processing and production of various non-metallic minerals. The industry includes cement production, coal cleaning and conversion, glass and glass fiber manufacture, lime manufacture, phosphate rock and taconite ore processing, as well as various other manufacturing processes.

(8) Petroleum related industries include oil and gas production, petroleum refining, and basic petrochemical production.

(9) Combustion sources include a large number of combustion units generally used to produce electricity, hot water, and process stream for industrial plants; or to provide space heating for industrial, commercial, or residential buildings. They include all utility, industrial, and residential combustion sources using coal, oil, gas, wood, or waste-derived fuels.

Although fugitive emissions are listed as a separate group, they can occur from storage and handling, reactor processes, and separation processes. Area fugitive sources include groups of valves, pressure relief devices, pumps and compressors, cooling towers, open-ended lines, and sampling systems. Process fugitive sources include hotwells, accumulators, and process drains from reactors, product recovery devices, and separation equipment. The emissions include:

(1) Volatile organic compounds (VOCs) from process vents, wastewater treatment units, and as fugitive emissions.

(2) Inorganic emissions from reaction systems, such as carbon monoxide, chlorine, hydrogen chloride, hydrogen fluoride, etc.; as well as from combustion sources, which would include sulfur dioxide, nitrogen oxides, etc.

(3) Particulate and dusts from combustion sources, storage piles, and solid products such as detergent granules, etc. (Shen et al., 1993).

3.2
Industrial wastes

Wastes may be classified into industrial, municipal, agricultural, and mining wastes, but this chapter covers only industrial wastes. Industrial wastes can be grouped into organic or inorganic wastes; and hazardous or non-hazardous wastes, depending upon their physical and chemical properties of the waste. There is no exact definition of industrial waste. Generally, the term is loosely used to signify a material stream without value to the industrial process concerned. This definition does not imply that the material in the waste stream would be totally worthless as such. On the contrary, many industrial waste streams are, or could be, valuable raw materials for other industrial processes. Problems tie in the rate of waste production, the means of transport, and very often, the lack of incentive because of negative attitudes or economic reasons. The purely technical problems, such as impurities or the form of the waste, often pose less significant barriers that can generally be overcome.

Hazardous waste may be defined as those wastes, because of its quantity, concentration, or physical or chemical characteristics, may cause or contribute to an increase in mortality or in serious irreversible, or incapacitating reversible illness; or pose a substantial present or potential hazard to human health or the environment when improperly treated, stored, transported, disposed of, or otherwise managed. Hazardous waste presents a real danger to human health and the environment. Its formation constitutes a problem of its own. Past neglect, improper waste deposition, and abandoned waste-dump sites containing hazardous wastes currently cause major pollution problems in many countries. Because of these problems, most industrialized countries have adopted laws regulating hazardous waste management.

In the United States, the 1976 Resource Conservation and Recovery Act (RCRA) has narrowly defined hazardous waste: "a solid waste, or combination of solid wastes, which because of its quantity, concentration, or physical, chemical, or infectious characteristics, may (1) cause, or significantly contribute to an increase in mortality or an increase in serious irreversible, or incapacitating reversible illness; or (2) pose a substantial present or potential hazard to human health or the environment when improperly treated, stored, transported, or disposed of, or otherwise managed." Hazardous wastes in a broader sense should include chemical, biological, flammable, explosive, and radioactive substances, which may all be in a solid, liquid, sludge, or contained gaseous state. There are also inconsistent definitions of hazardous waste under various federal laws such as the Water Pollution Control Act Amendment of 1972, the Hazardous materials Transportation Act of 1974, Toxic Substance Control Act of 1976, and the Solid Waste Disposal Act Amendment in 1976. To clarify the definition of hazardous waste, the U.S. Environmental Protection Agency (USEPA) lists four characteristics of hazardous Waste (USEPA, 1986):

(1) Ignitability, which identifies wastes that pose a fire hazard during routine management. Fires not only present immediate dangers of heat and smoke, but also can spread harmful particles and gases over wide areas.

(2) Corrosivity, which identifies wastes that require special containers because their ability to corrode standard materials, or requiring segregation from other wastes because of their ability to dissolve toxic contaminants.

(3) Reactivity (or explosiveness), which identifies wastes that, during routine management, tend to react spontaneously, react vigorously with air or water, are unstable to shock or heat, generate toxic gases, or explode.

(4) Toxicity, which identifies wastes that, when improperly managed, may release toxicants in sufficient quantities so as to pose a substantial hazard to human health or the environment. Toxic wastes are harmful and even fatal when ingested or absorbed. When toxic wastes are disposed of on land, contaminated liquid may drain (leach) from the waste and pollute groundwater. Toxicity is identified through a laboratory procedure called the Toxicity Characteristics Leaching Procedure (TCLP).

Today, although only one fifth of the earth's population enjoys the fruits of industrialization, we are already seeing the limitations of growth, depletion of natural resources, and degradation of our environmental quality. Some of the industrial development that have so successfully contributed to the economic improvement must be questioned. Presently, our knowledge of industrial waste and about environmental threats seems to grow faster than our ability cure them. Many of the consequences of overexploitation of the earth and generation of large amount of waste are resulting in global effects that threaten our future rather than in local or regional problems. In the United States, however, the industrial toxic releases to air, water, land as well as toxic transfers to public-owned treatment works and off-site locations have been decreased since 1987 (USEPA, 1992).

U.S. environmental laws regulate the generation, transport, storage, treatment, and disposal of hazardous wastes. These laws include the Resource Conservation and Recovery Act (RCRA); the Hazardous and Solid Waste Amendments of 1984; the Comprehensive Environmental Response, Compensation, and Liability Act (CERCLA) of 1980; the Superfund Amendments and Reauthorization Act (SARA) of 1986 including the Emergency Planning and Community Right-to-Know Act; the Clean Air Act Amendment of 1990; the Hazardous Materials Transportation Uniform Safety Act of 1990; and the Pollution Prevention Act of 1990. The Comprehensive Environmental Response, Compensation, and Liability Act (CERCLA) created the Superfund program to respond to a release or threatened release of hazardous substances, pollutants, and contaminants stemming from accidents or uncontrolled hazardous waste sites. Under the law, those responsible for the contamination are required to conduct the cleanup. Where enforcement is not successful, the federal government can clean up a site using the CERCLA Trust Fund (Superfund), which is supported by excise taxed on feedstock chemicals and petroleum and by a more broadly based corporate environmental tax.

Through November 1992, USEPA had identified 37592 potentially hazardous waste sites across the nation. Of these, 93% have undergone a preliminary assessment to determine the need for further action. Half of the sites was classified as no further federal action, as they could be handled by states and private parties. The EPA listed most serious sites on the Superfund National Priorities List (NPL), which consisted of 1252 sites. NPL listing made these sites eligible for Superfund cleanups. In fiscal 1992, permanent cleanup construction was completed at 86 sites – a total equaling more than the number completed in the first 11 years of Superfund. EPA targets call for 650 NPL sites to be cleaned up by the end of the year 2000. The expenditures associated with Superfund are forecast by EPA to increase from $2 billion per year (in annualized 1986 dollars) to over $8 billion per year in 2000 (CEQ, 1991 and 1992).

The TRI data collected and published annually demonstrate a steady decline in the volume of toxic chemicals released to the environment by the manufacturing sector. However, over the last several years, the total amount of wastes generated has been rising. TRI data for 1995 show a decline of 4.9% in releases of core chemicals reported in both 1994 and 1995. Overall, from the baseline year of 1988 until 1995, total releases (for chemicals reported in each of the years) decreased by 1.35 billion pounds, a 45.6% decline. However, total production-related waste generated in 1995 from all TRI chemicals was over 35 billion pounds, a 6.8% increase since 1991. The clearest measure of industrial pollutants in the manufacturing sector can be found in companies' annual reports of environmental releases of toxic chemicals to TRI. Correspondingly, one of the clearest indicators of corporate responsiveness to the need for reducing chemical releases and preventing pollution has been a company's participation in EPA's voluntary programs.

Although hazardous waste regulations differ from country to country, there are basic similarities: the scope of application is indicated by giving general information about the waste or by listing separately the wastes considered to be hazardous, using criteria such as the origin of the waste, toxic substances present, and their properties. There is still no consensus on a generally accepted list of toxic substances, but such a project has been undertaken. Many countries rely on the principle that the generator of the waste is responsible. Generally, the results in various procedures to ensure a cradle-to-grave management of the hazardous waste is an extended control from its generation to its proper disposal. Such legislation normally specifies that the waste generators must ensure that the waste produced is properly transported and disposed of, even if these tasks must be subcontracted out to someone else.

In the European Economic Community (EEC) as a whole, the total industrial and hazardous wastes released into the environment was estimated about 180 million tons in 1982. In 1988, the estimated quantities of industrial waste and hazardous waste produced by eight countries in EEC were 185.1 million tons and 20.2 million tons respectively (see Table 3-1).

Table 3-1 Estimated industrial and hazardous wastes released in the European Economic Community

Country	Industrial waste	Hazardous waste
Belgium	13.0	2.5
Denmark	0.8	0.1
Germany	52.0	4.5
France	38.0	2.0
Ireland	1.2	0.075
Italy	35.0	5.0
Netherlands	5.1	1.0
United Kingdom	40.0	5.0
Total (in million tons)	185.1	20.2

Source: Junger and Lefevre, 1988.

The first European Economic Community Directive on Toxic and Dangerous Waste was implemented in 1980. The Organization for Economic Cooperation and Development (OECD) adopted its first Decision on the Export of Hazardous Waste from OECD countries in 1986. The United Nations Environment Program (UNEP) in 1987 adopted the Cairo Guidelines and Principles for the Environmentally Sound Management of Hazardous Wastes. Many countries, however, still lack the legal framework to deal properly with hazardous wastes. Much current research and development activity is devoted to efforts aiming to reduce and reuse industrial waste streams. With the rising cost of waste handling, most companies have significantly decreased their industrial waste because of economic necessity. Virtually the only remaining option in industrial production is to decrease the formation of waste streams at the source by internal modification of process streams. These procedures may involve a change of raw materials, as well as ultimate redesigning of the product itself. Various industries have demonstrated that new preventive technologies can improve the environment (Johansson, 1992). Figure 3-2 shows a waste management diagram that illustrates increasing effectiveness of waste management.

3.3
Toxic chemicals

As many as 100,000 chemical substances are used commercially around the world, and for many produced in very high volumes – 10,000 metric tons or more annually – the public can get no toxicity data in most countries. The public even do not know the toxic chemicals in their houses. For example, some common household hazardous chemicals are such as:

- Hair dyes and bleaches are highly toxic if ingested.
- Ammonia is toxic and an irritant if inhaled or skin contacted.
- Drano contains sodium hydroxide.

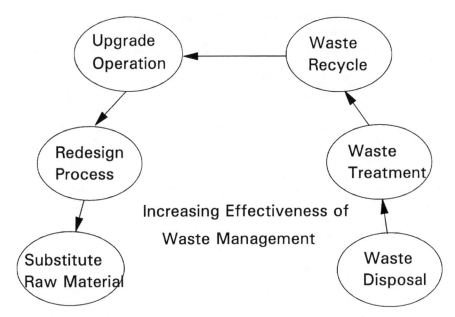

Fig. 3-2 A waste management diagram

- Fingernail polish remover contains acetate solvents.
- Polyurethane wood finish contains isocyanates.
- Fertilizer causes nitrate poisoning.
- Pesticides can be lethal to all living organisms.
- Others include carburetor cleaner, paint thinner, glues or expoxides, varnish, anti-freezer, brake fluid, rust remover, and lead paint.
- In addition, some common industrial hazardous chemicals are:
  - Chlorinated solvents such as trichloroethane.
- Non-chlorinated solvents such as benzene, acetone.
- Waste oils such as hydraulic oils, cutting oils and machine lubricants.
- Acids and alkalis such as sulfuric acid, caustic soda, lime.
- Developers, fixers, and inks contains formaldehyde, cyanide, acetic acid from textile manufacturing, print and photo and X-ray processing.
- PCB's contain polychlorinated biphenyls from insulating equipment.
- Asbestos used for insulting ducts and buildings.
- Paint wastes and lacquers from metal maintenance manufacturing processes.
- Resins such as urea formaldehyde, epoxy from chemical, rubber and plastic manufacturing.

In a long-needed step, the major chemical manufacturers in developed countries have agreed to share the responsibility for getting toxicity information. For example, the harmonization program of the European Community (EC) requires participating countries to negotiate common baseline standards, but not ceiling. The materials policies of the future are international in scope. Thus, more aggressive coordination across national boundaries will be essential (Geiser, 1993). International treaties and trade agreements become increasingly important vehicles for planning, guiding, and guaranteeing global economic integration that incorporates environmental considerations. The uses of chemicals already known to present hazards must be reduced and the manufacturing of substitutes for these chemical uses must be encouraged in both the developed and developing countries (Geiser and Irwin, 1993).

EPA compiles the quantities of toxic chemicals releasing to the environment and the amounts of waste managed on-site of transferred off-site for management elsewhere. EPA also asks questions of industry about source reduction, energy recovery, and treatment and disposal activities. By making toxic release inventory (TRI) data public, TRI provides a strong incentive for companies to reduce wastes. TRI milestone include:

- In 1994, EPA added 286 additional chemicals and chemical categories to the TRI, giving the public a broader picture of progress in preventing toxic waste generation and release, and began to collect toxic release data from federal agencies.
- In 1995, TRI data (announced in May 1997) showed that reported industrial releases declined by 45.6% (1.35 million pounds) from 1988. Of the 21,951 facilities, nearly 29% reported implementing at least one source reduction activity.
- In 1997, EPA expanded by 30% the number of industrial facilities required to report to TRI, to include the categories of metal mining, coal mining, electric utilities, commercial hazardous waste treatment, petroleum build terminals, chemical wholesalers, and solvent recovery services. In addition, 700 chemical manufacturing facilities which already report right-to-know information to the TRI, will also be required to report on additional types of pollution, such as hazardous waste treatment activities.

For obtaining TRI reports and data, readers may contact EPA's EPCRA Hotline 800-555-0202 or NCEPI 800-490-9198. Readers may also find data on Internet at ftp://ftp.epa.gov; or gopher://gopher.epa.gov; or http://www.epa.gov/opptin-tr/tri.

Although prevention can address many pollution problems, the U.S. emphasizes the reduction of industrial toxic chemicals. Major industrial sources of toxic releases and transfers in the order of quantity include chemicals, primary metals, paper, transportation, fabricated metals, plastics, electrical, petroleum, printing food, furniture, textile, machinery, printing, leather, photography, and stone/clay. Table 3-2 presents TRI facilities and forms by industry. In the U.S., the chemicals and allied products industry accounted for the largest share of total toxic waste generation. According to the U.S. TRI, their Toxic waste generated in 1988 and 1989, contributing 48% of the total. Facilities in the chemical industry reported TRI wastes totaling 2.8 billion pounds, more than three times the TRI wastes of the second-ranked industry, the primary metal industries, which accounted for 756.8 million pounds. In contrast to these high-impact industries, which played a large role in TRI totals, the eight industrial categories with the smallest among of toxic waste generation, each represented less than 1% of the TRI total as shown in Table 3-3 TRI Releases and Transfers by major U.S. industry, 1987–1989 (CEQ, 1991).

Individual industries have had very different experiences with TRI chemical releases and reductions. As Table 3-4 shows, several industries reported reductions of half of more to total releases since 1988, led by the electrical equipment industry (79.7%) and leather goods manufactures (77.8%). In 1995, the chemical manufacturing industry continued to rank in first place with the largest amount of chemicals released (36% of total releases), followed by the primary metals industry (15%), paper (11%), and plastics (5%) (USEPA, 1997).

The overall TRI patterns of chemical, environmental, and industrial distribution in the United States were deeply influenced by a small number of individual facilities reporting TRI wastes significantly larger than those reported by all others. The top ten parent companies operated a total of 410 facilities, or 2% of all TRI facilities, and accounted for a little more than one-fourth of all TRI wastes in 1989. With 88 facilities, Du Pont was the top-ranked parent company, reporting TRI wastes of 349 million pounds, 6% of the national total. Other top-ranked parent companies included Monsanto, American Cyanamid, BP America, and Renco Group as shown in Table 3-5. The Ten Parent Companies with the Largest TRI Releases and Transfers in 1989. The top 10 chemicals released into the environment (shown in Table 3-6) account for over half the total amount of releases of the expanded list of 643 TRI chemicals. The 10 compounds that reported the highest total releases of toxic chemicals in 1994 are shown in Table 3-7.

The TRI data collected and published annually demonstrate a steady decline in the volume of toxic chemicals released to the environment by the manufacturing sector. However, over the last several years, the total amount of wastes generated has been rising. TRI data for 1995 show a decline of 4.9% in releases of core chemicals reported in both 1994 and 1995. Overall, from the baseline year of 1988 until 1995, total releases (for chemicals reported in each of the years) decreased by 1.35 billion pounds, a 45.6% decline. However, total production-related waste generated in 1995 from all TRI chemicals was over 35 billion pounds, a 6.8% increase since 1991.

The parent companies operating the greatest number of TRI facilities did not necessarily report the largest TRI totals. For example, General Electric operated the largest number of facilities of any parent company, but its 138 facilities generated 38.9 million pounds of TRI wastes (an average of 28,1814 pounds per facility, well below the average for the top ten parent companies). The Cooper Industries, which operated 69 facilities, reported generating only 4.4 million pounds of wastes (an average of 63,403 pounds per facility). Altogether, the ten parent companies with the largest number of facilities accounted for 4% of TRI facilities, and contributed 11% to total TRI wastes in 1989 as shown in Table 3-8

Table 3-2 Toxic release inventory facilities and forms, 1989

TRI total rank	Industry	SIC code	Facilities	Forms	Air	Surface water	Land	Underground	Public sewage	Off-site total	Total releases and tranfers	%
			Number		Pounds							
11	Food	20	1,452	2,649	13,716,106	3,633,426	12,454,502	1,017,909	39,117,332	2,962,616	72,901,891	1.17
21	Tobacco	21	19	62	13,820,462	118,812	14,751	0	791,940	312,982	15,058,947	0.24
15	Textiles	22	401	890	35,617,602	4,693,758	191,896	0	15,274,101	3,547,714	59,325,071	0.95
22	Apparel	23	29	49	1,019,201	250	40,599	0	692,897	166,748	1,919,695	0.03
19	Lumber	24	616	1,719	27,813,245	101,014	198,009	0	952,592	4,303,183	33,368,043	0.53
13	Furniture	25	397	1,556	56,894,704	1,850	56,511	0	390,551	5,451,673	62,795,289	1.01
4	Paper	26	587	2,328	202,210,446	86,518,400	9,925,423	0	39,743,943	31,944,839	370,343,051	5.93
14	Printing	27	313	623	50,423,807	35,150	316	40,000	3,460,704	6,581,949	60,541,926	0.97
1	Chemicals	28	3,838	20,332	754,922,471	228,105,753	164,564,318	973,706,836	323,874,071	438,305,845	2,883,479,294	46.20
9	Petroleum	29	364	3,053	54,989,933	3,471,886	3,153,767	19,846,879	13,060,813	9,713,429	104,236,707	1.67
7	Plastics	30	1,293	3,213	158,832,600	733,598	155,662	3,004	4,957,837	24,443,360	189,126,061	3.03
17	Leather	31	132	372	14,255,347	680,505	353,215	0	20,340,615	2,176,552	37,806,234	0.61
16	Stone/clay	32	559	1,315	23,283,963	874,242	2,029,808	6,580,250	1,294,091	21,120,356	55,182,710	0.88
2	Primary metals	33	1,380	5,446	232,958,571	18,987,700	277,003,090	41,020,432	19,088,232	268,808,482	857,866,507	13.75
6	Fabricated metals	34	2,579	7,860	117.524.318	1,518,379	4,742,417	286,120	18,342,883	72,698,383	215,112,500	3.45
10	Machinery	35	870	2,315	51,754,691	548,021	257,368	52,800	2,993,295	19,725,887	75,332,062	1.21

8	Electrical	36	1,578	5,047	115,198,789	714,367	1,454,389	43,720	18,711,626	46,261,903	182,384,794	2.92
5	Transportation	37	1,054	4,447	201,297,144	331,889	1,256,795	21,384	7,502,020	50,770,863	261,180,095	4.18
12	Measure./photo.	38	344	1,000	45,076,520	685,863	490,993	25C	5,415,110	12,698,912	64,367,648	1.03
8	Miscellaneous	39	372	946	26,171,054	54,173	304,149	0	452,222	9,061,758	36,043,356	0.58
3	Mult, codes 20–39		1,303	5,133	217,850,529	8,547,184	80,320,516	172,724,324	30,813,971	69,259,997	579,516,521	9.29
20	No codes 20–39		282	776	11,938,600	1,238,018	2,588,388	0	3,280,462	4,096,876	23,142,344	0.37
	Total		19,762	71,131	2,427,570,103	361,594,238	561,556,882	1,215,343,908	570,551,308	1,104,414,307	6,241,030,746	100.0

*Toxics Release Inventory ranking by pounds of emissions, with 1=greatest volume of emissions.

**SIC=Standard Industrial Classification.

Source: U.S. Environmental Protection Agency, Toxics in the Community (Washington, DC: U S Government Printing Office, September 1990), pages 56–57.

(USEPA, 1991). During 1987–1990, the TRI shows a downward shift in total reported TRI releases and transfers into the environment as shown in Table 3-9(CEQ, 1992).

Table 3-3 TRI releases and transfers by industry, 1989

Industry	Total releases and transfers	Environmental distribution					
		Air	Surface water	Land	Under-ground	Public sewage	Off-site
	million lbs	% of total releases and transfers					
Food	67.80	27.77	4.15	6.41	1.63	55.26	4.79
Tobacco	1.49	93.14	3.04	0.10	0.00	1.11	2.61
Textiles	46.08	65.90	2.16	0.10	0.00	23.79	8.04
Apparel	2.06	65.78	0.01	0.02	0.00	21.42	12.76
Lumber	37.82	89.42	1.74	0.30	0.00	0.19	8.36
Furniture	65.37	92.51	0.00	0.03	0.00	0.93	6.62
Paper	313.25	61.94	13.46	3.17	0.00	14.68	6.75
Printing	60.92	90.85	0.01	0.00	0.00	1.23	7.91
Chemicals	2745.77	27.07	3.98	3.89	39.75	12.65	12.66
Petroleum	194.50	85.04	0.36	0.12	0.01	2.74	11.72
Plastics	194.50	85.04	0.36	0.12	0.01	2.74	11.72
Leather	24.86	49.83	0.92	1.03	0.00	37.15	11.06
Stone/clay	47.49	47.52	1.47	5.39	13.84	2.09	29.69
Primary metals	756.81	31.25	2.09	31.61	4.87	2.02	28.15
Fabr. metals	207.38	61.83	0.15	0.52	0.16	4.06	33.28
Machinery	74.92	72.30	0.53	0.41	0.00	3.79	22.96
Electrical	145.76	65.29	0.46	0.95	0.03	9.78	23.48
Transportation	245.32	80.50	0.05	0.07	0.00	3.18	16.20
Measure./photo.	69.54	67.64	0.62	0.14	0.00	4.36	27.24
Miscellaneous	38.89	64.74	0.09	0.14	0.00	1.20	33.83
Multiple codes	437.28	54.80	2.18	17.30	4.40	5.91	15.41
No codes	19.22	60.27	0.87	0.23	0.00	9.32	29.32
Total	5705.67	42.54	3.31	7.79	20.70	9.66	16.00

Source: U. S. Environmental Protection Agency, Office of Toxic Substances, Toxics in the Community: National and Local
Perspectives. The 1989 Toxics Release Inventory National Report, EPA 560/4-91-014, (Washington, DC: EPA, 1991).

Table 3-4 1995 toxic release inventory released by industry

Industry	Percentage change in releases, 1988–95	Total releases, 1995 (millions of pounds)
Electric equipment	−79.7%	30.5
Leather	−77.8%	3.1
Measure./photo.	−74.2%	16.9
Tobacco	−72.2%	1.7
Machinery	−67.6%	23.2
Textiles	−56.1%	17.8
Chemicals	−49.8%	787.7
Printing	−48.3%	31.6
Stone/clay/glass	−47.1%	36.0
Transportation	−44.4%	110.0
Petroleum	−40.6%	59.9
F'abricated metals	−40.1%	82.6
Primary metals	−38.2%	331.2
Furniture	−33.7%	41.0
Plastic	−31.1%	112.2
Food	−27.5%	86.0
Paper	−12.6%	233.2
l,umber	−5.0%	31.3
Apparel	+33.6%	1.3

Source: EPA, 1995 Toxics Release Inventory: Public Data Release
(EPA 745-R-97-005, April 1997), Tables 4-10, 5-5.

Table 3-5 Ten parent companies with the largest TRI total releases and transfers, 1989 (Source: USEPA, 1991)

TRI rank	Parent company	Facilities		Total releases and transfers		Average releases and transfers per facility
		Number	Percent	Pounds	Percent	Pounds
1	Du Pont	88	0.39	349,275,844	6.12	3,969,044
2	Monsanto Co.	33	0.15	293,833,577	5.15	8,904,048
3	American Cyanamid Co.	29	0.13	202,092,889	3.54	6,968,720
4	BP America	25	0.11	123,971,863	2.17	4,958,875
5	Renco Group	2	0.01	119,079,722	2.09	59,539,861
6	3M Co.	53	0.23	108,727,958	1.91	2,051,471
7	Vulcan Chemicals	1	0.00	92,349,716	1.62	92,349,716
8	General Motors Co.	136	0.60	90,279,073	1.58	663,817

Table 3-5 (continued)

TRI rank	Parent company	Facilities		Total releases and transfers		Average releases and transfers per facility
		Number	Percent	Pounds	Percent	Pounds
9	Eastman Kodak Co.	24	0.11	79,258,257	1.39	3,302,427
10	Phelps Dodge Corp.	19	0.08	77,423,843	1.36	4,074,939
Subtotal		410	1.82	1,536,292,742	26.93	3,747,055
Total for all other facilities		22159	98.18	4,169,377,638	73.07	189,157
Grand total		22569	100.00	5,705,670,380	100.00	252,810

Table 3-6 Top 10 chemicals released/disposed, 1995

Chemical	Number of pounds (millions)
Methanol	245.0
Ammonia	195.1
'l'oluene	145.9
Nitrate compounds	137.7
Xylene (mixed isomers)	95.7
Zinc compounds	87.6
Hydrochloric acid	85.3
Carbon disulfide	84.2
n-Hexane	77.4
Methyl ethyl ketone	70.0
'l'otal for top 10 chemicals	1,224.1
Total for all TRI chemicals	2,208.7

Source: EPA, 1995 Toxics Release Inventory: Public Data Release, Overview
(EPA 745-R-97-005, April 1997), Table 6.

Table 3-7 Top 10 companies based on total releases reported to TRI, 1994

Company	Total facilities	Releases (millions of pounds)
Du Pond	70	203.6
ASARCO Inc.	11	69.4
Renco Group Inc.	12	66.1
IMC Global Inc.	13	47.7
International Paper Co.	71	43.1

Table 3-7 (continued)

Company	Total facilities	Releases (millions of pounds)
General Motors Corp.	112	36.8
Courtaulds United States Inc.	9	34.5
Monsanto Co.	27	27.4
Arcadian Partners LP	8	26.4
Georgia-Pacific Corp.	90	26.2
Total for top IO companies	423	581.2
Total for all TRI facilities	22,744	2,260.2

Source: EPA, 1994 Toxics Release Inventory: Public Data Release, Executive Summary (EPA 745-S-96-001, June 1996), Tables E-1.

Table 3-8 Ten parent companies with the largest TRI total releases and transfers, 1989

TRI rank	Parent company	Facilities		Total releases and transfers		Average Release and transfers per facility
		Number	Percent	Pounds	Percent	Pounds
1	General Electric Co.	138	0.61	38,890,268	0.68	281,814
2	General Motors Co.	136	0.60	90,279,073	1.58	663,817
3	Du Pont	88	0.39	349,275,844	6.12	3,969,044
4	Allied-Signal Co.	86	0.38	52,921,041	0.93	615,361
5	Borden Inc.	70	0.31	8,806,357	0.15	125,805
6	Cooper Industries Inc.	69	0.31	4,374,775	0.08	63,403
7	Westinghouse Electric Corp.	67	0.30	13,654,595	0.24	203,800
8	United Technologies Corp.	66	0.29	10,487,983	0.18	158,909
9	International Paper	64	0.28	22,864,152	0.40	357,252
10	Georgia-Pacific Corp.	65	0.29	19,348,757	0.34	297,673
Subtotal		849	3.76	610,902,845	10.71	719,556
Total for all others		21,720	96.24	5,094,767,535	89.29	234,566
Grand total		22,569	100.00	5,705,670,380	100.00	252,810

Source: USEPA, 1991.

Table 3-9 Environmental distribution of TRI releases and transfers, 1987–1990 (billion pounds)

Type of release or transfer	1987	1988	1989	1990
Air emissions	2.710	2.630	2.550	2.200
Surface water discharges	0.412	0.311	0.193	0.197
Underground injection	1.330	1.340	1.170	0.723
Land disposal	0.728	0.531	0.454	0.441
Transfer to POTWs	0.610	0.574	0.557	0.447
Transfer to off-site location	1.240	1.080	0.932	0.804
Total releases and transfers	7.030	6.470	5.860	4.810

Sources: U. S. Environmental Protection Agency, Office of Toxic Substances, 1990 Taxic Release Inventory: Public Data Release, EPA 700-S-92-002, (Washington, DC: EPA, 1992).

The Organization for Economic Cooperation and Development's Pollution Prevention and Control Group (OECD/PPCG) is convening a series of five to seven workshops in 1994 and 1995 to develop guidance for governments on toxic chemical inventories as called for in Chapter19 of UNCED's Agenda 21. The Agenda 21's proposals on chemicals highlight the principle of the right of the communityand of workers to know about chemical risks. It calls on governments to adopt right-to-know programs and on international organizations to develop guidance for governments on establishing such programs. It asks businesses to report annual routine emissions of toxic chemicals even in the absence of host country requirements (WWF, 1994)

Some excerpts from Agenda's Chapter 19 include: The broadest possible awareness of chemical risks is a prerequisite for achieving chemical safety. The principle of the right of the community and of workers to know those risks should be recognized. However, the right to know the identity of hazardous ingredients should be balanced with industry's right to protect confidential business information (Section 19.8). Improve databases and information systems on toxic chemicals, such as emission inventory programs, through provision of training in the use of those systems as well as software, hardware and other facilities (Section 19.40b). Government should undertake concerted activities to reduce the risks of toxic chemicals, taking into account the entire life-cycle of the chemicals. These activities could encompass both regulatory and non-regulatory measures...(Section 19.49b). Industries should be encouraged to adopt, on a voluntary basis, community right-to-know programs based on international guidelines, including sharing of information on causes of accidental and potential releases and means of preventing them, and reporting on annual routine emissions of toxic chemicals to the environment in the absence of host country requirements (Section 19.50c). Government should direct information campaigns such as programs providing information about chemical stockpiles, environmentally safe alternatives and emission inventories that could also be a tool for risk reduction to the general public to increase the awareness of problems of chemical safety (Section 19.60a). Government should consider adoption of community right-to-know or other public information dissemination programs... Appropriate international organization, in particular UNEP, OECD, the Economic Commission for Europe (ECE) and other interested parties, should consider the possibility of developing a guidance document on the establishment of such programs for use by interested Governments. The document should build on existing work on accidents and include new guidance on toxic emission inventories and risk communication. Such guidance should include harmonization of requirements, definitions and data elements to promote uniformity and allow sharing of data internationally...(Section 19.61c).

3.4
Waste characterization

Physical-chemical properties of industrial waste are fundamental and critical to effectively manage waste . Waste characteristics is a major factor in assessing the feasibility of applying waste minimization and clean technologies. It affects the operating and maintenance practices of a specific facility. In order to minimize waste generation, it is necessary to know: (a) whether the waste is gaseous, liquid or solid; (b) whether the waste is all or partially organic, and its concentration in the waste; (c) whether any toxic metals and halogenated compounds in it and how much; and (d) is it classified as hazardous waste and to what degree (Shen, 1993).

For priority toxic pollutants, USEPA has identified twenty-four toxic substances from its Toxic Release Inventory, based on health and environmental effects, potential for exposure, production volume. The Consumer Products Safety Commission (CPSC), the Federal Drug Administration (FDA), and the Occupational Safety and Health Administration (OSHA) are also concerned with the control of toxic substances. Such substances include acrylonitrile, arsenic, asbestos, benzene, beryllium, cadmium, chlorinated solvents, (trichloroethylene, perchloroethylene, methylchloroform, and chloroform), chlorofluorocarbons, chromates, coke oven emissions, diethylstilbestrol (DES), dibromochloropropane (DBCP), ethylene dibromide, ethylene oxide, lead, mercury and mercury compounds, nitrosamines, ozone, polybrominated biphenyls (PBBs), polychorinated biphenyls (PCBs), radiation, sulfur dioxide, vinyl chloride and polyvinyl chlorine, and toxic waste disposals that may enter the food chain.

For hazardous air pollutants, key waste stream properties for organic and inorganic vapor emissions are: hazardous air pollutant contents, organic content, heat content, oxygen content, halogen/metal content, moisture content, flow rate, temperature and pressure. Key waste stream properties for particulate emissions are: hazardous air pollutant content, particulate content, moisture content, sulfur trioxide content, flow rate, temperature, particle mean diameter, particle size distribution, drift velocity, and particle resistivity.

References

CEQ (1990) Environmental Quality. 21st Annual Report of the Council on Environmental
    Quality, p. 56–56. The Superintendent of Documents, U.S. Government Printing Office,
    Washington, D.C.
CEQ (1991) Environmental Quality. 22nd Annual Report of the Council on Environmental
    Quality, p. 111
CEQ (1992) Environmental Quality. 23rd Annual Report of the Council on Environmental
    Quality, p. 127 and p. 417
Geiser G and FH Irwin (1993) Rethinking the Materials We Use: A New Focus for Pollution
    Policy. The World Wildlife Fund Publications, Washington D.C., pp. 33–35.
Irwin FH (1991) Environmental Law. World Wildlife Fund Publications Vol. 22:1. Washing-
    ton, D.C.
Johansson A (1992) Clean Technology. Lewis Publishers, London
Junger JM and Lefevre B (1988) UNEP Ind. Environs., 11(1), 15
Shen TT (1991) "Safety and Control For Toxic Chemical and Hazardous Wastes," presented
    at the National Workshop on Safety and Control For Toxic Chemicals and Hazardous
    Wastes in Beijing, China, August 27–30, 1991
Shen TT, C Schmidt and T Card (1993) Assessment and Control of VOC Emissions From
    Waste Treatment and Disposal Facilities, Chapter2: Waste Characterization. Van Nos-
    trand Reinhold, New York
U.S. EPA (1978) "Solid Waste Facts" SW-694, p. 5. Washington, D.C.
U.S. EPA (1986) "Solving the Hazardous Waste Problem". Office of Solid Waste, Washington,
    D.C., EPA/530-sw-86-037, November
U.S. EPA (1991) Control Technologies for Hazardous Air Emissions. Office of R&D, Center
    for Environmental Research Information, Washington, D.C.
U.S. EPA (1997) Toxics Release Inventory: Public Data Release (EPA 745-R-97-005, April
    1997), Tables 4-10, 5-5.
WWF (1994) The Right-To-Know: The Promise of Low-Cost Public Inventories of Toxic
    Chemicals. World Widelife Fund Publications, Baltimore, Md.

4
Pollution prevention technology

The influence of science and technology on the modern society has rapidly increased over the whole period of industrialization and there is every reason to believe that this trend will continue. However, it is only relatively recently that we have been make aware of the negative effects of man's dominance over the Nature. Today, environmental questions, quite rightfully so, are a major concern of industrialists as well as politicians, not to forget those who are responsible for the development of new technology, the scientists and engineers. It seems fairly evident that we need new technical solutions, and in some cases even new attitudes towards the role of technical development, in order to prevent the emerging environmental pollution.

This Section presents (1) major technology perspective such as industrial toxics, energy efficiency, renewable energy, material recycling, biodegradable materials, system approach; and (2) preventive technologies in industrial processes such as improved plant operations, in-process modification, materials and product substitutions, materials separation, and solvent alternative; and (3) energy technology such as energy efficiency and renewable energy.

4.1
Major technology perspective

For the past decade, new preventive technologies have been directing to cope with changing waste characteristics. Listed below are the major themes of current and future technology development.

4.1.1
Focus on industrial toxics

Although opportunities for pollution prevention range across many economic sectors, current efforts have focused largely on toxic industrial wastes, which government regulations consider extremely hazardous. Methods of controlling the use of toxic chemicals include:

- Increasing knowledge of chemicals;
- Improving information;

- Substituting toxic chemicals by less toxic ones; and
- For those chemicals that cannot be replaced, ensuring that handling does not result in exposure that may damage public health or the environment.

In 1992 USEPA launched the "design for the environment" (DfE) program to apply pollution prevention principles to the design of chemical processes and products. By disseminating information on the industrial toxics, comparative risk and performance of chemicals, DfE helps industries make informed, environmentally responsible design choice. The program includes:

- Chemical design grants program;
- National center for pollution prevention;
- Demonstration projects;
- Insurance incentive; and
- Full cost accounting (CEQ, 1992).

4.1.2
Material recycling

Encouraging examples show that rationally remanufactured goods, if properly designed originally, open up new possibilities for economic benefits in waste reduction and the recycling of materials. There already exists a rather well-developed market for remanufactured and reconditioned products such as engines, gearboxes, and generators in the automobile industry, for word- and data-processing equipment in the office machine sector, and for some household devices and electric tools.

4.1.3
Biodegradable materials

Interest in biodegradable materials has increased considerably during the last decade, as they offer potential for reducing the cost of waste handling and the increasing problem of littering, due to faster decomposition of the disposed materials. Degradation mechanisms include biodegradation, photodegradation, chemical degradation, and environmental erosion.

4.1.4
Engineering design

Many progressive companies are looking beyond traditional pollution control and prevention strategies to make their companies leaders in an ever more competitive global marketplace. A number of companies have adopted "design for the environment" or "cleaner production" concept. Basically both design the environment and cleaner production concepts are looking at the raw materials in one generation of product becoming the raw materials of the second generation of products. The two concepts require that a large amount of forethought be given to the products and processes to ultimately conserve valuable materials and resources.

Design for the Environment (DfE) promotes the design of safer products and processes in areas such as dry cleaning, screen printing and electronics, and harnesses environmental information to advance new prevention approaches and technologies among business and industry. We must utilize expertise and leadership to facilitate information exchange and research on pollution prevention efforts. DfE works with businesses, trade associations, and other stakeholder industries to evaluate the risks, costs, and performance of alternative chemicals, processes, and technologies. In addition, DfE helps individual businesses apply specific tools and methods to undertake environmental design efforts and is a key in distinguishing their processes, products, and services.

Cleaner Production (CP) seeks continuous improvement of industrial processes, products and services to reduce the use of natural resources, to prevent the pollution of air, water and land, and to reduce waste generation in order to minimize risks to the human pollution and the environment. In general, the concept, methodology and evaluation frameworks of DfE and CP are not much different from technical viewpoint. In a pollution prevention project, the terms DfE and CP are used interchangeably to mean any traditional or novel chemical product, technology, or process that can be used to perform a particular function. Most environmental professionals in industry and business prefer using the term of cleaner production, while professionals in governments and academic institutes prefer using term of design for environment.

4.1.5
System approach

Preventive technologies such as design for environment and cleaner production are necessary as part of a systematic approach that focuses not just on the parts, but also on the internal consistent as a whole. Preventive technology includes process waste minimization, recycling and reuse; while cleaner production emphasizes on input materials and energy, and also the output products that are environmentally acceptable. Some are now viewing pollution prevention as a more overriding concept that includes waste minimization, cleaner production, design for the environment, and life-cycle analysis as tools to move us toward a more sustainable future. Pollution prevention is a broad view in which one seeks not only to reduce pollutants and wastes, but also to optimize the total materials cycle from virgin material, to finished material, to component, to product, to obsolete product, to ultimate disposal, and also to various services as well. The system approach recognizes the many elements of economic competitiveness and the need to incorporate both existing and emerging technologies.

In fact, pollution prevention technology is a combination of efficient product designs and operational procedures that use materials with minimum waste and toxic by-products, and technical processes that are precisely sensed, monitored, and controlled. Much of the technology to prevent pollution is available now, and is frequently not complex. Among current techniques, much headway has been taken in the area of solvents substitution in coating/painting, water and energy conservation, minimizing hazardous waste generation and recycling.

While capital investments of pollution prevention can generate large savings, smaller process and maintenance actions still significantly add to overall savings. Likewise, pollution prevention techniques apply to small businesses just well as to large ones. However, since preventive techniques often use existing techniques, small businesses face the special challenge of gathering information and financing first stage costs.

4.2
Applications in industrial processes

Industry involves thousands of products and production processes, resulting in a decentralized enterprise system. Technical progress toward pollution prevention is likewise decentralized, since it is driven by economic consideration, and frequently specific to one of thousands of different industrial processes. Current pollution prevention technology generally employs conventional engineering approaches. Even as priority shifts from waste treatment and control to prevention, engineers will still employ available technology at first to achieve their objectives. In the future, however, process modifications and friendly product designs will become more innovative for the environment.

Pollution prevention techniques for industrial manufacturing facilities such as waste minimization and source reduction can be understood by observing the path of material as it passes through an industrial site. Even before materials arrive at the site, we could avoid toxic materials when less toxic substitutes exist. Pollution prevention technologies for industries can be generalized into five groups: Improved plant operations, in-process recycling, process modification, materials and product substitutions, and materials separations (Hunt, 1986; OTA, 1987; CEQ, 1990).

4.2.1
Improved plant operations

Manufacturers could implement a variety of improved management or "housekeeping" procedures that would aid pollution reduction; they could conduct environmental audits, establish regular preventive maintenance, specify proper material handling procedures, implement employee training, as well as record and report data.

(A) Environmental audits
Environmental audits may be conducted in many different settings by individuals with varied backgrounds and skills, but each audit tends to contain certain

common elements. It is better to identify and correct problems associated with plant operation to minimize waste generation. One aspect of improved plant operation is cost saving. Production costs and disposal costs can be cut simultaneously by improving plant operation. The practice of environmental auditing also examines critically the operations on a site and, if necessary, identifying areas for improvement to assist the management to meet requirements. The essential steps include (1) collecting information and facts, (2) evaluating that information and facts, (3) drawing conclusions concerning the status of the programs audited with respect to specific criteria, and (4) identifying aspects that need improvement, and (5) reporting the conclusions to appropriate management (ICC, 1989).

Audits enable manufacturers to inventory and trace input chemicals and to identify how much waste is generated through specific processes. Consequently, they can effectively target the areas where waste can be reduced and formulate additional strategies to achieve reductions. The audit consists of a careful review of the plant's operations and waste streams and the selection of specific streams and/or operations to assess. It is an extremely useful tool in diagnosing how a facility can reduce or eliminate hazardous and non-hazardous wastes. It focuses on regulatory compliance and environmental protection (Detailed description of facility environmental audit will be presented in Section 5.2).

(B) Regular preventive maintenance

Preventive maintenance involves regular inspection and maintenance of plant equipment, including lubrication, testing, measuring, replacement of worn or broken part, and operational conveyance systems. Equipment such as seals, gaskets should be replaced periodically to prevent leaks. The benefits of preventive maintenance are increased efficiency and longevity of equipment, fewer shutdowns and slow-downs due to equipment failure, and less waste from rejected, off-specification products. Maintenance can directly affect and reduce the likelihood of spills, leaks, and fires. An effective maintenance program includes identification of equipment for inspection, periodic inspection, appropriate and timely equipment repairs or replacement, and maintenance of inspection records.

Corrective maintenance is needed when the design levels of a process change and adjustments to indirect factors are required. This type of maintenance includes recognizing the signs of equipment failure and anticipating what repairs or adjustments need to be make to fix the problem or improve the overall efficiency of the machinery. Visual inspection ensures that all of the elements of the process system are working properly. However, routine inspections are not a substitute for the more thorough annual compliance inspections. After each visual inspection, it is important to document the results and evaluate the effectiveness of corrective previous actions. Any necessary future corrective action should also be identified.

In reducing fugitive emissions, conscientious leak detection and repair programs have proven to be extremely effective at a fraction of the cost of replacing conventional equipment with leafless technology components. Besides being expensive, changing to leafless technology is not always feasible, and reduces emissions over well-maintained, high-quality conventional equipment only marginally.

(C) Material handling and storage

Material handling and storage operations can cause two types of fugitive emissions: (1) low-level leaks from process equipment, and (2) episodic fugitive emissions, which an event such as equipment failure results in a sudden large release. Often, methods for reducing low-level equipment leaks result in fewer episodes, and vice versa. Methods for reducing or eliminating both types of fugitive emissions can be divided into two groups: (1) leak detection and repair and (2) equipment modification. Such emissions can be prevented by good practices. Proper materials handling and storage ensures that raw materials reach a process without spills, leaks, or other types of loses which could result in waste generation. Some basic guidelines for good operation practices are suggested to reduce wastes by:

- Space containers to facilitate inspection;
- Labeling all containers with material identification, health hazards, and first aid recommendations;
- Stacking containers according to manufacturers' instructions to prevent cracking and tearing from improper weight distribution;
- Separating different hazardous substances to prevent cross-contamination and facilitate inventory control; and
- Raising containers off the floor to inhibit corrosion from "sweating" concrete.

Another good operating procedure is the use of larger containers for storage, if this does not increase wastes from spoilage. Large containers reduce the ratio of container surface to volume and, therefore, reduce the area that has to be cleaned and the waste generated from the process. An alternative to the standard 55 gallon drums are 300 gallon polyethylene containers supported by wire mesh. These containers are portable, reusable, allow bottom or top access, and can be cleaned easily. They are lockable and designed for easy trucking transfer. Additional simple practices, such as keeping containers sealed tightly and using the best, cost-effective means of cleaning-up small spills, help to reduce overall waste generation in a facility (NYSDEC, 1989).

Spills and leaks are major sources of pollutants in industrial processes and material handling. When material arrives at a facility, it is handled and stored prior to use; material may also be stored during stages of the production process. It is important to prevent spillage, evaporation, leakage from containers or conduits, and shelf-life expirations. Standard operating procedures to eliminate and minimize spills and leaks should take place regularly. Better technology might consist of tighter inventory practices, sealless pumps, welded rather than flanged joints, bellows seal valves, floating roofs on storage tanks, and rolling covers versus hinged covers on openings. While these techniques are not novel, they still could lead to large replacement costs if a company has many locations where leakage can occur. Conversely, they could provide large economic benefits by reducing the loss of valuable materials.

Waste from storage vessels takes many forms, from emissions due to vapor displacement during loading and unloading of storage tanks to wastes formed during storage to the storage containers themselves if they are discarded. Reducing waste from storage vessels therefore consists of a variety of activities. Storage tanks for storing organic liquids are found at petroleum refineries, organic chemical manufacturing facilities, bulk storage and transport facilities, and other facilities handling organic liquids. They are used to dampen fluctuation in input and out put flow. Storage tanks can be disastrous source of waste when weakly active undesired reactions run away, and is important to monitor temperatures where this can occur and to design tanks so that heat dissipation effects dominate. Inadequate heat dissipation is of particular concern in the storage of bulk solids and viscous liquids. Other aspects of pollution prevention for storage tanks involving tank bottoms, standing and breathing losses, and emissions due to the loading and unloading of storage tanks.

Vapors that are displaced in loading and off-loading operations can be a significant source of VOC emissions from storage containers. Vapor recovery devices that trap and condense displaced gases reduce losses due to loading and unloading of fixed-roof storage tanks by 90–98%. Vapor balance, where vapors from the container being filled are fed to the container being emptied, is another technique that can be applied in some cases to reduce emissions. Spills due to overfilling of storage containers are another source of emissions that occur during loading and unloading operations. These spills can be prevented through the use of appropriate overflow control equipment and/or overflow alarms.

(D) Employee training

Employee training is paramount to successful implementation of any industrial pollution prevention program. All the plant operations staff should be trained according to the objectives and the elements of the program. Training should address, among other things, spill prevention, response, and reporting procedures; good housekeeping practices; material management practices; and proper fueling and storage procedures. Properly trained employees can more effectively prevent spills and reduce emission of pollutants.

Well-informed employees are better able to make valuable waste reduction suggestions. Plant personnel should comprehend fully the costs and liabilities incurs in generating wastes and understand what the wastes are and where they come from in relation to their specific tasks. They should have a basic idea of why and where waste is produced and whether the waste is planned or unplanned.

Employee training can take place in three stages: prior to job assignment, during job training, and on-going throughout employment. Before beginning an assignment, employees should become familiar with toxic properties and health risks associated with exposures to all hazardous substances that they will be

handling. In addition, employees should learn the consequences of fire and explosion involving these substances. Finally, they should learn what protective
clothing or gear is required and how to use it. During job training, employees
should learn how to operate the equipment safely, the methods and signs of material releases, and what procedures to follow when a spill or leak occurs. On-going education includes regular drill, updates on operating and clean-up practices, and safety meetings with other personnel (NYSDEC, 1989).

The challenge of education and training today is how to integrate air-water-
land pollution management through integrated waste prevention prior to the
application of waste treatment and disposal technologies. A multi-media plan
would help to implement pollution prevention. Cross-disciplinary education
and training will enable trainees to understand the importance of multi-media
pollution prevention principles and strategies so that they can carry out such
principles and strategies for pollution prevention (Shen, 1990).

(E) Operating manual and record keeping

Over the past decade, governments and industry trade organizations, have developed guides and handbooks for reducing wastes. These materials are useful
for analyses of individual facilities, although some guides attempt to be more
general. Good facility documentation can have many benefits for the plant, including waste reduction. Facility documentation of process procedures, control
parameters, operator responsibilities, and hazards in a manual or set of guidelines will contribute to safe and efficient operation. It also promotes consistency,
thereby lessening the likelihood of producing an unacceptable product which
must be discarded. An facility operating manual will assist the operators in monitoring waste generation and identifying unplanned waste releases and will also
assist operators in responding to equipment failure.

Diligent record-keeping with regard to waste generation, waste handling and
disposal costs, and spills and leaks helps to identify areas where operating practices might be improved and later will help in assessing the results of those improved practices. Record-keeping can be instrumental in assuring compliance
with environmental regulations and is a sign of concern and good faith on the
part of the company. The industrial pollution prevention program should include a record of spills or other discharges, the quality and quantity of accidental
releases, site inspections, maintenance activities and any other information that
would enhance the effectiveness of the program. In addition, all records should
be retained for at least three years.

4.2.2
In-process recycling

Materials are processed, frequently in the presence of heat, pressure, and/or catalysts, to form products. As materials are reacted, combined, shaped, painted,
plated, and polished, excess materials not required for subsequent stages become waste, frequently in combination with toxic solvents used to cleanse the

excess from the product. The industry disposes of these wastes either by recycling them into productive reuse or by discharging them as wastes into the air, water, or land. Often costly treatment is required to reduce the toxicity and pollutants in the waste discharge before final disposal. These liquid, solid, or gaseous wastes at each stage of the production process are the source of pollution problems. On-site recycling of process waste back into the production process will often allow manufacturers to reduce pollution and save costs for less waste treatment and disposal.

For example, solvents are being recycled in many industrial processes. The current goal of solvent recycling is to recover and refine its purity similar to virgin solvent for reuse in the same process, or of sufficient purity to be used in another process application. Recycling activities may be performed either on-site or off-site. On-site recycling activities are: (1) Direct use or reuse of the waste material in a process. It differs from closed-loop recycling, in that wastes are allowed to accumulate before reuse; and (2) Reclamation by recovering secondary materials for a separate end-use or by removing impurities so that the waste may be reused.

(1) Advantages of on-site recycling include:
   – Less waste leaving the facility;
   – Control of reclaimed solvent's purity;
   – Reduced liability and cost of transporting waste off site;
   – Reduced reporting (manifesting); and
   – Possible lower unit cost of reclaimed solvent.

(2) Disadvantages of on-site recycling must also be considered, including:
   – Capital outlay for recycling equipment;
   – Liabilities for worker health, fires, explosions, leaks, spills, and other risks as a result of improper equipment operation;
   – Possible need for operator training; and additional operating costs (NYS-DEC, 1989).

Off-site commercial recycling services are well suited for small quantity generators of waste since they do not have sufficient volume of waste solvent to justify on-site recycling. Commercial recycling facilities are privately owned companies that offer a variety of services ranging from operating a waste recycling unit on the generator's property to accepting and recycling batches of solvent waste at a central facility.

4.2.3
Process modification

Pollution can be prevented in many ways specific to particular processes. Many industrial plants have prevented pollution successfully by modifying production processes. Such modifications include adopting more advanced technology through process variable controls, changing cleaning processes, chemical catalysts, segregating and separating wastes as follows (USEPA, 1992):

(A) Process variable controls
Temperature and pressure applications are critical variables as materials are reacted and handled in industrial processes. They can significantly alter the formation of toxins. Improvements include better control mechanisms to meter materials into mixtures; better sensors to measure reactions; more precise methods, such as lasers, to apply heat; and computer assists to automate the activity.

(B) Changing cleaning processes
The cleaning of parts, equipment, and storage containers is a significant source of contamination. Toxic deposits are common on equipment walls. The use of solvents to remove such contamination creates two problems: disposal of the contaminants and emissions from the cleaning process itself. Some changes include the use of water-based cleansers versus toxic solvents, non-stick liners on equipment walls, nitrogen blankets to inhibit oxidation induced corrosion, and such solvent-minimizing techniques as high pressure nozzles for water rinse-out.

For example, the Sandia Laboratory has embarked on a program to reduce hazardous liquid waste by-products of cleaning processes used in the manufacture of electronic assemblies and precious machine parts by: (1) alternative solvents used to remove solder flux residues during electronics assembly manufacture, (2) alternative solvents used in ceramic header fabrication, and (3) alternative manufacturing processes that eliminate the need for solvent cleaning of precision optical components prior to mounting (Oborny, et al.,1990). Alternative solvents technologies for cleaning processes will be further discussed in Section4.2.6.

(C) Chemical catalysts
Since catalysts facilitate chemical reactions, they welcome pollution prevention research. Better catalysts and better ways to replenish or recycle them would induce more complete reactions and less waste. Substitution of feedstock materials that interact better with existing catalysts can accomplish the same objective.

(D) Coating and painting
The paints and coatings industry will have to change technologies to accommodate environmental preventive goals. Manufacturers of architectural coatings under increasing environmental regulations will reduce the volatile organic compounds contained in their coatings by displacing oil based products with water based coatings. In particular, the paint industry will center its research upon re-formulations and increasing the efficiency of coating applications via water based paints, powder coatings, high-solids enamels, reactive diluents, and radiation curable coatings. For the common source of toxic waste, technical improvements include better spray equipment, such as electrostatic systems and robots; and alternatives to solvents, such as bead blasting.

(E) Segregating and separating wastes

A drop of pollutant in a pure solution creates a container of pollution. Segregating wastes and non-wastes reduces the quantity of waste that must be handled. Various technical changes and modifications provide more precise and reliable separation of materials unavoidably mixed together in a waste stream by taking advantage of different characteristics of materials, such as boiling or freezing points, density, and solubility. Separation techniques such as distillation, super-critical extraction, membranes, reverse osmosis, ultra-filtration, electrodialysis, adsorption, separate pollutants or mixed wastes back to their constituent parts. Although simple in principle, these processes become high-tech in the precision with which they are applied to facilitate other options in the hierarchy such as recycling, treatment, and disposal.

(F) Support activities

Garages, motor pools, powerhouses, boilers, and laboratories – all can produce wastes that we must address. We can significantly reduce their sources of pollution through improvement of operation practices as described previously in Section 4.1.

4.2.4
Materials and product substitutions

The issues involved materials and product substitutions are complex and include economic and consumer preferences as well as technological considerations. Obviously, the use of less toxic materials in production can effectively prevent pollution in a decentralized society. Scientists and engineers are actively evaluating and measuring material toxicity and developing safer materials. Likewise, the life-cycle approach requires that products be designed with an awareness of implication from the raw material stage through final disposal stage. Examples of the product life-cycle applications include substitutes for fast-food packaging, disposal diapers, plastic containers, certain drugs and pesticides.

(A) Materials substitution

Industrial plants could use less hazardous materials and/or more efficient inputs to decrease pollution. Input substitution has been especially successful in material coating processes, with many companies substituting water-based for solvent-based coatings. Water-based coatings decrease volatile organic compound emissions, while conserving energy. Substitutes, however, may take a more exotic form, such as oil derived from the seed of a native African plant, Vernonia galamensis, to substitute for traditional solvents in alkyd resin paints (Reisch, 1989).

(B) Product substitution

Manufacturers also could reduce pollution by redesigning or reformulating end products to be less hazardous. For example, chemical products could be produced as pellets instead of powder, decreasing the amount of waste dust lost during packaging. Unbleached paper products could replace bleached alternatives. With uncertain consumer acceptance, redesigning products could be one of the most challenging avenues for preventing pollution in the industrial sector. Moreover, product redesign may require substantial alterations in production technology and inputs, but refined market research and consumer education strategies, such as product labeling, will encourage consumer support.

Changes in end products could involve reformulation and a rearrangement of the products' requirements to incorporate environmental considerations. For example, the end product could be made from renewable resources, have an energy-efficient manufacturing process, have a long life, and be non-toxic as well as easy to reuse or recycle. In the design of a new product, these environmental considerations could become an integral part of the program of requirements.

In both the redesign of existing products and the design of new products, additional environmental requirements will affect the methods applied and procedures followed. These new environmental criteria will be added to the list of traditional criteria. Environmental criteria for product design include:

- Using renewable natural resource materials.
- Using recycled materials.
- Using fewer toxic solvents or replacing solvent with less toxics.
- Reusing scrap and excess material.
- Use water-based inks instead of solvent-based ones.
- Reducing packaging requirements.
- Producing more replaceable component parts.
- Minimizing product filter.
- Producing more durable products.
- Producing goods and packaging that can be reused by consumers.
- Manufacturing recyclable final products.

Figure 4-1 shows a diagram of material and product substitution which can be accomplished through redesigning or reformulating the product to meet new environmental requirements.

4.2.5
Materials separation

In the chemical process industry, separation processes account for a significant portion of investments and energy consumption. For example, distillation of liquids is the dominant separation process in the chemical industry. Pollution preventive technology aims to find methods that provide a sharper separation than distillation, thus reducing the amounts of waste, improving the use of raw materials, and yielding better energy economy. The relationship of various separation technologies to particle size is given in Fig. 4-2. The figure describes the physical selection parameter with respect to particle size for various separation techniques.

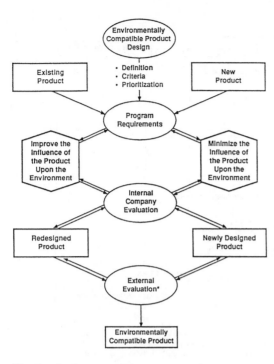

Fig. 4-1 Material and product substitution

Adapted From: Dr. J. C. van Weenan, IDES, University of Amsterdam, February 18, 1991.

In examining separation equipment for waste reduction, three levels of analysis can be considered. One level of analysis involves minimizing the wastes and emissions that are routinely generated in the operation of the equipment. A second level of analysis seeks to control excursions in operating conditions. The third level of analysis seeks to improve the design efficiency of the separation units. Waste reduction opportunities derived from each of these levels of analysis are presented below.

Distillation columns produce wastes by inefficiently separating materials, through off-normal operation, and by generating sludge in heating equipment. The following solutions to these waste problems have been proposed (Nelson, 1990):

– Increase the reflux ratio, add a section to the column, retray/repack the column, or improve feed distribution to increase column efficiency.
– Insulate or preheat the column feed to reduce the load on the reboiler. A higher reboiler load results in higher temperatures and more sludge generation.
– Reduce the pressure drop in the column, which lowers the load on the reboiler.
– In addition, vacuum distillation reduces reboiler requirements, which reduces sludge formation.

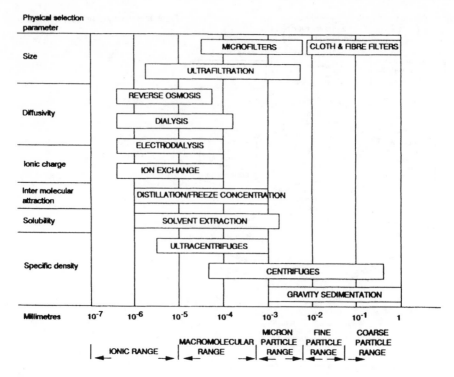

Fig. 4-2 The relationship of various separation technologies to particle size (Source: UNEP Incl. Env. 12(1), 15, 1989)

- Changes in tray configurations or tower packing may prevent pollution from distillation processes.
- Another method for preventing pollution from distillation columns involves reboiler redesign.

(A) Supercritical extraction

Supercritical extraction is essentially a liquid extraction process employing compressed gases instead of solvents under supercritical conditions. The extraction characteristics are based on the solvent properties of the compressed gases or mixtures. We have known about the solvent power of supercritical gases or liquids for more than 100 years, but the first industrial application did not begin until the late 1970s.

From an environmental point of view, the choice of extraction gas is critical, and to date, only the use of carbon dioxide would qualify as an environmentally benign solution. From a chemical engineering point of view, the advantage offered by supercritical extraction is that it combines the positive properties of both gases and liquids, i.e., low viscosity with high density, which results in good transport properties and high solvent capacity. In addition, under supercritical conditions, solvent characteristics can be varied over a wide range by means of pressure and temperature changes.

(B) Membranes

Membrane technology offers other new techniques for combining reaction and separation activities when the product molecules are smaller than the reactant molecules. Removal of product also makes membrane reactor advantageous if the product can react with a reactant to form a waste. Membranes are important in modern separation processes, because they work on continuous flows, are easily automated, and can be adapted to work on several physical parameters, such as: molecular size, ionic character of compounds, polarity, and hydrophilic/hydrophobic character of components.

Micro-filtration, ultra-filtration, and reverse osmosis differ mainly in the size of the molecules and particles that can be separated by the membrane. Liquid membrane technology offers a novel membrane separation method in that separation is affected by the solubility of the component to separate into a liquid membrane rather than by its permeation through pores, as is the case in conventional membrane processes, such as ultra-filtration and reverse osmosis. The component to be separated is extracted from the continuous phase to the surface of the liquid membrane, through which it diffuses into the interior liquid phase. Promising results have been reported for a variety of applications, and it is claimed to offer distinct advantages over alternative methods, but liquid membrane extraction is not yet widely available (Saari, 1987).

(C) Ultra-filtration

Ultra-filtration separates two components of different molecular mass. The size of the membrane pores constitutes the sieve mesh covering a range on the order of 0.002 to 0.05 micron meter. The permeability of the membrane to the solvent is generally quite high, which may cause an accumulation of the molecular phase close to the surface of the membrane, resulting in increased filtration resistance, i.e., membrane polarization and back diffusion. However, the application of trans-membrane feed flow is being used effectively to reduce membrane polarization.

(D) Reverse osmosis

Reverse osmosis is generally based on the use of membranes that are permeable only to the solvent component, which in most applications is water. The osmotic pressure due to the concentration gradient between the solutions on both sides of the membrane must be counteracted by an external pressure applied on the side of the concentrate in order to create a solvent flux through the membrane. Desalting of water is one area where reverse osmosis already is an established technique. The major field for future work will be increasing the membrane flux and lowering the operating pressure currently required in demineralization and desalination by reverse osmosis.

(E) Electrodialysis

Electrodialysis is used to separate ionic components in an electric field in the presence of semi-permeable membranes, permeable only to anions or cations. Applications are demineralization and desalination of brackish water or recuperation of ionic components such as hydrofluoric acid.

(F) Adsorption

Adsorption involves intermolecular attraction forces between the molecules in gas or liquid that are weaker than the attractive forces between these molecules and those of a solid surface. Adsorption is the removal of a pollutant from a gas or liquid stream to become attached to a solid surface. Gas adsorption processes for example can be used to separate a wide range of materials from process gas streams. Normally adsorption processes are considered for use when the pollutant is fairly dilute in the gas stream. The magnitude of adsorption force, which determines the efficiency, depends on the molecular properties of the solid surface and the surrounding conditions.

Adsorbents may be polar or non-polar, however, polar sorbents will have a high affinity for water vapor and will be ineffective in gas streams that have any appreciable humidity. Gas stream associated with industrial processes will be humid or even saturated with water vapor. Activated carbon, a non-polar adsorbent, is effective at removing most volatile organic compounds. Examples of adsorbents for separating gaseous pollutants include activated carbon, activated alumina, silica gel, molecular sieves, charcoal, and Zeolite.

4.2.6
Solvent alternative technologies

The following alternative technologies are being investigated and implemented at the Douglas Aircraft Company (DAC):

(1) High solids topcoats. It is an alternative technology to painting aircraft with conventional topcoats which contain high levels of solvents for sprayability and drying. While it was recently acceptable to simply substitute exempt solvents, such as 1,1,1 trichloroethane, in order to comply with regulations.

(2) Chromium elimination. It covers a variety of processes, such as painting, sealing, plating, and chemical processing. Chromium has long been a main ingredient of many airframe processes because of it's excellent corrosion and wear resistant properties. In 1990 DAC expanded the use of a thin film sulfuric acid anodize process to include some commercial work, thus reducing the use of the popular chromic acid anodize. Chrome-free aircraft sealers, alternative plating technologies, and non-chromated deoxidizes are also being researched by DAC engineers.

(3) Alkaline/aqueous degreasing technologies. It is a cleaning process that uses solvent vapors alone to effectively remove a variety of contaminants from the workpiece. It is a relatively simple, one step process that provides a clean, dry part ready for subsequent processing. Tests are presently being conducted on various immersion type cleaners to replace solvent vapor degreasing. Some of the candidates are aqueous cleaners, terpene based cleaners, and the use of ultrasonic technology with immersion cleaners.

(4) Aqueous cleaners are typically alkaline in nature, their pH being in the range of 9–11. Many chemical suppliers already provide alkaline cleaners on the open market. Aldaline/aqueous cleaning is presently the leading contender to replace vapor degreasing, but the implementation of this will require some change in process and equipment which will require some operator training and/or familiarity.

(5) Alternative handwipe solvents cleaners. They are used extensively for clean up and repair during the manufacture and assembly of transport aircraft. For example, the solvents used for vapor degreasing, these solvents may be ozone depletors and/or carcinogens. The Douglas Aircraft Company's engineers are working with their suppliers to develop cleaners that will work effectively at ambient temperatures to remove the common aircraft industry contaminants.

(6) CFC elimination. Chlorofluorocarbons (CFCs) are used as cleaners of small electronic parts such as printed circuit boards (PCBs), in wire assembly areas, and in many maintenance tasks. They are also used in air conditioners and machine tool chillers, and as propellants in aerosol can applications. These are several lubricants and mold release compounds at DAC that are comprised of CFCs. These are not used in high volume, but do provide an opportunity for reduction of CFC emissions. DAC has begun the process of substituting the propellants used in some of the aerosol lubricants.

(7) Resource recovery and waste minimization. It is another broad category that encompasses many technologies: chemical processing, waste disposal, recycling and housekeeping are just a few. There are many opportunities for improvement under this category as environmental technology continues to advance.

Waste minimization was highlighted during DAC's recent implementation of the Total Quality Management philosophy. This effort enlightened and encouraged every employee to consider his/her impact on the environment and the workplace. An effective example of waste minimization was accomplished by simply reducing the size of their vendor-provided wipe rags. An on-site survey conducted to evaluate the usage of wipe rags discovered that the three foot square rags were too large for convenient wipe operations. The supplier agreed to provide smaller rags at less cost to DAC, thus reducing both the volume and weight of rag-generated wastes. Because of the wide variety of uses for wipe rags, they are liable to become contaminated with many products including hazardous substances that require the disposal of these rags as hazardous waste.

Each of these projects contributes to eliminate the negative impact of manufacturing processes upon the environment. At Douglas Aircraft Company, they are working diligently with their suppliers and subcontractors to develop, test and implement new alternative technologies. Alternative technologies are becoming increasingly necessary to meet the ever-tighting demands of an aware public when it comes to environmental legislation. It is noteworthy that environmental professionals tend to share technological developments and breakthroughs to lead the society a better, cleaner, and healthier place to live (Locklin, 1993).

Chlorinated hydrocarbon solvents and Chlorofluorocarbons are used extensively in cleaning operations in the Department of Energy (DOE) defense program, the nuclear weapons complex, the Department of Defense (DOD) weapons refurbishment facilities, and in industry. A Solvent Utilization Handbook has been published by their joint task force to provide guidelines for the selection of nontoxic environmentally safe substitute solvents for these operations. The information contained will include cleaning performance, corrosion testing, treatability operations, recycle/recovery techniques, volatile organic compound emissions and control techniques as well as other information. The Handbook will be updated on an annual basis with information on new solvent substitutes that appear in the marketplace. The handbook database is under revision. Toxicological information, handling and disposal, and economics of solvent usage will also be included in the updated handbook (Chavez and Herd, 1993)

4.3
Energy technology

Energy is essential for economic production and the high standard of living. In 1992 the United States expanded energy initiatives to improve environmental quality and energy security and to stimulate economic growth. The National Energy Strategy Act of 1992 allows implementation of a balanced, pro-environment, pro-growth national energy strategy. During 1991 and 1992 the U.S. government introduced over 90 National Energy Strategy initiatives, including expanded energy efficiency and renewable energy programs, and formed partnerships among federal agencies, industries, and states to reduce energy and water use.

4.3.1
Energy efficiency

Energy efficiency is the generic term which, to some degree, embodies energy intensity, energy savings, and energy rationalization. (1) Energy intensity is the consumption of energy per unit of product output, or GDP. (2) Energy savings represent a reduction in this intensity. (3) Energy rationalization is a combination of energy savings and switching from higher cost to lower cost fuels. In the process of producing, delivering and consuming energy, there is a continuum of activities – investment in energy supply capacity; production and transformations; transmission and distribution processes; and final use. The efficiency concept applies to all of these activities.

The concept of efficiency in energy supply covers all the processes embodied in energy sector management: investment, production/generation, transmission, distribution, and final supply for each of the energy sources – power, coal, petroleum, gas, and biofuels. The overriding objective of energy supply decisions should be to secure high quality supplies at the lowest possible investment and operating cost to the economy, and to price these supplies to consumers at a level which reflects their long run marginal costs. All nations, especially developing nations, clearly need to strengthen efficiency in all three areas: investment planning, operational procedures, and pricing.

The United States uses 10% more energy in 1992 than it did in 1973, yet there are 30% more homes, 50% more vehicles, and the gross domestic product is 50% higher. The challenge is to produce and use energy more efficiently and cleanly. The U.S. produced approximately 65 quadrillion Btu in 1991, industry accounts for about 36% of end-use energy consumption, relying on a mix of fuels, and uses 70% of the energy it consumes to provide heat and power for manufacturing (CEQ, 1992). U.S. domestic energy resources are extensive and diverse, with coal, oil, natural gas, and uranium found in significant quantities within U.S. borders. Unconventional sources, such as coalbed methane, are also potential future energy sources. Renewable energy sources, such as geothermal, solar, and wind, are avail-able and limited only by the cost-effectiveness of technologies to harness them.

In the energy sector, for example, preventive technology and management can reduce environmental damages from extraction, processing, transport, and combustion of fuels. Preventive technology and management approaches include:

– increasing efficiency in energy production, distribution and use;
– substituting environmentally benign fuel sources; and
– design changes that reduce the demand for energy.

In 1991, the National Industrial Competitiveness through Environment, Energy, and Economics program was established jointly by the Department of Commerce, the Environmental Protection Agency, and the Department of Energy to fund development of generic, pre-competitive technology to enhance industrial competitiveness through improved energy efficiency. Grants are awarded to state and industry consortia to develop energy efficient technologies with promise of commercial viability for U.S. industry.

Increasingly more developing countries realize that the traditional approaches used by industrialized countries to spur economic growth will no longer work, because such approaches cost too much economically and environmentally. Since social and economic developments of these developing countries still depend on providing more and more energy services, the global community must identify, explore, and implement alternative paths that are better for the environment, and the economy (Levine, et al, 1991).

4.3.2
Renewable energy

Renewable energy resources will be an increasingly important part of the power generation mix over the next several decades. Not only do these technologies help reduce global carbon emissions, but they also add some much needed flexibility to the energy resource mix by decreasing our dependence on limited reserves of fossil fuels. Experts concur that solar energy, wind power, photovoltaics (PV), hydropower and biomass are on track to become strong players in the energy market of the next century.

The solar energy makes possible a number of promising renewable energy technologies, including solar thermal, solar photovoltaic, wind, ocean thermal, small-scale hydropower, biomass, and hydrogen produced from renewable energy. Geothermal energy is another renewable source that draws on the earth's interior heat. Some of these sources are now nearly competitive in cost with fossil-fuel energy. Renewable technologies could be competitive in meeting 80% of U.S. energy needs with 15 to 20 years if given adequate support through economic incentives and funding of applied research, development, and demonstration projects.

Wind power is the fastest growing electricity technology currently available. Wind-generated electricity is already competitive with fossil-fuel based electricity in some locations, and installed wind power capacity now exceeds 7000 MW worldwide. Meanwhile, PV electricity is seeing impressive growth worldwide. PV although currently costs three to four times more than conventional delivery electricity, is particularly attractive for applications not served by the power grid. Many of the developing countries are attracted to the PV technologies' modular nature, located close to the user, the units are far cheaper and quicker to install than central station power plants and their extensive lengths of transmission line (EPRI, 1998).

U.S. Windpower is the world's leading manufacturer of wind turbines and the largest supplier of wind-generated electricity. The Aotanmont-Solano project includes more than 3000 turbines located in the Altammont Pass and Montezuma Hills of California. In 1989, the project produced 600 million kilowatts, enough electricity to power 100,000 homes. Windpower is a promising energy source for the North Central and Northwest United States, and is already supplying significant amounts of energy in northern Europe and Asia (Renew America, 1990).

4.3.3
Fuel cells

Fuel cells offer compact, modular packaging, high efficiency, fuel and siting flexibility, and pollution free operation. They could become widely used as distributed premium power sources at industrial sites and in manufacturing plants, office buildings, institutional settings, and perhaps eventually homes. In many areas, fuel cells are expected to provide strong competition to commercial and industrial electricity rates at the point of end use. As part of an integrated ecological design, for example, a major new skyscraper under construction at Times Square in New York City will include fuel cells for powering external lighting and serving some of the buildings heating requirements.

Advanced fuel cell technologies also are expected to find many off-grid applications as lightweight, compact, remote portable power generators. Several electric power companies are already positioned in joint ventures with fuel cell developers to market the systems to customers that have special power quality and reliability requirements. Other energy service providers are planning to market fuel cells as part of premium-power offerings. Because of their high efficiency in converting natural gas, methanol, hydrogen, and even gasoline into electricity, fuel cells offer the lowest carbon dioxide emissions of any fossil power system (EPRI, 1998).

4.3.4
Advanced power electronics

Power electronics based on silicon semiconductor switching and converter devices are transforming the ability to manage the power delivery system in real time. They are analogous to the low-power transistors and integrated circuits that brought about the computer age, but they operate at multi-megawatt power levels. They can switch electricity to a wide range of voltages, frequencies, and phases with minimal electrical loss and component wear.

Advanced high-power electronics based on new types of semiconductor materials are expected to enable precise control and turning of all circuits – even gigawatt-scale power systems. And packaged devices make with these new materials could be as much as 100 times smaller and lighter than today's silicon devices. As a result, advanced power electronics promise unprecedented increases in the efficiency and cost-effectiveness of devices for a wide range of electricity production, delivery, and end-use applications. Such megawatt electronics are also critical for a variety of electric propulsion, control, and weapons applications for defense systems.

For electricity providers, power electronics represent the critical enabling technology for improving power system performance, offering value-added services to customers, and succeeding in a competitive marketplace. For semiconductor manufacturers, the new pollution-free technology of advanced power electronics could trigger a second electronics revolution and unlock an entirely new multi-billion-dollar market (EPRI, 1998).

References

CEQ (1990) Environmental Quality, The 21st Annual Report of the Council on Environmental Quality, pp. 92–94.

CEQ (1992) Environmental Quality, The 23rd Annual Report of the Council on Environmental Quality, pp. 71–75, 181, and 208.

Chavez AA and MD Herd (1993) "DOE/DOD Solvent Utilization Handbook", in: Solvent Substitution For Pollution Prevention, Noyes Data Corporation, Park Ridge, NJ, pp. 116–117.

EPRI (1998) "Technologies for Tomorrow", EPRI Journal, January/February 1998, p. 37–40.

Levine MD, SP Meyer and T Wilbands (1991) "Energy Efficiency and Developing Countries," Environ. Sci & Technol, Vol. 25, No. 4, pp. 485–489.

Locklin JM (1993) "Alternative Technologies For Environmental Compliance" in: Solvent Substitution For Pollution Prevention, Noyes Data Corporation, Park Ridge, NJ, pp. 92–96.

NYSDEC (1989) Waste Reduction Guidance Manual. Prepared by ICF Technology Inc. for New York State Department of Environmental Conservation, pp. 3–3 to 3–11.

Oborny MC et al. (1990) "Sandia's Search for Environmental Sound Cleaning Processes for the Manufacture of Electronic Assemblies and Precision Machined Parts," in: Environmental Challenge of the 1990's Proceedings, International Conference on Pollution Prevention: Clean Technologies and Clean Products, USEPA/600/9–90/039, September.

OTA (1987) Serious Reduction, Report published by Congressional Office of Technology Assessment, Washington, D.C., pp. 78–82.

Reisch MS (1989) "Demand puts paint sales at record levels," Chemical and Engineering News, October 30, p. 41.

Renew America (1990) "Searching for Success", Washington, D.C.

Saari M (1987) Prosessiteollisuuden Erotusmenetelmat, VTT Research Note, Technical Centre of Finland, p. 730.

Shen TT (1990) "Educational Aspects of Multimedia Pollution Prevention", in: Environmental Challenge of the 1990's Proceedings, International Conference on Pollution Prevention: Clean Technologies and Clean Products, USEPA/600/9–90/039, September.

USEPA (1991) "Pollution Prevention 1991: Progress on Reducing Industrial Pollutants", EPA 21P-3003, October.

USEPA (1992) Facility Pollution Prevention Guide. Office of R&D, Washington, D.C., EPA/600/R-92/088, May.

van Weenan JC (1990) Waste Prevention: Theory and Practice. The Haque: CIP-Gegecans Koninklije Bibliostheek.

5
Total environmental quality management

Pollution prevention is a concept and practice that has quickly taken us beyond our traditional "command and control" approach to controlling waste and toxic emissions. Because the concept focuses on not generating waste in the first place, it has forced companies to look at the flow of chemicals in the workplace and to look at where and why wastes are generated. Decisions related to pollution prevention have to be made before waste is ever generated from those involved in product and process design and operation, to those making decisions about materials use.

Industrial pollution prevention requires new blood with creative and probing personalities. The future industrial manufacturing facility will spend considerable time bench-marking the competition. This refers to a thorough evaluation of products and services that are made by a competitor and are considered by the consumer to be better. The facility will consider the environmental implications of a new product and service. It will plan pollution prevention into its research experiments. It will insist on pollution prevention in the acquisition of new components or systems from others so that waste generation in the life cycle of its product will be minimized. The future facility will constantly use total quality management tools to find problems and seek solutions. It will inculcate the total quality management culture into every person in the company.

This Chapter consists of four sections: (1) risk assessment and risk management, (2) environmental auditing program, (3) design for the environment, and (4) voluntary environmental labeling program.

- Risk is a measure of the likelihood and magnitude of adverse effects, including injury, disease, or economic loss (Kolluru, 1996). Risk assessment is a systematic method of characterizing the quantitative or qualitative process. The methods characterize the different types of risks to environment and to human health, safety and welfare vary, but share some common elements. The results of risk assessments provide input to risk management decisions, which take into account political, economic, and social considerations for assessing competing risks and allocating resources toward their management.
- Environmental auditing program ranges from the desire to determine compliance with specific regulations, standards, or policies, to the goal of potentially pollution conditions and also to identify potentially pollution preven-

tion opportunities. It provides an opportunity to derive neutral sets of data and compare the magnitudes of problems in terms of human and environmental health and costs of alternatives.

– Design for the environment is a process which aims at minimizing or preventing environmental damage as a design objective in the first place. This involves not only waste and pollution created during development and manufacture of a product and service, but also product use and disposal of the used-product by the consumer. In other words, it considers the environmental effect of a product through its entire life-cycle. The word "product" can be any manufacturing or constructing products such as airplane, automobile, bridge, building, park, land-use, city-plan, etc. Design for the environment is a new and challenging frontier for engineering professionals to learn and use the product life-cycle assessment method to design environmentally compatible products and services.

– Voluntary environmental labeling program has the potential to improve the quality of our environment by producing goods or products which met certain environmental criteria to label with a recognized symbol, such as "green product."

5.1
Risk assessment and risk management

In parallel with heightened public sensitivities to multiple hazards in the environment, and perceptions that the hazards are getting out of control, there has been growing interest in the United States and worldwide to use risk-based approaches for environmental quality management. A variety of approaches are being used to characterize diverse risks, and to deploy resources effectively for protecting public health and the environment.

This section describes risk concepts and definitions, different types of risks and assessment methods, risk assessment paradigm, and potential applications to industrial pollution prevention.

5.1.1
Concepts and definitions

Risk is inherent in everyday living and in all decision-making. Most of us have an intuitive sense of hazards and risks – that somehow they are undesirable and are to be avoided. Because risk has many dimensions with scientific, political, social, and economic overtones, there are many definitions. The words risk assessment and risk analysis are often used synonymously, but risk analysis is sometimes used more broadly to include risk management aspects as well.

In risk assessment and management, we look at an existing or a potential situation and ask such questions as: What can go wrong and why, how likely is it, how bad can it be, and what can we do about it? Risk assessment and management, in essence, is a matter of answering these questions in a scientific, systematic, and transparent manner. Fig. 5-1 illustrates a risk assessment and risk management model which is based on the guidelines issued by the National Academy of Sciences and National Research Council and adopted by EPA.

Risk assessment, then, is the process of characterizing the likelihood of occurrence of an event and the likely magnitude of consequences. Risk management is a decision-making framework that integrates the "science" of risk assessment with "public values" to avoid or reduce identified risks. This is accomplished through the implementation of policy options by, for example, prioritizing the different types of risk using comparative risk analysis.

5.1.2
Types of risks and assessment methods

Some of the major types of risks and "end points" are shown below (Kolluru, 1994 and 1996):
- Human health, safety, and welfare risks

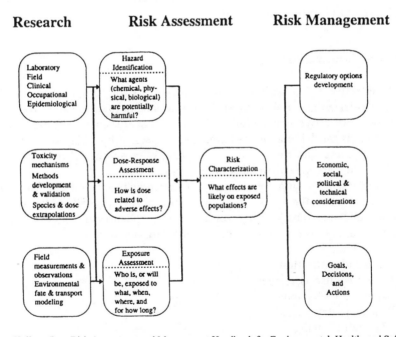

Source: Kollure, Rao, Risk Assessment and Management Handbook for Environmental, Health, and Safety Professional, McGraw-Hill, New York, 1996. Adopted from National Research Council risk assessment and management paradigm.

Fig. 5-1 Risk assessment process

- Ecological/environmental risks
- Financial risks
- Occupational/worker risks
- Environmental/public risks
- Consumer/residential risks
- Acute, sub-chronic, and chronic risks
- Cancer cases; non-cancer systemic effects.

The methodologies for assessing these different types of risks vary to a considerable degree. (It is beyond the scope of this book to go into the details of the different types of risks and their assessment. Readers may consult a book such as: Rao Kolluru: Risk Assessment and Management Handbook for Environmental, Health, and Safety Professionals, McGraw-Hill, 1996). In general, risk assessments can be classified as qualitative or quantitative. They can also be "deterministic" "point" estimates, or probabilistic methods. The former uses single values for input variables that are representative of, for example, 50th, 90th, or 95th percentile values, depending on the degree of conservatism desired in the risk estimates. The probabilistic methods, on the other hand, use a range of values from probability distributions of input variables. Since probabilities are inherent to the concept of risk, the probabilistic methods are more in keeping with the science and spirit of risk ssessment.

5.1.3
Risk assessment process and steps

To be effective, risk assessments must always start with proper planning. Even before undertaking a risk assessment, it is important to define and agree on the scope and objectives. For example: Who are the constituencies or interested parties? What is the driving force – human, ecological, or financial concerns? What are the uses and who are the users of the data? What are the deliverables and the timing? What are the data quality objectives (DQOs)?

In the case of pollutant emissions from an industrial facility, or from a hazardous waste site, the focus is on long-term exposures, usually to low environmental concentrations. In other words, chronic exposures and effects. The basic risk assessment process consists of the following steps:

- Hazard identification
- Dose-response assessment
- Exposure assessment
- Risk characterization
- Uncertainty analysis

Risk assessments usually start off with an initial phase of research and data collection. In the context of an industrial facility, the collection phase may include data from historical operating records; data on discharges to environmental media including air, water, and land; pollutant concentrations onsite and in surrounding media from site assessment sampling; survey of existing or potential receptors, and so on.

The hazard identification step involves data review and identification of sources of possible risk. The sources are prioritized in terms of the potential to cause harm. An example is the selection of chemicals of concern based on their concentration, toxicity, and exposure potential to perform detailed quantitative analysis.

The purpose of dose-response or toxicity assessment is to portray the relationship between the extent of exposure (or dose) and the likelihood or extent of adverse effects in exposed populations. The responses range from no-observable-effect to temporary and reversible to permanent injury to death. In human health, the responses are generally categorized as threshold effects and zero-threshold cancer effects.

Exposure is coming into contact with a hazard, or access to a hazard. Exposure is what bridges the gap between a hazard and a risk. Exposure to a hazard, such as pollutant emissions from a facility, can occur either directly or indirectly via inhalation of air, incidental ingestion of contaminated soil, ingestion of contaminated food and water, or absorption through the skin via dermal contact. Contaminant sources, release mechanisms, transport and transformation characteristics, as well as the nature, location, and activity patterns of "receptors" are all part of exposure assessment.

In risk characterization, the results of hazard, dose-response, and exposure assessments are all integrated to arrive at quantitative estimates of risks. In safety risk assessment this may involve estimates of likelihood of fatalities or injuries. It may also include property damage and financial losses. In human health risk assessment, potential effects are generally presented in terms of cancer cases, and non-cancer hazard index.

Since uncertainty is inherent in all risk assessments, sources of uncertainty and the tendency to over- or under-estimate the risks should be clearly described. Risks should be presented in understandable language in proper context, rather than as isolated numbers. It is often helpful to include sensitivity analysis in the report to inform the decision makers about the input variables that wield the most (and the least) influence on the risk projections. Uncertainty and sensitivity analysis can be considered a part of risk characterization, or a separate step.

5.1.4
Risk assessment applications and advantages

Many federal agencies including the Environmental Protection Agency (EPA), Occupational Safety and Health Administration (OSHA), Food and Drug Administration (FDA), and Department of Transportation (DOT) routinely use risk criteria in their decisions. Risk assessments are applicable in a variety of situations, such as to:

- Conduct baseline assessment of a site to determine the need for remedial action and the extent of cleanup required for acceptable residual risks
- Analyze current or potential emissions to assess risks and achieve cost-effective reductions
- Assess process safety and transport hazards in facility siting decisions
- Evaluate existing and new technologies for effective prevention, control, or mitigation of hazards and risks
- Develop a set of criteria for closing or decommissioning facilities and minimizing future liability
- Address public health and safety concerns within a consistent and scientific framework.

As an example of risks from an operating industrial facility, the risks to potentially exposed populations can be calculated by estimating the different types of pollutant emissions from the facility. In addition, permissible emissions from a facility can be back-calculated to meet a set of external regulatory guidelines or internal company policy guidelines.

Risk assessments offer both economic and non-economic benefits, as suggested by the preceding applications. One advantage is that the different types of risks – environmental, health, safety – can all be addressed by means of a common language. A common framework can be used to allocate resources in an optimal manner to remedy existing problems or avoid future problems. Risk assessments afford control over costs and timing of implementation. They also yield insights into time-sensitive hazards and risks, and the timing of remedial action.

Among the limitations of risk assessments are the speculative nature of many of the assumptions and the high degree of variability in the results of assessments conducted by different professionals. There are unrealistic expectations on the part of some sponsors that risk assessments represent a panacea, a substitute for action. Moreover, few professionals have the necessary inter-disciplinary qualifications to conduct risk assessments. There is also the pressing need to strengthen the scientific component of risk assessments.

5.1.5
Risk management

Application of risk assessment to industrial pollution prevention provides an opportunity to derive neutral sets of data and compare the magnitudes of problems in terms of human and environmental health and costs of alternatives. The U.S. EPA and other federal agencies have emphasized risk assessment as a basic tool to determine priority of environmental problems, rather than being pushed into action by the perception of risks (USEPA 1990).

A particular industrial facility can apply risk assessment at various points of production. Risks due to transportation of raw materials, their input into process, the production process itself, toxic releases and products leaving the production process, and the transport of products to distributors and consumers can be evaluated using historical data, and measures could be taken to decrease risks of accidents and resulting pollution. Within the plant, one can evaluate process safety (and probability of spill and/or accident that would result in pollution) by applying probabilistic risk analysis. Based on probable severity of release resulting in air, water, and/or land contamination, one can perform risk assessment of effects of pollutants on human and ecological health. Similarly, routine releases can be evaluated using chemical risk analysis or carcinogen risk analysis.

From the risk assessment report, one can perform comparative risk assessments for each particular chemical or process, and rank them according to the magnitude of risk (either probability of cancer, or magnitude by which releases exceed environmental release standard). Numerous computer programs exist to simulate and evaluate risks from chemicals. A good starting point for performing chemical risk analysis may be toxic release inventory (TRI) reports, since they represent an inventory of pollution, even if incomplete. Chemical risk assessment and economic risk assessment can be applied to evaluate various pollution prevention and control practices. Based on such assessment one can set priorities in pollution prevention projects (Molak 1995).

Since the chemical industry deals with the largest number of toxic chemicals and processes, the application of risk assessment and risk management is most appropriate. An attempt should be made to minimize loss or release of chemicals during the process and establish safety measures to minimize risk of other types of injuries (e.g., from explosions or falls). Releases of toxic chemicals and hazardous wastes will cause serious environmental pollution and must be handle with special caution and carefully prevented to avoid major social and economic losses.

5.2
Environmental audit program

One of the best ways to prevent environmental pollution is to ensure environmentally sound company's operations and management. For example, environmental management audit uncovers small pollution control or waste management problems before they become large plant or environmental liabilities. The audit aims not only at minimizing potential negative impacts of the company on the environment, but also maximizing the positive impact of an environmentally sound system on the company's other activities.

A growing number of industrial companies are instituting environment management audits at their facilities. Such audits stimulate environmentally sound technology that helps reduce waste and resource expenditures and thereby increases profits. Environmentally oriented management will broaden managerial vision in concepts for the firm, markets, and growth opportunities.

Regulatory agencies consider such auditing as an important corporate management technique because it ensures compliance with environmental requirements and related corporate policies. The regulatory agencies have accordingly encouraged more companies to evaluate and implement individualized environmental auditing programs, since private-sector auditing can increase environmental compliance while conserving scarce government resources.

5.2.1
Why audit?

Motivations for developing an environmental auditing program range from the desire to measure compliance with specific regulations, standards, or policies, to the goal of identifying potentially hazardous conditions. Thus, while auditing may appear to serve the universal need of evaluating and verifying environmental compliance, in practice, auditing programs are designed to meet a broad range of objectives, depending on the corporate culture, management philosophy, and size.

Many reasons exist for companies to initiate audit programs. For example, auditing systems can help corporate management to (ADL, 1983):

- Determine and document compliance status;
- Improve overall environmental performance at the operation facilities;
- Increase the overall level of environmental awareness;
- Improve the environmental risk management system;
- Develop a basis for optimizing environmental resources;
- Project positive concern about environmental effects and commitment to take steps necessary to correct identified problems;
- Avoid substantial tort liability arising from personal injury, property damage, or "toxic tort" claims;
- Avoid the surprise, disruption, and unplanned costs of sudden enforcement actions;
- Develop better relations with government agencies through the presence of an affirmative program designed to find and correct problems before they become dangers;
- Build closer links with community and government not only to minimize the chance of environmental conflict, such as enforced shutdowns and product bans, but in particular, but also help better the image of the company and industry in general, in terms of the positive role and contribution they provide; and
- Realize savings through process changes which reduce the amount of raw materials needed or create less pollution to be civil or criminal liability under the current environmental laws and regulations.

Private-sector environmental auditing systems and plant-level compliance assurance programs are available to offer chief executive officers and other corporate stakeholders both more effective environmental management and increased comfort or security that audit procedures to assure compliance with environmental requirements and corporate policies are not only in place, but are operating as planned.

5.2.2
Why is waste generated?

Industrial waste is generated because less than 100% of the raw materials entering a process end up as product. Historically, maximizing the product yield, or maximizing raw material utilization, always has been a primary goal of any process design or operation. Therefore, a key question at the outset of any pollution prevention program is whether or not this goal has been successfully achieved in light of high waste disposal costs and the possibility of a ban on land disposal.

To characterize major causes of waste generation, along with the associated operational and design factors for typical sources of waste, the chemical processing and metal finishing industries are selected as illustrative example.

5.2.2.1
Chemical reactions

(A) Typical Causes of waste generation:
- Incomplete reactant conversion
- By-product formation
- Spent catalyst – deactivation due to poisoning, sintering, etc.

(B) Operational factors:
- Inadequate temperature control
- Inadequate mixing
- Poor feed flow control
- Poor feed purity control

(C) Design factors:
- Inadequate reactor design
- Catalyst design or selection
- Choice of process path
- Choice of reaction conditions
- Fast quench
- Inadequate instrumentation or controls design
- Poor heat transfer

5.2.2.2
Contact between aqueous and organic phases

(A) Typical causes of waste generation:
- Vacuum production via steam jets
- Presence of water as a reaction by-product
- Use of water for protective rinse
- Equipment cleaning
- Cleaning of spills

(B) Operational factors:
- Indiscriminate use of water for cleaning or washing

(C) Design factors:
- Choice of process route
- Vacuum production via vacuum purge
- Use of reboilers instead of steam stripping
- Clinger reduction

5.2.2.3
Disposal of unusable raw materials and off-spec products

(A) Typical causes:
- Off-spec product generation caused by contamination, temperature or pressure excursions, improper reactants proportioning, inadequate pre-cleaning of equipment or workpiece
- Obsolete raw material inventories

(B) Operational factors:
- Poor operator training and supervision
- Inadequate quality control
- Inadequate production planning and inventorying of feedstocks

(C) Design factors:
- Extension use of automation
- Maximizing dedication of equipment to a single process function

5.2.2.4
Process equipment cleaning

(A) Typical causes:
- Presence of clinger (process side) and scale (cooling water side)
- Deposit formation
- Use of filter aids
- Use of chemical cleaning agents

(B) Operational factors:
- Insufficient drainage prior to cleaning
- Inadequate cooling water treatment
- Excessive cooling water temperatures

(C) Design factors:
- Design for lower film temperature in heat exchangers
- Design of reactors or mixing tanks with wall wiper blades
- Controls to prevent cleaning water from overheating

5.2.2.5
Metal parts cleaning

(A) Typical causes:
- Disposal of spent solvent, cleaning sludge or spent cleaning solution

(B) Operational factors:
  – Indiscriminate use of solvents and water

(C) Design factors:
  – Choice between cold dip tank or vapor degreasing
  – Choice between solvent vs. aqueous cleaning solution

5.2.2.6
Metal surface treatment

(A) Typical causes:
  – Dropout
  – Disposal of spent treatment solutions

(B) Operational factors:
  – Poor rack maintenance
  – Indiscriminate rinsing with water
  – Fast withdrawal of workplace

(C) Design factors:
  – Counter-current rinsing
  – Fog rinsing
  – Dropout collection tanks

5.2.2.7
Spills and leaks cleanup

(A) Typical causes:
  – Manual material transfer and handling operations
  – Leaking pump seals
  – Loading flange gaskets

(B) Operational factors:
  – Inadequate maintenance
  – Poor operator training
  – Lack of operator attention
  – Indiscriminate use of water in cleaning

(C) Design factors:
  – Choice of gasketing material
  – Choice of seals
  – Use of welded or seal-welded construction.

It is useful to distinguish between the operational and design factors affecting waste generation. While operational factors usually can be addressed and implemented with a very short time period, the design factors typically require more insight and, in many cases, extensive research (e.g., a search for a better, more selective catalyst).

Early recognition of waste generation aspects during the process or product development stage is extremely important, as evidenced by the experience at 3M Corporation. At 3M, the formulation process of new products involves an evaluation of the waste generation associated with the future manufacturing process. The company thus has been able to reduce the generation of waste before new products reach the production stage (Ling, 1994).

5.2.3
Environmental management audit

The environmental management audit places special emphasis on the pollution prevention at the source. It helps corporate management to achieve all activities as described previously in Section 5.2.1 Why Audit. To this end, the managerial personnel share most responsibilities. However, the production personnel are responsible to audit the existing production procedures and examine opportunities for process or plant modification on a continual basis.

The management audit examines the organization of the environmental function particularly as it relates to pollution prevention. Staffing and organizational structure are reviewed as well as the links between the activities of pollution prevention and control, risk and hazard prevention, product safety and occupational health and safety. Particular attention should also be given to the links between the environmental group and other functional and staff groups and especially top management.

Over the past several years industry has increasingly come to realize that sound environmental management can be equated with good management. For much of the world, a manufacturer can no longer disregard the environment in the pursuit of profits. Today's manufacturers are increasingly obliged to take stock of the environment in which they operate and to make their activities conform to the standards and norms required by the public and government. Industry could benefit from a critical examination of its processes and technologies through environmental auditing to determine where problems might arise, particularly with regard to human health (ADL, 1983).

The environmental management audit begins with an assessment of the degree of compliance with environmental legislation and the assessment of the potential environmental risk associated with the operations. The evidence is sought for compliance with prevailing legislation on air pollution, water pollution, solid waste disposal, hazardous waste disposal, in-plant hazards etc.

In addition, an assessment is made of the general state of equipment, plant, materials storage areas, together with evidence for the reliability and the performance of processes, the quality of operations and operating procedures, and the state of records and documentation of all emissions, spills, effluents, and waste, together with details of off-site disposal and of any infringements or conflicts with the authorities or the local communities. This phase would be accompanied by a list of recommendations for the modification of equipment, processes, operating procedures, materials storage etc. (Royston and McCarthy, 1988).

5.2.3.1
Technology

Technology can turn many of the environmental problems to opportunities and benefit for the corporation. The management audit therefore covers the technology used and the management process by which decisions are taken on technological issues and the way in which the environmental implications of these decisions are treated.

During the process of technology acquisition, development and deployment, evidence would be gathered regarding the impact of the technology on the natural environment, the health and safety of workers and the community. At the same time, attitudes and procedures will be assessed which allow the needs of the environment to be met positively by product, process and service technologies. In addition, it might also include the identification of environmental technologies which could be offered to other organizations.

Specifically, the environmental management audit will look for both evidence of management practices in terms of environment quality, new and future technologies, as well as of the company's main technologies in terms of their being environmental assets or liabilities. The management audit will also cover the organizational aspects of technological decision-making and the structural links between top management, R&D, engineering, production, marketing and the environmental group in ensuring the necessary conditions.

5.2.3.2
External affairs

An inadequate external relations program can nullify an excellent performance in pollution prevention and technology selection. The management audit should therefore examine the enterprise's organization for external and public affairs and evaluate how it performs.

Internally the environmental management audit examines how the function is structured, staffed and managed. It evaluates the links with top management and other corporate functions. It also evaluates how the enterprise communicates with its employees on environmental issues and how this important source of information and support is being managed.

Externally the audit reviews the degree and the quality of the enterprise involvement with the outside world. This is a critical area because there are so many external contacts, links and activities that the effective use of limited management resources needs objective evaluation.

A key objective of auditing external affairs is to assess the liabilities and risks that exist in the enterprise's relations with the communities, local and national government and environmental pressure groups. Opportunities to reduce such risks and liabilities are identified and appropriate corrective measures should be recommended. In essence, the audit assesses whether the enterprise has the optimum level and quality of communication with the society, both in terms of its involvement and participation in discussions on environmental issues and effective feedback all parts of the enterprise.

5.2.3.3

Management

A very important aspect of the environmental management audit is the assessment of the total management system in terms of its being an asset or a liability for the company's environmental performance. Thus evidence would be collected in relation to environmental policies and attitudes, knowledge and skills at all levels and all divisions of the organization; environmental training activities; environmental organization; and environmental successes and failures which have occurred.

The environmental management audit could focus on any of the above areas. Alternatively, a comprehensive audit could be carried out in a series of stages. In the first stage, for example, the audit could focus on the corporation's management systems and the role of top management in establishing and maintaining an environmentally-oriented culture. Interviews could be carried out with the chief executive officer and other members of top management to assess the state of environmental policies and practices in the company and to map out the environmental organization and the allocation of environmental responsibilities. This could also be the opportunity to make the first inventory of recent environmental successes and failures.

The next stage could get involvement of senior line managers and those staff managers directly responsible for environment, external affairs and technology. In this stage of the environmental management audit, each area could be assessed to ascertain whether and how it represented an asset or a liability for the environmental performance of the enterprise. Each of its areas of responsibilities would be listed with its activities and its decision areas. For each would be determined the extent of the impact on the environment and what specific management steps and management resources were being allocated to minimize the negative and maximize the positive impact on the environment. Again, the balance could be drawn between recent environmental successes and failures and the present organization, procedures, staffing and attitudes representing an asset or a liability for the enterprise.

The final stage of the audit would involve an investigation of each production facility and product in order to assess the extent to which they conform to all regulatory requirements or represent risks for the environment and hence the enterprise. Each stage of the audit would be accompanied by a report which could set out not only the present situation, but also managerial proposals to correct any deficiencies and to exploit any strengths.

5.2.3.4

Confidentiality and disclosure

Some managers considering whether to conduct environmental audits of their facilities have expressed concern that information generated by environmental audits, if made public, could be embarrassing or trigger inspection or enforcement actions that would not otherwise occur. This perception is perhaps the greatest deterrent to wider corporate implementation of environmental auditing. Information developed by an audit can expose a company and its managers to civil and criminal liability, and is potentially subject to pretrial discovery by an adversary party during litigation. However, research and the experience of individuals responsible for ongoing corporate audit programs suggest that this threat of disclosure has neither restricted program effectiveness nor created adverse consequences (UNEP, 1988).

Two basic auditing premises need to be re-stated. First, companies without serious commitment to environmental goals rarely implement an environmental auditing program. In companies with auditing programs, top managers seem to analyze their situation and conclude that the potential benefits from auditing are greater than the potential costs involved, including any costs from possible information disclosures. This may be true either because the risk of disclosure is low, or because the firm has little to lose even if data is disclosed. Second, the audit is a working, evolving management tool used to analyze and evaluate internal corporate procedures, and should not be regarded as primarily a generator of legal documents designed to function in legal areas.

Industry representatives have noted that it is becoming increasingly difficult for corporate management to ignore environmental problems, whether actual or potential, at the plant level. Deliberate or structured ignorance occurs more and more at the managers' professional and personal peril. Environmental auditing can help reduce these dangers by instituting procedures for managing compliance more effectively and ensure that problems are promptly identified and remedied before the arrival of a government inspector, a reporter, or third party complaint. A properly structured audit program can achieve these benefits without substantial confidentiality concerns.

5.2.4
Facility environmental auditing

A facility environmental audit is defined as a basic in-plant management tool comprising of a systematic, documented, periodic and objective evaluation of how well facility environmental management systems and equipment are performing. The aim of the audit is to facilitate effective control of environmental practices, and to enable the facility to assess compliance with company policies, including meeting regulatory requirements. Facility environment auditing is seen as an internal process that should become a necessary and routine part of most, if not all, industrial management whatever the size of the company (UNEP, 1988).

The concept of facility environmental auditing came into being during the early 1970 s, but under the guise of a number of different approaches and names: environmental reviews, surveys, assessments and quality controls. Today the term "environmental audit" is widely used. In practice, such audits are carried out by a small, qualified, independent team of people who visit a particular site to check the environmental program and performance of a plant. Although there is no set way in which an environmental audit is performed, the procedures have become more formalized.

One great advantage of regular facility environmental auditing is that it provides the company with a greater overall awareness of its workers and processes, identifying compliance problems and areas of risk, pinpointing both strengths and weaknesses, and encouraging continual improvement. In that regard facility environmental auditing encourages the use of low-waste technologies, prudent utilization of resources, and identification of potential hazards and risks.

The scope and requirements for facility environmental audits will undoubtedly vary from one plant to another and there is not necessarily any one single approach. Different approaches and attitudes to environmental audits may accommodate each company's special needs. The decision is as much a question of how as it is why. Any decision to conduct an environmental audit must be based on an adequate assessment of the strategies available. This is particularly important in the context of smaller companies with limited financial resources. It has also been observed that, depending on existing circumstances and corporate organizational resources, it may not be desirable to conduct an extensive and formal environmental audit. At the same time, companies which forego any method of extraordinary self-study should recognize a continuing necessity to reevaluate this decision periodically in light of changing circumstances, particularly when regulatory agencies establish specific programs.

5.2.4.1
Company-wide vs. plant specific
To a large extent, the issue of company-wide vs. plant specific audits may be decided on the basis of a risk assessment among other additional practical considerations. It is hard to imagine that a company would voluntarily initiate an audit program on a multi-plant basis from the outset. Too little is known about how to apply an abstract auditing approaching to the specific operations of a particular company, particularly in the case of highly diversified companies. To complicate matters further, even companies in which divisions are involved in the same general business, such as foundries, may challenge a single auditing approach. This practical reality calls for a prototype framework based on representative information about the operations previously studied. This framework would establish detailed procedures for a company-wide audit or perhaps a series of distinct plant audits, each with its own unique characteristics.

In any case, it is important to determine initially whether any company-wide strategy would be cost effective in terms of the potential usefulness of data it would generate. If it yields only an unmanageable mass of data which memorializes noncompliance but does not assist in management decisions, it will not be valuable.

5.2.4.2
One facility vs. another
If a particular company opts to audit on a selective basis, it will then need to decide at which facilities it will concentrate efforts and for which environmental programs. Some companies may choose to audit all locations for a selected

group of programs or one or two locations for numerous programs (including air, water, solid and hazardous wastes, and underground injection). As mentioned above, the selection of particular environmental requirements for study is largely a function of the type of manufacturing process being conducted. Where the type of process does not afford sound decision criteria, the selection may reflect on a balancing of known or perceived exposures.

Where the decision involves the selection of one plant over another, a number of factors may be considered: the quality of existing information at each plant, age, size, number of past enforcement actions or noncompliance problems, locations, type of process conducted, quality of plant management and personnel control, expenditures for pollution control equipment, possible expansion of operations, possible sale, and the degree of past scrutiny by corporate headquarters (Guida, 1982).

5.2.4.3
Audit report format

The audit report should concentrate factual, objective observation. It is both a comfort statement and an exception list. Recommendations and good practices at the facility should both be noted. An audit report format is described below:

PART I:Executive summary

PART II:Detailed audit report
1. Scope
2. Audit participants
3. Exit interview
4. Definitions:
   a. Significant findings
   b. Noncompliance findings
   c. Notable findings
5. Audit results
   a. Air
      (1) Significant findings
      (2) Noncompliance findings
      (3) Notable findings
   b. Water
      Additional Sections dependent on the scope of the audit.
   c. Wastes

5.2.5
Development and implementation of environmental auditing

This sub-section provides basic information to develop and implement an environmental auditing. There are four essential phases of an environmental auditing: (1) starting an audit; (2) pre-audit activities; (3) audit inspection; and (4) post-audit activities (NYSDEC, 1989 and ICC, 1989).

5.2.5.1
Starting an audit

It should begin with a commitment from the top management and a statement of goals, usually in the form of a policy statement. The statement should make a firm commitment to pollution prevention and establish some broad goals for the audit such as protecting the environment and reducing the cost of treating and disposing wastes.

The policy statement will be stronger if it establishes quantifiable goals, such as reducing waste by 10% in two years or eliminating all hazardous waste in ten years. It is very important to be realistic in the establishment of these goals. Naming specific goals and clearly stating the commitment of the company to the pollution prevention program helps to overcome various barriers such as a lack of awareness of the benefits of pollution prevention, limitations of technical staff, and concern over tampering with an established process.

5.2.5.2
Pre-audit activities

Preparation for each audit covers a number of activities including (1) selecting the review site and audit team; (2) developing an audit plan which defines the technical, geographic and time scope; (3) obtaining background information on the plant (for example by means of a questionnaire) and the criteria to be used in evaluating programs. The intent of these activities is to minimize time spent at the site and to prepare the audit team to operate at maximum productivity throughout the on-site portion of the audit.

Depending on the nature and size of the facility, much of the data needed for a pollution prevention program may already have been collected as a normal part of plant operations or as part of a response to existing regulatory requirements. A number of information sources to consider for facility include: (1) Regulatory information such as waste shipment manifests, emission inventories, biennial hazardous waste reports, permits and/or permit applications; (2) Process information such as process flow diagrams, design and actual material and heat balances; (3) Raw material/production information such as product composition and batch sheets, material application diagrams, material safety data sheets, product and raw material inventory records, operator data logs, operating procedures, and production schedules; and (4) Accounting information such as waste treatment and disposal costs, water and sewer costs, costs for non-hazardous waste disposal, and costs of energy and material.

With respect to the composition of the audit team, there are both advantages and disadvantages in including a member from the site being audited. Advantages include: (1) the insider's knowledge of the specifics of the plant as regards both physical installations and organizational patterns; and (2) associating a local employee with the audit report may make it appear more credible to the plant's workforce. The main disadvantage is that the insider may have difficulty in taking or expressing an objective view, especially if this might be seen as criticism of his superior or immediate colleagues. Independent consultants may provide assistance especially to smaller companies, in the event of a lack of internal expertise.

5.2.5.3
Audit inspection
Audit inspection at the site typically include five basic steps:

1. Identifying and understanding management control systems. Internal controls are incorporated in the facility's environmental management system. They include the organizational monitoring and record-keeping procedures, formal planning documents such as plans for prevention and control of accidental release, internal inspection programs, physical controls such as containment of released material, and a variety of other control system elements. The audit team gains information from numerous sources through use of formal questionnaires, observations, and interviews.

2. Assessing management control systems. The second step involves evaluating the effectiveness of management control systems in achieving their objectives. In some cases, regulations specify the design of the control system. For example, regulations may list specific elements to be included in plans for responding to accidental releases. More commonly, team members must rely on their own professional judgment to assess adequate control.

3. In this step the team gathers evidence required to verify that the controls do in practice provide the result intended. Team members follow testing sequences outlined in the audit protocol which have been modified to consider special conditions at the site. Examples of typical tests include review of a sample of effluent monitoring data to confirm compliance with limits, of training records to confirm that appropriate people have been trained, or of purchasing department records to verify that only approved waste disposal contractors have been used. All of the information gathered is recorded for ease of analysis and as a record of conditions at the time of the audit. Where a control element is in some way deficient, the condition is recorded as a "Finding".

4. Evaluating audit findings. After the individual controls have been tested and team members have reached conclusions concerning individual elements of the control system, the team meets to integrate and evaluate the findings and to assess the significance of each deficiency of pattern of deficiencies in the overall functioning of the control system. In evaluating the audit findings, the team confirms that there is sufficient evidence to support the findings and summarizes related findings in a way that most clearly communicates their significance.

5. Reporting audit findings. Findings are normally discussed individually with facility personnel in the course of the audit. At the conclusion of the audit, a formal exit meeting is held with facility management to report fully all findings and their significance in the operation of the control system.

The team may provide a written summary to management which serves as an interim report prior to preparation of the final report. Some of the information can be gathered during site inspection by asking typical questions such as:

- What is the composition of the wastestreams and emissions generated in the facility? What is their quantity?
- From which production processes or treatments do these wastestreams and emissions originate?
- Which waste materials and emissions fall under environmental regulations?
- What raw materials and input materials or production process generate these wastestreams and emissions?
- How much of a specific raw or input material is found in each wastestream?
- What quantity of materials is lost in the form of volatile emissions?
- How efficient is the production process and the various steps of that process?
- Does mixing materials produce any unnecessary waste materials or emissions which could otherwise be reused with other waste materials?
- Which good housekeeping practices are already in place to limit waste generation (Evers, 1995)?

5.2.5.4
Post-audit activities

At the conclusion of the on-site audit two important activities remain: preparation of the final report with implementation plan and development of a corrective action program.

(1) Final audit report

The final audit report is generally prepared by the team leader and, after review in draft by those in a position to evaluate its accuracy, it is provided to appropriate management.

(2) Action plan preparation and implementation

Facility personnel, sometimes assisted by the audit team or outside experts, develop a plan to address all findings. This action plan serves as a mechanism for obtaining management approval and for tracking progress toward its completion. It is imperative that this activity take place as soon as possible so that management can be assured that appropriate corrective action is planned. A primary benefit of the audit is lost, of course, if corrective action is not taken promptly.

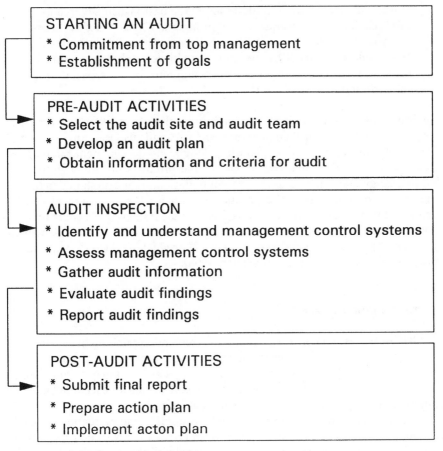

Fig. 5-2 Basic steps of an environmental audit

Follow-up to ensure that the corrective action plan is carried out and all necessary corrective action is taken as an important step. This may be done by an audit team, by internal environmental experts, or by management. A summary of basic steps of an environmental audit is shown in Fig. 5-2.

5.3
Designing environmentally compatible products

Designing environmentally compatible processes and products is a new and challenging frontier for engineering professionals. As stated in the National Research Council report on Frontiers in Chemical Engineering (NRC, 1988):

"Chemical plants have been designed in the past principally to maximize reliability, product quality and profitability. Such issues as chronic emissions, waste disposal and process safety have often been treated as secondary factors. It has become clear, however, that these considerations are as important as the others and must be addressed during the earliest design stages of the plant."

Environmentally compatible products minimize the adverse effects on the environment resulting from their manufacture, use, and disposal. Environmental considerations during product planning, design, and development can help industries minimize the negative impact of the products on the environment. While changing product design to prevent pollution, engineers should maintain the quality or function of the product. Design for the environment can be achieved by the people directly involved, within the framework of company policy and with support from company management. It can be implemented through: (1) product life-cycle assessment, (2) product life-cycle design, and (3) product life-cycle costing analysis (Shen, 1996).

5.3.1
Product life-cycle assessment

A full life-cycle assessment (LCA) requires a tremendous amount of data to complete even the inventory component for all of the inputs and outputs of each component through out the life-cycle. For some purposes, it may be sufficient to use simplified LCA, which does not attempt to collect data for each input and out put of each component of product in each state of the life-cycle.

The life-cycle concept is simply the holistic approach to evaluating the environmental impacts of a product system from cradle to grave. Life-cycle assessment in its fully quantitative form is the rigorous analytical application of the life- cycle concept. It is a tool for providing certain types of information, which can, in turn, lead to research and development of improved materials and technologies. Life-cycle assessment can make manufacturers and individuals more aware of the environmental consequences of the products they buy and use. It can help foster development of other tools that better integrate the needs of the policy and science-based communities. It can also complement, not substitute, for efforts directed at eliminating uses of substances posing unacceptable risk (WWF, 1993).

The definition and description of product life-cycle assessment, developed internationally by the Society for Environmental Toxicology and Chemistry (SETAC), is as follows:

"Life-cycle assessment is an objective process to evaluate the environmental burdens associated with a product, process or activity by identifying and quantifying energy and materials used and wastes released to the environment, to assess the impact of those energy and material uses and releases to the environment, and to evaluate and implement opportunities to affect environmental improvements. The assessment includes the entire life-cycle of the product, process or activity, encompassing extracting and processing raw materials; manufacturing, transportation and distribution; use, re-use, maintenance; recycling and final disposal." (SETAC, 1993).

Environmentally compatible products are designed for durable, disposable, disassemblable, recyclable, re-usable. Product and process designs employ a systems analysis and decision-making process where goals depend upon the type of products and processes as well as the specific environmental concerns to be addressed at each stage in the product life-cycle. Capital goods, consumer goods, commodity-type products, and buildings have very different life-cycles, materials, environmental impacts, economics, and thus very different design solutions. Table 5-1 shows how product differences can affect the approach and outcome of life-cycle analysis and the contribution that product design or redesign can make (Henn and Fava, 1993).

Product life-cycle assessment is an objective tool or method to identify and evaluate all aspects of environmental effects associated with a specific product of any given activity from the initial gathering of raw material from the earth, until the point at which all residuals are returned to the earth. The method can also be used to evaluate the types and quantities of product inputs, such as raw materials, energy, and water, and as product outputs, by-products, and waste that affect various resources management options to create sustainable systems (Vigon et al., 1993).

Table 5-1 Product differentiation in life-cycle analysis

Consumable products and packages	Durable goods and large capital projects
Low technology	High technology
All product sizes	Mostly large size
Large volumes of solid waste	Lesser volumes of solid waste
Widespread wastes	More concentrated wastes
Waste release in manufacturing	Waste release in life-cycle
Moderate products recycling	High product recycling
High packages recycling	High recycling when applicable
Short life-cycle duration	Long life-cycle duration
Low purchase price of item	High purchase price of item
Small cost of ownership	Large cost of ownership
Little after-sale operation	Extensive after-sale operation
Little after-sale service	Extensive after-sale service

Source: Henn and Fava, 1993, p. 544 with modification

Life-cycle assessment (LCA), as shown in Fig. 5-3, is composed of four components: (1) goal definition and scope; (2) inventory analysis; (3) impact analysis; and (4) improvement analysis. A brief discussion of these four components follows:

5.3.1.1
LCA goal definition and scope

Goal definition and scope take place throughout the life-cycle assessment. Life-cycle assessment is the evaluation of the effects of all inputs and outputs of a product during its life-cycle. The scope of the assessment defines the system boundaries that are necessary to ensure that the analysis addresses the purpose of the assessment. The scope also includes definition of assumptions, data requirements, and limitations of the assessment (SETAC, 1993).

5.3.1.2
Inventory analysis

Inventory analysis is a technical, data-based process which quantifies energy and raw material requirements, atmospheric emissions, waterborne effluents, solid wastes, and other releases for the entire life-cycle of a product, package, process, material, or activity (USEPA, 1993b). Inventory analysis begins with raw material

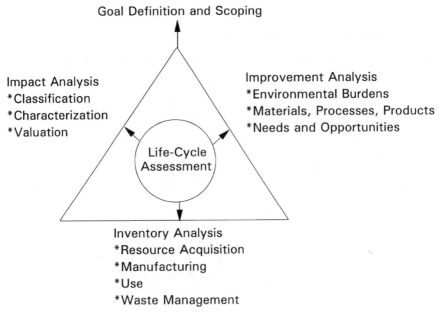

Fig. 5-3 Product life-cycle assessment framework

extraction and continues through final product consumption and disposal. The steps of performing a life-cycle inventory include (Vigon et al., 1993):

(1) Defining the purpose and scope of the inventory – It is based on one or several objectives such as (a) establishing a baseline of information on a system's overall resource use, energy consumption, and environmental loading; (b) identifying stages within the life-cycle of a product or process where a reduction in resource use and emissions might be achieved; (c) comparing the system inputs and outputs associated with alternative products, processes, or activities; (d) guiding the development of new products, processes, or activities toward a net reduction of resource requirements and emissions; and (e) identifying areas to be addressed during life-cycle impact analysis.

(2) Defining the system boundaries – A complete life-cycle inventory will set the boundaries of the total system broadly to quantify resource and energy use and environmental releases throughout the entire life-cycle of a product or process, as shown in Fig. 5-4.

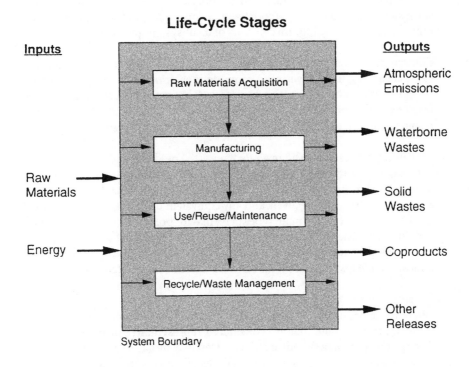

Source: EPA/600/R-92/245, February, 1993.

Fig. 5-4 Define system boundaries

(3) Devising an inventory checklist – A checklist can be prepared to guide data collection and validation and to enable construction of the computational model. Figure 5-5 shows a generic example of an inventory checklist and an accompanying data worksheet.

(4) Instituting a peer review process – A peer review process should address the four areas: (a) scope/boundaries methodology; (b) data acquisition/compilation; (c) validity of key assumptions and results; and (d) communication of results. The peer review panel should consist of a diverse group of 3 to 5 individuals representing various sectors: such as government, academia, industry, environmental or consumer groups, and life-cycle practitioners.

(5) Gathering data – Data collected for an inventory should always be associated with a quality measure. Formal data quality monitors such as accuracy, precision, representativeness, and completeness are strongly preferred.

(6) Developing stand-alone data – Stand-alone data must be developed for each subsystem to fit the subsystem into a single system. There are two goals to achieve in this step: (1) presenting data for each subsystem consistently; (2) developing the data in terms of the life-cycle of only the product being examined in the inventory. A standard unit of output must be determined for each subsystem. All data could be reported in terms of the production of a certain number of pounds, kilograms, or tons of a subsystem product.

(7) Constructing a computational model – It consists of incorporating the normalized data and material flows into a computational framework using a computer spreadsheet or other accounting technique. The systems accounting data that result from the computations of the model give the total results for energy and resource use and environmental releases from the overall system.

(8) Presenting the results – The inventory report should explicitly define the systems analyzed and the boundaries that were set. All assumptions made in performing the inventory should be clearly explained. The basis for comparison among systems should be given, and any equivalent usage ratios that were used should be explained. Two main types of format for presenting results are tabular and graphical. The choice among the various ways of creating tables and graphs varies, based on the purpose and scope of the study. Graphical presentation of information helps to augment tabular data and can aid in interpretation.

(9) Interpreting and communicating the results – An important criterion in understanding or interpreting the results is data accuracy. Many life-cycle inventories present data that are considered representative of the profiled industry or group. Data for one particular step may be gathered from a number of manufacturers and production facilities. Final conclusions about the results of inventory studies may involve making value judgements regarding

the relative importance of air and water quality, solid waste issues, resource depletion, and energy use. Based upon each individual's locale, background, and life style, value judgements vary (Vigon et. al, 1993).

The methodology for inventory analysis is considered to be well-defined and understood, but there are some key issues that are still being discussed in the practitioner community. These include the sources, availability and quality of data, the allocation procedures for co-products, the inclusion of waste management impacts, and the allocation of inputs and outputs in a system involving recycling (SETAC, 1991).

A product life-cycle inventory quantifies three categories of environmental releases: atmospheric emissions, wastewater effluents and solid and hazardous wastes. Products and by-products also are quantified. Most inventories consider environmental releases to be actual discharges of pollutants or other materials from a process or operation under evaluation.

(1) Atmospheric emissions – Emissions of air pollutants mostly come from the process, production and combustion of fuel for process or transportation. Emissions may also be released from material and waste storage facilities. A life-cycle inventory should include these emissions, reported in units of weight. Air emissions include all substances classified as pollutants per unit weight of product output.

(2) Wastewater effluents – As with atmospheric emissions, wastewater effluents result mostly form the process, production and combustion of fuels. A life-cycle inventory should also include these effluents. They are reported in units of weight and include all substances generally regarded as pollutants per unit of product output. The discharge of wastes and spills to the ocean or some other body of water, should be reported as waste.

(3) Solid and hazardous wastes – They should include discarded materials, by-products, off-spec products, process solid waste, slugs and residues from waste treatment facilities. They are generally reported in units of weight per day or per unit of product output.

Most analysts prefer using a process flow diagram which guides them to direct the collection of data and to construct inputs and outputs of a given industrial process. The concept and practice can be effectively used for product life-cycle inventory. Fig. 5-5_provides a life-cycle inventory checklist of tow parts: Part I covers the inventory scope and procedures and Part II covers the inventory module inputs and outputs.

LIFE-CYCLE INVENTORY CHECKLIST PART I—SCOPE AND PROCEDURES
INVENTORY OF: _____

Purpose of Inventory: (check all that apply)

Private Sector Use

Internal Evaluation and Decision Making
- ☐ Comparison of Materials, Products, or Activities
- ☐ Resource Use and Release Comparison with Other Manufacturer's Data
- ☐ Personnel Training for Product and Process Design
- ☐ Baseline Information for Full LCA

External Evaluation and Decision Making
- ☐ Provide Information on Resource Use and Releases
- ☐ Substantiate Statements of Reductions in Resource Use and Releases

Public Sector Use

Evaluation and Policy-making
- ☐ Support Information for Policy and Regulatory Evaluation
- ☐ Information Gap Identification
- ☐ Help Evaluate Statements of Reductions in Resource Use and Releases

Public Education
- ☐ Develop Support Materials for Public Education
- ☐ Assist in Curriculum Design

Systems Analyzed
List the product/process systems analyzed in this inventory: _____
_____

Key Assumptions: (list and describe)
_____
_____
_____
_____

Define the Boundaries
For each system analyzed, define the boundaries by life-cycle stage, geographic scope, primary processes, and ancillary inputs included in the system boundaries.

Postconsumer Solid Waste Management Options: Mark and describe the options analyzed for each system.
- ☐ Landfill _____
- ☐ Combustion _____
- ☐ Composting _____
- ☐ Open-loop Recycling _____
- ☐ Closed-loop Recycling _____
- ☐ Other _____

Basis for Comparison
- ☐ This is not a comparative study.   ☐ This is a comparative study.

State basis for comparison between systems: *(Example: 1000 units, 1,000 uses)* _____

If products or processes are not normally used on a one-to-one basis, state how equivalent function was established.
_____

Computational Model Construction
- ☐ System calculations are made using computer spreadsheets that relate each system component to the total system.
- ☐ System calculations are made using another technique. Describe: _____

Describe how inputs to and outputs from postconsumer solid waste management are handled. _____
_____

Quality Assurance: (state specific activities and initials of reviewer)
Review performed on:
- ☐ Data Gathering Techniques _____
- ☐ Coproduct Allocation _____
- ☐ Input Data _____
- ☐ Model Calculations and Formulas _____
- ☐ Results and Reporting _____

Peer Review: (state specific activities and initials of reviewer)
Review performed on:
- ☐ Scope and Boundary _____
- ☐ Data Gathering Techniques _____
- ☐ Coproduct Allocation _____
- ☐ Input Data _____
- ☐ Model Calculations and Formulas _____
- ☐ Results and Reporting _____

Results Presentation
- ☐ Methodology is fully described.
- ☐ Individual pollutants are reported.
- ☐ Emissions are reported as aggregrated totals only.
  Explain why: _____
  _____
- ☐ Report is sufficiently detailed for its defined purpose.
- ☐ Report may need more detail for additional use beyond defined purpose.
- ☐ Sensitivity analyses are included in the report. List: _____
- ☐ Sensitivity analyses have been performed but are not included in the report. List: _____

Fig. 5-5 Life-cycle inventory checklist

LIFE-CYCLE INVENTORY CHECKLIST PART II—MODULE WORKSHEET

Inventory of: _____ Preparer: _____

Life-Cycle Stage Description: _____

Date: _____ Quality Assurance Approval: _____

MODULE DESCRIPTION: _____

	Data Value[a]	Type[b]	Data[c] Age/Scope	Quality Measures[d]
MODULE INPUTS				
Materials				
Process				
Other[e]				
Energy				
Process				
Precombustion				
Water Usage				
Process				
Fuel-related				
MODULE OUTPUTS				
Product				
Coproducts[f]				
Air Emissions				
Process				
Fuel-related				
Water Effluents				
Process				
Fuel-related				
Solid Waste				
Process				
Fuel-related				
Capital Repl.				
Transportation				
Personnel				

(a) Include units.

(b) Indicate whether data are actual measurements, engineering estimates, or theoretical or published values and whether the numbers are from a specific manufacturer or facility, or whether they represent industry-average values. List a specific source if pertinent, e.g., "obtained from Atlanta facility wastewater permit monitoring data."

(c) Indicate whether emissions are all available, regulated only, or selected. Designate data as to geographic specificity, e.g., North America, and indicate the period covered, e.g., average of monthly for 1991.

(d) List measures of data quality available for the data item, e.g., accuracy, precision, representativeness, consistency-checked, other, or none.

(e) Include nontraditional inputs, e.g., land use, when appropriate and necessary.

(f) If coproduct allocation method was applied, indicate basis in quality measures column, e.g., weight.

Fig. 5-5 (continued)

5.3.1.3
Impact analysis
The purpose of impact analysis is to evaluate impacts and risks associated with the material and energy transfers and transformations quantified in the inventory data into environmental effects is achieved through a wide range of impact analysis models, including hazard and risk analysis models. The final result of an impact analysis is an environmental profile of the product system.

Impact analysis is a systematic process which identifies, characterizes, and values potential ecosystem, human health, and natural resource impacts associated with the inputs and outs of a product or process system. Impact analysis is a technical, quantitative, and/or qualitative process. The analysis addresses both ecological and human health impacts, resource depletion, and possibly social welfare. The impact analysis component covers other effects that an inventory cannot easily quantify, such as habitat modification and heat and noise pollution (USEPA, 1993a).

Impact analysis evaluates impacts caused by design activities. The final result of an impact analysis is an environmental profile of the product system. Inventory data can be translated into environmental impacts through many different models. Most impact models are centered on hazard and risk assessment. Environmental impacts can be organized into resource depletion, ecological degradation, human health effects, and other human welfare effects.

(A) Resource depletion
Acquisition has two basic environmental consequences. One is ecological degradation from habitat disruption (e.g., physical disruption from mining). The other is reduction in the global resource base.

(B) Ecological degradation
Risk assessment includes many of the elements of human health risk assessment but is much more complex. The ecological stress agents, as well as the potentially affected ecosystem must be identified. Ecological stress agents can be categorized as chemical, physical, or biological.

(C) Human health effects
Health risk assessment includes hazard identification, risk assessment, exposure assessment, and risk characterization. Human health and safety risks can also be assessed using models that evaluate process system reliability.

Impact analysis represents one of the most challenging analysis functions of product systems development. Although current methods for evaluating environmental impacts are incomplete, impact assessment is important, because it enables designers and planners to understand the environmental consequences of a design more fully. Even when models exist, they may rely on many assumptions or require considerable data. Despite these problems, some form of impact analysis helps designers and planners better understand the environmental consequences of a design.

The conceptual framework for life-cycle impact analysis that has received some degree of consensus includes three steps: classification, characterization, and valuation. Classification is the assignment of inventory items to impact categories based upon an elaboration of impact networks, which are the potential cause-and-effect linkages between inventory items and ultimate impacts on human health, the ecosystem, and resources.

Characterization is the process of aggregating and quantifying impacts within the impact categories which involves an understanding of the environmental processes that lead from the inventory item to the ultimate impact of concern. Valuation is the step where the different impact categories are considered in relation to each other in order to further aggregate the results of the impact assessment or to allow for further interpretation of the results. The valuation step is not an objective process, but depends upon social and cultural values and preferences. For instance, the importance ascribed to human health impacts versus ecological impacts is a cultural, ethical, and political issue (USEPA, 1993b; SETAC 1993).

Front-end environmental impact analysis can pay big dividends in many cases through redesign efforts, but a cost-risk-benefit analysis of most proposed design changes at each stage in the life-cycle, and in total, is also essential to a sound environmental strategy. Table 5-2 is a summation of the multiplicity of impacts on almost every type of business activity by "green product design". Optimizing the many life-cycle impacts of products, packages, and processes on producers, consumers, and the environment is a future challenge (Henn and Fava, 1993).

5.3.1.4
Improvement analysis

The results of both inventory and impact assessments are valuable in understanding (1) the relative environmental burdens resulting from evolutionary changes in given processes, products, or packaging over time; (2) the relative environmental burdens between alternative processes or materials used to make, distribute, or use the same product; and (3) the comparison of the environmental aspects among alternative products that serve the same use.

Improvement analysis systematically evaluates needs and opportunities to reduce the environmental burden associated with energy and raw material use and waste releases throughout the life-cycle of a product, process, or activity. This analysis may include both quantitative and qualitative measures of improvements.

The analytical information from one phase can complement the other two phases. For example, the inventory alone may identify opportunities to reduce waste releases, energy consumption, or material use. Impact analysis typically identifies the activities with greater and lesser environmental effects, while improvement analysis helps ensure that any potential reduction strategies are optimized and that improvement programs do not produce additional and unanticipated adverse impacts on human health and the environment (Vigon et al., 1993).

5.3.1.5
Limitations of life-cycle assessment

Experience indicates that quantitative life-cycle assessments will be difficult becoming the decision-making documents that their advocates hope. Four factors restrict the use of quantitative life-cycle assessments as decision-making documents:

(1) Cost limitations – Society can hardly afford it because life-cycle assessment requires a large amount of data and answers many endless questions.

(2) Data limitations – Much of the critical data on product production is considered proprietary business information. Access to such data is severely restricted even to government regulators.

(3) Lack of common unit – No common unit exists for comparing different types of environmental impacts. Some researchers have gravitated toward dollars as the common unit for certain studies, but the values assigned to environmental costs and benefits are often wholly subjective. Others use pounds of pollutants or Btus of energy to report impacts.

(4) Inability to quantify important environmental impacts – Even if agreement could be reached on a common unit and differing impacts could be converted to it, the problem of quantifying certain impacts would still exist. This is true especially in terms of the value of such impacts such as human life versus ecological damages.

5.3.2
Product life-cycle design

Environmental criteria often are not considered at the beginning of the design phase when it is easiest to eliminate potential negative effects through a product's life-cycle. As a result, most industries continue to spend more resources to fix problems, rather than prevent them. Although the most innovative firms are adopting ambitious environmental policies to shift their focus to prevention, translating these policies into successful action is a major challenge. In particular, without proper support, environmental design programs, which are vital to this process, may be launched without specific objectives, definitions, or principles (Keoleian et al., 1993).

Products of high quality have little environmental impact and less cost to produce may sound like a far-off fantasy, but researchers and designers are making them a reality through a concept call green design. Green design means waste prevention: by substituting less hazardous materials for more hazardous materials; by designing materials and products that use materials more effectively; or by designing products that can be recycled or remanufactured more easily. The Congressional Office of Technology Assessment, which did a report "Green

Table 5-2 A summation of the multiplicity of impacts by green product design

Key development teams	Opportunity	Feasibility	Design	Preproduction	Production	Sales & distribution	Education & regulation	Promotion
Research & development	Identify alternative products and/or applications	Assess existing and alternative products' environmental impact (LCA)	Design for: Least impact, Long life, Secondary use, Recycling, Maintenance and repair, Disposability	Weigh animal testing vs. alternative testing	Develop packaging recycling infrastructures	Develop product recycling infrastructures	Establish stakeholder alliances and third party endorsement	Certify green products and processes
Manufacturing and technology	Identify alternative materials, energy sources, processing methods	Assess extraction, transporting, processing, and waste disposal environmental impact	Design for pollution prevention and zero waste	Comply with federal and state regulations and local environmental ordinances	Practice quality environmental management	Green audit packing materials, transport vehicles, storage facilities, etc.	Establish stakeholder alliances and third party endorsement	Transfer green technology
Finance and economics	Compare the cost of managing for conformance vs. managing for assurance	Project the cost of impact on: Human and environmental health, Liability, Resource supply	Institute sustainable development planning	Investigate energy savings and resource conservation incentives	Consider product development partnerships and joint business ventures	Lobby for free market environmentalism	Establish stakeholder alliances and third party endorsement	Budget for environmental education, purchase incentives and stakeholder alliance programs
Marketing	Conduct consumer and customer environmental marketing research	Conduct product purchase, usage and performance research	Satisfy consumers' and customers' basic needs such as value, price, and performance	Engage support of green suppliers, ecopreneurs, and environmentalists	Be attentive to community relations and community right-to-know	Support trade with environmental education and community outreach programs	Promote consumer environmental education and empowerment	Comply with state green labeling regulations and FTC guidelines

Source: Coddington Environmental Marketing, Inc., New York.

Products by Design: Choices for a Cleaner Environment," states that green design should incorporate environmental objectives while maintaining product performance, useful life and functionality (Chemecology, 1993).

5.3.2.1
LCD objectives and scopes

The primary objective of life-cycle design (LCD) is to reduce total environmental impacts and health risks caused by product development and use. This objective can only be achieved in concert with other life-cycle design goals. Life-cycle design seeks to:

– Conserve resources;
– Prevent pollution;
– Support environmental equity;
– Preserve diverse, sustainable ecosystems; and
– Maintain long-term, viable economic systems.

Life-cycle design addresses the entire product system, not just isolated components. In life-cycle design, we must focus on the four components of a product that include: (1) product, (2) process, (3) distribution network, and (4) management responsibilities. We then incorporate these elements in the product design phase and assure that every activity related to making and using products is considered.

(1) Product consists of all materials in the final product and includes all forms of these materials from acquisition to disposal.

(2) Processing transforms materials and energy into intermediary and final products.

(3) Distribution network consists of packaging systems and transportation networks used to contain, protect, and transport items.

(4) Management responsibilities include administrative services, financial management, personnel, purchasing, marketing, customer services, and training and educational programs. The management component also develops information and conveys it to others.

The process, distribution, and management components can be further classified into several subcomponents: facility or plant, operations or process steps, equipment and tools, labor, secondary material inputs, and energy (Keoleian and Meanery, 1993).

5.3.2.2
Managing design

Design actions translate life-cycle goals into high-quality, low-impact products. Corporate management must exert a major influence on all phases of product life-cycle design. Commitment from all levels of management is a vital part of life-cycle design. The progress of LCD programs should be monitored and assessed using clearly established environmental and financial measures. Appropriate measures of success are necessary to motivate individuals to pursue environmental impact and health-risk reductions.

Life-cycle design can be applied to products in any number of ways:

(1) Improvement, or minor modifications of existing products and processes;

(2) New features associated with developing the next generation of an existing product or process; and

(3) Innovations characteristic of new designs.

No single design method or set of rules applies to all types of products. Designers should use the available information and references to develop tools best suited to their specific projects.

Companies that look beyond quick profits to focus on customers and cooperation with suppliers have the best chance of succeeding with life-cycle design. Achieving this depends on cross-disciplinary teams. These teams may include any of the following participants: accounting, advertising, community, customers, distribution, packaging, environmental resources staff, government regulators, industrial designers, lawyers, management, marketing and sales, process designers and engineers, purchasing, production workers, research and development staff, and service personnel. Effectively coordinating these teams, however, and balancing the diverse interests of all participants present a significant challenge.

5.3.2.3
Design requirements

Formulating design requirements is one of the most critical activities in life-cycle design. Requirements define products in terms of functions, attributes, and constraints. Functions describe what a successful design does, not how it is accomplished. Attributes are additional details that provide useful descriptions of functions. Constraints are conditions that the design must meet to satisfy project goals. Constraints provide limits on functions that restrict the design search to manageable areas. Considerable research and analysis are needed to develop proper requirements. Too few requirements usually indicate that the design is ambiguous (Keoleian et al., 1993).

A well-conceived set of requirements translates project objectives into a specific solution. In life-cycle design, environmental functions are critical to overall system quality. Thus all requirements must be balanced in successful designs. A product that fails in the marketplace benefits no one. For this reason, environmental requirements should be developed at the same time as performance, cost, cultural, and legal criteria.

Product design requires a system approach which is based on the product life-cycle analysis (PLA) as discussed previously. PLA includes raw materials acquisition and processing, manufacturing, use/service, resource recovery, and disposal. A life-cycle design framework can provide guidance to more effectively conserve resources and energy, prevent pollution, and reduce the aggregate environmental impacts and health risks associated with a product system. The framework addresses the product, process, distribution, and management or information components of each product system (Keoleian and Menerey, 1993).

First, we must identify the aspects of a product that adversely impact the environment. Second, we need to define goals by modifying aspects of its performance that are environmentally unacceptable and can be improved. Factors that should be considered include whether it uses a scarce material, contains toxic substance, uses too much energy and water, or is not readily reused or recycled. These environmental criteria can be added to the initial program of requirements for the product, such as quality, customer acceptance, and production price.

The goals of new product design can be reformulation and a rearrangement of the products' requirements to incorporate environmental considerations. For example, the new product can be made out of renewable resources, have an energy-efficient manufacturing process, have a long life, be non-toxic and be easy to reuse or recycle. In the design of a new product, these environmental considerations can become an integral part of the program of requirements (USEPA, 1992). Environmental criteria for designing products are to:

- Use renewable natural resource materials.
- Use recycled materials.
- Use fewer toxic solvents or replace solvents with an alternative material.
- Reuse scrap and excess materials.
- Use water-based inks instead of solvent-based ones.
- Produce combined or condensed products that reduce packaging requirements.
- Minimize product filter and packaging.
- Produce more durable products.
- Produce goods and packaging reusable by the consumer.
- Manufacture recyclable final products.

5.3.2.4
Design strategies

Effective design strategies can only be adopted after project objectives are defined by requirements. Deciding on a course of action before the destination is known means to invite disaster. Thus, a successful strategy satisfies the entire set of design requirement. No single strategy, however, should be expected to satisfy all project requirements. Most development projects should adopt a range of strategies. Possible life-cycle design strategies are:

– Product system life extension – durable, adaptable, reliable, serviceable, re-manufacturable, reusable
– Material life extension – recycling
– Material selection – substitution, reformulation
– Reduced material intensiveness
– Process management – process substitution, process control, improved layout, inventory control and material handling, facilities planning
– Efficient distribution – transportation, packaging
– Improved business management – office management, information provision, labeling, advertising (Keoleian et al., 1993)

5.3.2.5
Research and development
Research and development discovers new approaches for reducing environmental impacts. In life-cycle design, current and future environmental needs are translated into appropriate designs. A typical design project begins with a needs analysis, then proceeds through requirements formulation, conceptual design, preliminary design, detailed design, and implementation.

Comparative analysis referred to as benchmarking is also necessary to demonstrate that a new design or modification is an improvement over competitive or alternative designs. The design team continuously evaluates alternatives throughout design. If assessments show that requirements cannot be met or reasonably modified, the project should end.

5.3.2.6
Limitations
Limitations of life-cycle product design must be considered such as lack of data and models for determining life-cycle impacts, as well as lack of motivation. When the scope of design is broadened from that portion of the life-cycle controlled by individual players to other participants, interest in life-cycle design can dwindle. In particular, it can be difficult for one party to take actions that mainly benefit others.

Incorporating environmental constraints into all plant design stages will require a new generation of design procedures that must be incorporated into formal education and continuous training. The essential elements of these new design procedures are:

– Identification and prioritization of waste streams from processes and products;
– Procedures for selecting environmentally compatible materials;
– Design of unit operations that minimize waste;
– Economics of pollution prevention; and
– Process flowsheeting for minimizing wastes (Allen et al., 1992).

5.3.3
Product life-cycle cost analysis

Life-cycle cost analysis is based on product-specific costs that occur within the life-cycle framework. The purpose is to measure or describe costs that can help prevention pollution in product-related decisions. Life-cycle cost will helpfully enable us to make the difficult choices for pollution prevention. Less waste means more pollution prevention. Recycling and reclamation seem clearly better than disposal of wastes in the environment, but substitutions are much more difficult to evaluate. For example PCBs were originally used in transformers as substitutes for materials of less stability. Asbestos likewise appeared to be a fiber of miraculous properties. Currently, advocates of pollution prevention encourage the substitution of water-based cleaners for solvent-based ones; but many aqueous cleaners can have adverse impacts on wastewater treatment plants and water bodies due to high BOD and metals.

For example, in the paper versus plastic bag question, many of the classic life-cycle costing elements such as maintenance and repairs are not applicable. Assuming equivalent performance (e.g., carrying capacity) of the bags, a supermarket chain or government procurement officer might consider the cost differential resulting from the greater bulkiness of paper bags -more storage space needed or more frequent refilling of storage bins. And, if paper bag storage areas attract insects, costs of pesticide application or monitoring of infestation will arise.

Another example, in deciding between stocking disposable or washable diapers in hospitals, a procuring agency will need to consider the costs of collecting and cleaning washable diapers, wastewater treatment and/or disposal, and the diapers' useful lifetime (and ultimate disposal) versus the costs of using and disposing non-measurable diapers for an equivalent population of children (Bailey, 1991).

Products must have an attractive price to be successful. Fortunately, some strategies for reducing environmental impacts can also lower costs while meeting all other critical requirements. To assist in life-cycle design, cost analysis practices need to be modified to reflect the actual costs of development. An accurate estimate of costs to develop and use a product are central to life-cycle design. Material and energy flows identified during the inventory analysis provide a detailed template for assigning costs to individual products. In an effort to be more complete, life-cycle costing analysis also uses an extended time scale (Keoleian and Menerey, 1993).

By adding the appropriate missing costs to the life-cycle cost analysis, we can achieve "full-cost accounting", probably necessary to evaluate pollution prevention projects. Life-cycle cost analysis should include costs involved in hazardous materials and hazardous waste management with cost savings of pollution prevention of hazardous materials and wastes. Unless liability costs and less tangible benefits are considered, the economic justification for change toward prevention is insufficient. Full-cost accounting includes (1) usual costs, (2) hidden and regulatory costs, and (3) liability and other costs (Keoleian and Menerey, 1993).

5.3.3.1
Usual costs

Life-cycle costing analysis first identifies standard capital and operating expenses and revenues for product systems. Many low-impact designs offer benefits when evaluated solely by usual costs. These savings result from eliminating or reducing pollution control equipment, non-hazardous and hazardous waste disposal costs, and labor costs. In addition, pollution prevention and resource conservation design strategies can reduce material and energy costs. Some examples of usual costs are:

– Capital costs – buildings and equipment;
– Expenses – disposal, utilities, raw materials, supplies, and labor;
– Revenue – primary products and marketable by-products.

5.3.3.2
Hidden and regulatory costs

Many hidden costs for entire plants or business units are assigned to general overhead. Hidden costs are mainly related to regulation associated with product development. Design projects based on pollution prevention and resource conservation can reduce such regulatory costs. Some examples of hidden and regulatory costs are:

– Hidden capital costs such as monitoring equipment, preparedness and protective equipment, additional technology, and others; and
– Expenses such as notification, reporting, monitoring/testing, record keeping, planning, studies, modeling, training, inspections, manifesting, labeling, preparedness and protective equipment, closure/post closure care, medical surveillance, and insurance/special taxes.

5.3.3.3
Liability and other costs

Present liability costs include fines from non-compliance while future liabilities include forced cleanup, personal injury, and property damage. Poor design may cause damage to workers, consumers, the community, or the ecosystem. In addition to private lawsuits, industries face public liability, that is, government penalties for various regulatory violations. Liability costs include hiring legal staff or consultants; penalties and fines; future liabilities for customer injury and from hazardous waste sites.

Liability costs involve monetizing the potential liabilities associated with waste disposal. The monetization methodology has been used to illustrate the "true costs" of waste disposal directly in landfills compared to other alternatives. Liability costs tend to favor adequate waste treatment and waste reduction at the source, over options that rely more on land disposal of untreated waste or improperly treated wastes. In recent years, the monetization methodology has expanded to include a more complete range of legal liabilities and better data on cleanup and compensation costs.

Avoiding liability through design and redesign is the best industrial policy. However, when potential environmental problems do occur, industries should disclose this information in their financial statements. Because estimating potential environmental liability costs is difficult, these costs are often understated. The accounting staff must work closely with other members of the development team to estimate liability costs.

In addition to legal liabilities, pollution may impose other social costs on society that manufacturers, sellers, and/or purchasers of goods do not internalize well. These can include, for example, the difficult-to-measure costs of environmental degradation and diminished quality of life from air pollution; cleanup or compensation costs not adequately covered by liability laws; costs incurred by taxpayers to managing solid waste by public agencies; and other social costs not covered in potential legal liabilities because the legal system cannot or does not render them recoverable liabilities.

5.4
Environmental labeling

The past several years have seen a marked increased in the public's awareness of and concern for a range of environmental issues worldwide. Consumer goods have become a large and important part of the daily lives of our society. The variety and number of these goods increase daily and choices about which product to buy are becoming increasingly difficult to make. A growing awareness of the damaging consequences of the manufacture, use and disposal of many of these products on our environment has led to a significant public demand for clear and credible guidance on "green" products. One way in which the public, as consumers, seeks to lessen the environmental impacts of daily activities is by purchasing and using products perceived to be less environmentally harmful.

The labeling of environmentally friendly goods has the potential to improve the quality of our environment. It would be the result of voluntary actions by individual consumers and companies. A labeling program would authorize those companies producing goods which met certain environmental criteria to label them as "green product" with a recognized symbol. If these goods enjoyed an increased market share through the positive choices made by consumers, there would be an incentive for more companies to change their products to meet the criteria. This incentive would lead to environmental improvements, such as an increase in the efficiency of resource and energy use amongst the producers of consumer goods, and a reduction in pollution and waste.

All labeling programs are conducted by groups independent from marketers, and are considered "third-party" as opposed to "first party" environmental claims made by marketers themselves. Participation in these programs can be voluntary or mandatory. Labeling programs can be positive, neutral, or negative; that is, they can promote positive attributes of products, they can require disclosure of information that is inherently neither good nor bad, or they can require warnings about the hazards of products.

The lack of standardized usage, and consumer misunderstanding of environmental terms, often lead to specific environmental claims perceived as beneficial to the environment, even though virtually all products are associated with some adverse environmental impacts. In fact, surveys measuring consumer understanding of environmental marketing terms indicate that consumers do not comprehend many terms. Terms which are commonly used by marketers, such as recycled and biodegradable, are better understood by consumers, suggesting that consumers can learn about environmental issues if exposed to such information over time. Educating consumers may therefore be as important as developing environmental marketing guidelines or other regulatory programs to enable them to make informed purchasing decisions (Cude, 1991).

This section begins with a description of environmental marketing with emphasis on third-party environmental certification programs. Environmental labeling will be discussed in three approaches: report card, seal-of-approval, and single attribute certification. Four representative international environmental labeling programs, methods and procedures are selected for illustration.

5.4.1
Environmental marketing

Traditionally, marketers have depended largely on market research, quality and financial controls, trade relations, and promotion to sell their products. Now, environmental considerations have entered the picture. Events such as Earth Day have emphasized choices that individuals can make to decrease their impact on the environment; a large number of consumers have responded by purchasing and using products they perceive to be less environmentally harmful. Several surveys indicate that a majority of Americans consider themselves to be environmentalists and would prefer to buy products with a lessened environmental impact when quality and cost are comparable (Abt, 1990). Environmental labeling seeks to allow credible third-party organizations to pass judgment on the environmental performance of products and packages rather than leave the assertions to the product manufacturers themselves, who have their own vested interests.

Environmental marketing may be grouped into "first party" and "third party" activities. First-party environmental claims are made by marketers themselves which include claims, cause-related marketing, and other activities. Such activities are designed to promote generally the environmental attributes of either claims of specific products or the company-related promotion of environmental performance and cause-related marketing. Third-party activities are conducted by groups which exclude marketers. Participation in these programs can be voluntary or mandatory. Environmental labeling is an important part of environmental marketing. Fig. 5-6 shows a diagram of environmental marketing.

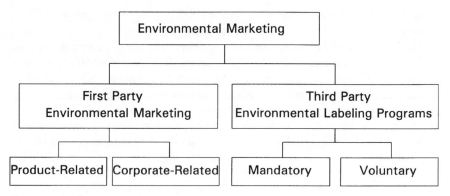

Fig. 5-6 Types of environmental marketing

Marketers have responded to consumer demand for "green" products by advertising the environmental attributes of their products, introducing new products, and/or redesigning existing products and packaging to enhance their environmental characteristics. The environmental marketing claims used to describe products and packaging range from vague, general terms such as earth-friendly or natural, to more specific claims such as "contains no chlorofluorocarbons" or "made with recycled material". Government and private parties alike have acknowledged that this trend offers an opportunity not only to decrease the environmental impacts of the consumption patterns of the public, but also to increase consumer education and sustain interest in environmental issues (USEPA, 1993a).

The lack of standardized usage, and consumer misunderstanding of environmental terms, often leads to specific environmental claims being perceived to mean "good for the environment," even though virtually all products are associated with some adverse environmental impacts. In fact, surveys measuring consumer understanding of environmental marketing terms indicate that consumer comprehension, which varies among terms, is for the most part quite low. Hurried point-of-purchase situations are likely to further decrease this comprehension. Many of the terms most commonly used by marketers, such as recycled and biodegradable, are also the most widely understood by consumers, suggesting that consumers learn about environmental issues if exposed to such information over time.

The government guidelines do not define terms, but give examples of more or less acceptable ways to present the following claims: composable; recyclable; recycled content; source reduction; refillable; and ozone safe and ozone friendly. Educating consumers may therefore be as important as developing environmental marketing guidelines or other regulatory programs to enable them to make informed purchasing decisions (Cude, 1991).

Consumer confusion over environmental marketing terms is exacerbated by several factors characteristic of environmental attributes. Advertising claims that cover easily-discernible attributes of a product, such as soft or tasty, are readily evaluated by a consumer who can judge the validity of claims against his or her own experience. However, environmental claims such as ozone-friendly pertain to product characteristics with which a consumer generally has little or no experience or cannot physically perceive; consumers cannot therefore evaluate the credibility or value of the claim. In addition, environmental claims such as recyclable and composable relate to more than just the inherent qualities of the product promoted; they also reflect the context in which a product is used or disposed (USEPA, 1990).

The success of market-driven environmental initiatives depends in large part on consumer awareness and knowledge of environmental issues. To use the market effectively as an environmental policy tool, there must be some assurance that environmental claims made about products are truthful and result in real environmental quality improvements. When consumers are misled by trivial or false advertising, environmental policy goals driven by those concerns are undermined. The effectiveness of all environmental marketing activities as policy tools can be ensured through several means: industry self-regulation, governmental intervention by regulation or voluntary initiatives, or third-party certification by public or private groups (USEPA, 1993).

Among countries involved in environmental marketing, the U.S. is virtually alone in its emphasis on defining individual terms rather than developing a national program. Germany, Canada, Japan, the European Community and the Nordic Council, among others, have government-run or government-associated environmental certification programs (ECPs) in place. With heightened interest in trade issues due to the General Agreement on Tariffs and Trade (GATT) and North American Free Trade Agreement (NAFTA) negotiations, ECPs are being examined both as potential barriers to trade and as an option to avoid barriers to trade. An ECP, for example, can effectively export its award criteria to foreign manufacturers wishing to enter its market (USEPA, 1993d).

The recent rush to market green products has achieved some success because of environmentally conscious consumers who translate their environmental values and fears into purchase decisions. A host of global and local environmental threats are increasing the number of those consumers. Developing countries need green products also and companies need green products to make their products green. Opportunities abound!

5.4.2
Environmental certification programs (ECPs)

Most of the environmental certification programs (ECPs) share three main objectives: (1) preventing misleading environmental advertising by providing an objective, expert assessment of the relative environmental impacts of products; (2) raising the awareness of consumers and encouraging them to take environmental considerations into account in their purchasing decisions by providing them with accurate information on the environmental consequences of products; and (3) providing manufacturers with market-based incentives to develop new products and processes with fewer environmental impacts.

A recent development in the marketplace is the rise of positive environmental certification programs, which are now actively reviewing products and issuing awards in more than 20 countries. The programs strive to make credible, unbiased, and independent judgments in certifying a claim or product. They are expected to provide the consumers with information and/or assessments that are often not apparent or not available to the consumer, that can help the consumer make purchasing decisions based on the environmental impacts of products, and that individuals can make to decrease their impact on the environment; a large number of consumers have responded by purchasing and using products they perceive to be less environmentally harmful. Several surveys indicate that a majority of Americans consider themselves to be environmentalists and would prefer to buy products with a lessened environmental impact when quality and cost are comparable (Abt, 1990).

Third-party ECP's consumer product labeling can serve three functions in the marketplace: (1) as an independent evaluation and endorsement of a product, (2) as a consumer protection tool, and (3) as a method of achieving specific policy goals. As an independent endorsement of a product, a program can offer companies a selling point that is more credible than claims made on their own behalf. For consumer protection, labeling can provide product information that is not readily apparent or easily discerned, or is not a positive selling point and thus would not necessarily be supplied by the marketer. As a policy instrument, labeling can influence marketplace behavior, guiding consumers and producers to act toward public policy goals. A variety of approaches to product labeling have been developed, prompted by health and safety concerns, concerns about hidden operational costs and, more recently, about the environmental impacts of products (USEPA, 1993).

All ECPs are voluntary programs and they are conducted under third-party certification processes of "Environmental Labeling Program" as shown in Fig. 5-7. ECPs can be positive or neutral labels; that is, they act either as a positive selling point in encouraging the sale of the product, or as a neutral disclosure of the environmental impacts of the product. Most of the ECPs share three main objectives: (1) to prevent misleading environmental advertising by providing an objective, expert assessment of the relative environmental impacts of products; (2) to raise the awareness of consumers and to encourage them to take environmental considerations into account in their purchasing decisions by providing them with accurate information on the environmental consequences of products; and (3) to provide manufacturers with market-based incentives to develop new products and processes with fewer environmental impacts. The current environmental certification programs (ECPs) mainly focus on three programs: (1) report card, (2) single attribute certification, and (3) seal-of-approval.

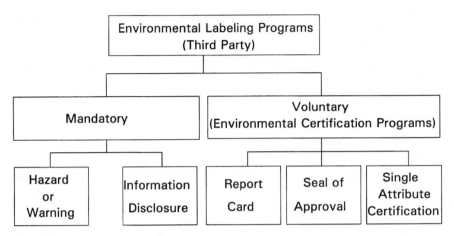

Fig. 5-7 Third party environmental labeling programs

5.4.2.1
Report card

The environmental report card approach to labeling involves categorizing and quantifying various impacts that a product has on the environment. The major advantage of the report card approach is that it provides manufacturers and consumers with a large amount of information. Criticisms of the report card fall into two basic categories: concern over the difficulty of obtaining information, and the difficulty in displaying information clearly and simply.

At this time, only Scientific Certification Systems (SCS) is using this method of labeling, although the Council on Economic Priorities has published a book ranking companies on environmental as well as various social criteria. SCS's Environmental Report Card is based on a " "life-cycle inventory" (LCI). LCI is the first step of the life-cycle assessment procedure, which involves a cradle-to-grave evaluation of the environmental burdens associated with the raw material extraction, manufacture, transportation, use, and disposal of a product. Under SCSE's LCI, inputs and outputs are quantified for each stage of material extraction, manufacturing, use, and disposal such as resource depletion, energy use, air emissions, wastewater effluents, and solid waste generation.

SCSE's Report Card Certification is a multi-step process, involving identification and quantification of inputs and outputs for every stage of a product's life-cycle, site inspections, record audits, pollutants sampling and testing, and quarterly monitoring. The certification process has nine stages as follows:

(1) Define project – Identify products to be evaluated, define systems of inputs and outputs, define project objectives, write proposal, and identify project manager and client technical teams.

(2) Prepare flow diagrams – Collect existing flow diagrams from client, prepare flow diagrams for each component of system.

(3) Prepare inventory data sheets – Generate inventory data sheets for each component identified in flow diagrams, determine required documentation for initial documentation review, determine which sites should be physically inspected.

(4) Site inspections (2–3 visits/site) – Review inventory sheets, assist in completion of inventory sheets where additional data are needed, examine technical processes to ensure full consideration of factors, confirm flow diagrams and make necessary changes, conduct preliminary review of relevant documentation.

(5) Data management – Prepare data input sheets and input data, conduct data quality review, perform necessary calculations and revisions.

(6) Presentation of findings to client – Submit detailed life-cycle inventory report, review report with client.

(7) Final auditing and testing – Full audit of records, invoices, and support documentation; sampling and testing to confirm emissions data; final inspection; development of quarterly update certification plan.

(8) Peer review

(9) Certification – Issuance of full report, with executive summary describing data quality, and corresponding Environmental Report Card. Companies are free to publish their Report Cards, provided that the more in-depth study reports are made available upon request to the public (OTA, 1990; USEPA, 1993d).

5.4.2.2
Single attribute certification programs
Single attribute certification programs certify that claims made for products meet a specified definition. Such programs define specific terms and accept applications from marketers for the use of those terms. If the programs verify that the product attributes meet their definitions, they award the use of a logo to the marketer.

Environmental Choice, Australia's verification process, gives government approval to product environmental claims that can be tested and quantified. If a product passes the required tests, its manufacturer may display "an agreed form of words" on the product. The Environmental Choice program categorizes environmental claims as follows:

– Claims that can be quantified;
– Claims dependent upon common understanding of terms used;

- Meaningless claims; and
- Misplaced or misleading claims.

Manufacturers applying for verification are not allowed to use claims that are meaningless or misleading. Random testing of products and services is conducted to ensure that providers of products and services remain in compliance with the program's requirements. Fines will be levied on those parties who misuse the Environmental Choice Logo (USEPA, 1993a).

5.4.2.3
Seal-of-approval

Seal-of-approval programs identify products or services as being less harmful to the environment than similar products or services with the same function. Single attribute certification programs typically indicate that an independent third party has validated a particular environmental claim made by the manufacturer. Unlike report cards offering consumers neutral information about a product, seal-of- approval programs are required by law. They use a logo for products judged to be less environmentally harmful than comparable products, based on a specified set of award criteria. First, product categories are defined based on similar use or other relevant characteristics; award criteria are then developed for a product category. All products within a product category are compared against the same set of award criteria. How these product categories and evaluation award criteria are set defines the most important differences among the seal-of-approval programs currently in existence. It is a complex task requiring the consideration of many factors, including environmental policy goals, consumer awareness of environmental issues, and economic effects on industry.

In general, seal-of-approval programs tend to have similar administrative structures. In a typical program, the government's environmental agency is involved to some extent, ranging from actually administering the program to simply providing advice or funding. The bulk of the responsibility rests in a central decision-making board composed of environmental groups, academics and scientists, business and trade representatives, consumer groups, and/or government representatives. Such board members usually serve for fixed terms (two to five years). Technical expertise is provided by the government, standards-setting organizations, consultants, expert panels, and/or task forces established for specific product categories.

Virtually all seal-of-approval programs follow the same overall certification process, with some minor variations. First, product categories are defined and chosen. Award criteria for that product group are then set to reduce those impacts considered to be the most important or relevant. Once the award criteria have been finalized and published, manufacturers are invited to submit products for testing and then, if accepted, apply for a license to use the logo (USEPA, 1993d).

In September 1991, the United Nations Environmental Program (UNEP) defined the following basic features common to all ECPs' seal of approval:

- Determination of award criteria based on life-cycle review of a product category;
- Voluntary participation of potential licensees;
- Run by a not-for-profit organization without commercial interests;
- Recommendations for product categories and environmental award criteria;
- A legally protected symbol or logo;
- Open access to potential licensees from all countries;
- Endorsement from government;
- Award criteria levels established to encourage the development of products and services that are significantly less damaging to the environment; and
- Periodic review and, if necessary, an update of both environmental award criteria and categories, taking into account technological and marketplace developments (UNEP, 1991).

Third-party product certification programs are well advanced in Germany, Canada and Japan. Germany's Blue Angel program has certified over 300 products in some 60 product categories. Canada has issued its EcoLogo license to 120 companies offering some 700 products in 34 product categories. Japan has certified so much that it has threatened the legitimacy of the program. U.S. consumers strongly support the idea of a product certification authority. The recent U.S. survey shows that over 80% of consumers prefer a product certification program on which to base purchase decisions.

The leaders of several of the U.S. major environmental and consumer organizations have formed non-profit organizations to assist consumers who want to make purchasing decisions based on the environmental impacts of products. These organizations label products that are truly environmentally acceptable. They began by certifying recycled contents of a package. More extensive "life-cycle analyses" were considered not technically feasible. In 1991, however, the Scientific Certification System in the United States started to develop methods and procedures for life-cycle analysis. Meanwhile, product certification program proved to have valuable functions in auditing the environmental performance of their products and in better conveying how to improve the environmental performance of their products and packages (USEPA, 1991a).

5.4.3
Methodologies and processes

Environmental labeling methodologies have received attention in the last ten years as more and more programs have been developed. The following consensus statement resulted from an environmental labeling conference held in Berlin in 1990:

> "Objective environment-related product labeling demands that the products and/or product groups be looked at in a comprehensive and technically sound way. The products to be labeled are therefore to un-

dergo a thorough assessment taking the form, for example, of an ecological balance sheet, where possible comprising the entire life-cycle of
a product and the relevant environmental aspects which apply, and depending on the nature of the product, the suitability for use and safety."
(German Ministry for Environment, 1990).

The United Nations Environmental Program, Industry and Environment Office,
and the University of Lund, Sweden, sponsored an expert seminar on environmental labeling in 1991, resulting in the following declaration on methodologies
for criteria development (set out in part). The characteristics of environmental
labeling programs include: (1) Determination of criteria based on life-cycle review of a product category; and (2) Criteria levels established to encourage the
development of products/services that are significantly less damaging to the environment (UNEP, 1991).

Possible process for the operation of an environmental labeling program for
"green products" include:

(1) Establishing an administrative body to decide on product category;
(2) Appointing an expert committee to review environmental criteria and re-
    quirements;
(3) Drafting criteria for industry and public comments;
(4) Finalizing criteria chosen;
(5) Publishing the criteria;
(6) Applying for labeling by manufactures or retailers;
(7) Assessing factory or product;
(8) Assessing and deciding the application;
(9) Issuing the environmental label if the application is approved.

5.4.4
Selected environmental labeling programs

Figure 5-8_displays 16 selected eco-labels around the world (USEPA, 1993d).
Eco-labels serve an effective way of helping consumers find environmentally
preferred products.

5.4.4.1
The Green Seal program
Led by the Earth Day 1990 Chair Denis Hayes, Green Seal provides analysis of
consumer products' environmental impacts. Green Seal is an independent, non-
profit environmental labeling program which operates throughout the United
States. Green Seal has established environmental standards for major categories
of consumer products. Suppliers of products including manufacturers, retailers,
and importers who meet or exceed those standards are eligible to license the use
of the Green Seal on their products and in their advertising. Consumers are able
to buy those products with the confidence that an independent group of scien-

Source: EPA 742-R-9-93-001. September 1993.

Fig. 5-8 Eco-labeling around the world

tists and other experts has found them environmentally acceptable (Hayes, 1991; USEPA, 1990).

The Environmental Standards Council, composed of independent scientists, academics, and other experts, acts as an appeals board for manufacturers and others who disagree with Green Seal's technical judgments. In some cases, advisory panels composed of representatives from businesses, government, academia, and the public interest community may be formed to assist in the development of specific standards.

Green Seal uses an "environmental impact evaluation", which attempts to identify the most significant environmental impacts in each stage of the product's life-cycle. The intent is to "reduce to the extent technologically and economically feasible, the environmental impacts associated with the manufacture, use, and disposal of products". Environmental impact evaluations have utilized published information on products and their production processes and information provided by product manufacturers. Both quantitative and qualitative information is used in the evaluations (Green Seal, 1991).

Award process includes four stages: selecting product categories; setting standards; evaluating products; and awarding the seal. The Green Seal program intends to achieve three specific objectives: technical accuracy, public credibility, and openness.

(1) Technical accuracy. The Green Seal program primarily aims to rely on the best available technical skills and methods to assess the environmental impacts of consumer products through their entire life-cycle. To ensure first-rate technical work, Green Seal convenes a diverse, high powered and balanced group of scientists and other experts to supervise this environmental impact assessment and standard setting process.

(2) Public credibility. In order to assure unbiased decisions and thereby inspire consumer and business trust in Green Seal, the organization conducts its operations under a strict Code of Ethics. This code prohibits anyone with a direct financial interest in firms that might benefit from the Green Seal, from participating in the decision-making process.

(3) Openness. To ensure that its decisions are fully informed, Green Seal conducts its decision making process in the open and invites the participation of industry, government agencies, consumers, and environmental organizations. Proposed environmental criteria for categories of consumer products are published for the public for comments, and public hearings and meetings are held on important issues affecting the labeling program.

The general criteria used to select product categories for labeling are as follows:

- Minimal environmental impact from use;
- Significant potential for improvement in the environment by using the product;
- Minimal environmental impact from disposal after use; and
- Other significant contributions to the environment.
- Additional criteria that are required for approval include:
- Appropriate pollution control measures at the production stage;
- Ease of treatment for disposal of product;
- Energy conservation during use of the product;
- Compliance with regulations and standards for quality and safety; and
- Price not extraordinarily higher than comparable products (Hashimoto, 1990).

The program seeks to label products that call attention to an "environmental" life style, rather than to reduce the environmental impacts of general consumer products. Activities considered as part of this environmental life style may also receive the label (USEPA, 1993a and 1993d).

5.4.4.2
European Communities Eco-labeling program
The European Communities (EC) Eco-labeling program operates through offi-
cial environmental labeling bodies in the member states and an approval process by the EC Commission, the governing political body of the EC. Proposals for
product categories and criteria are made by member states to the Commission,
which engages in consultation with a Consultation Forum of interest groups in
choosing categories. A category is assigned to the participating Competent Body
in a member state, which drafts criteria for certification. If approved, they are
adopted by the Commission (UK, 1993).

German "Blue Angel" Program, for example, has been in operation since
1978. The decision-making process for establishing criteria for environmental
labeling is a joint effort by the German Federal Ministry of the Environment, including the Federal Environmental Agency, the German Institute for Quality
Control and Labeling, and by a representative Environmental Labeling Jury.
While the Environmental Labeling Jury and the Institute for Quality Control,
which make decisions on criteria and administer the program, are non-governmental, the government does participate in the process (German Ministry for
Environment, 1990).

The Federal Environmental Agency provides the initial scientific review of the
product category and drafts the labeling criteria. Then, the Institute for Quality
Control and Labeling convenes expert hearings to discuss the criteria proposed
by the Environmental Agency. The representative Environmental Labeling Jury,
which has members from industry, environmental organizations, consumer associations, trade unions, environmental professional associations, and the federal states, determines the criteria for product categories based upon the proposal by the Environmental Agency and the comments in the expert hearings
(German Ministry of Environment, 1990).

The Blue Angel Program has issued more than 75 sets of labeling criteria,
many of which focus on a limited number of attributes in the product class. The
Program states, however, that it has always considered the life-cycle of the product in establishing criteria, usually by using a simple qualitative matrix of environmental impacts in each of the life-cycle stages. The criteria that are ultimately
developed focus on state-of-the-art technology and are intended to create more
significant environmental benefits than those created by generally used technologies (Neltzel, 1992).

The current procedure has been described as stepwise: it first uses the checklist or matrix approach to identify important parts of the life-cycle and the most
significant environmental attributes of the product class. This is supplemented
by the use of expert panels. If there is insufficient information to develop criteria
from this screening approach, an existing LCA will be done using available information with caution.

Most of the Blue Angel standards do not deal directly with the production process. The Blue Angel is one of the few programs that does not require producers to
demonstrate that they meet national environmental standards in the production

process. Two reasons are given for this omission: (1) analytic methods are frequently unavailable to separate the impacts of the product being considered from the whole manufacturing facility's impacts; and (2) such manufacturing standards would penalize countries with less stringent environmental standards and may constitute a trade barrier contrary to GATT regulations (Neitzel, 1992; USEPA, 1993a).

5.4.4.3
The Green Cross and Scientific Certification System

The Green Cross and Scientific Certification System was formerly known as the Green Cross. The certification program is operated by the non-profit Green Cross Certification System Company, a division of Scientific Certification Systems, Inc. (CSC). CSC's president, Stanley Rhodes, is a toxicologist by training. CSC claims that they are more scientific than activist.

In April 1990, four retail and supermarket chains serving the western United States announced a comprehensive "Green Cross" program for dealing with environmental product claims. Participating chains are ABCO Market, Inc. (75 supermarkets in Arizona); Fred Meyer, Inc. (125 stores in seven western states); Raley's (58 stores in northern California and northern Nevada); and Ralphs (143 stores in southern California.

Under the Green Cross program, products containing "the maximum practical, state-of-the-art level of recycled content" will receive the Green Cross Recycling Seal of Approval. Upon approval, CSC will certify manufacturers' claims. The Green Cross definition of recycling includes consumer and industrial waste, but excludes industrial scrap. Green Cross has also reached agreement with the paper company to remove environmental claims from their grocery bags relating to biodegradability, recyclability, and the non-toxicity of the ink used. Green Cross and participating stores developed an Environmental Performance Ranking program to rate products on different factors in environmental performance. Product rankings help spur manufacturers to upgrade their environmental practices. Buyers use them in retail stores in choosing products. The Environmental Seal of Approval is awarded based on a product's total environmental impact. Criteria for the seal include at least 50% sustainable or recycled materials content in the product and its packaging; a solid waste disposal plan that demonstrates zero environmental burden; and zero tolerance (no detected residues) for cancer-causing chemicals and reproductive toxins in all effluents and emissions associated with the product and packaging (USEPA, 1990).

5.4.5
Effectiveness of ECPs

The effectiveness of an environmental certification programs (ECPs) can be examined in terms of its impacts on consumer awareness, consumer acceptance, consumer behavior, manufacturer behavior, and environmental benefits. Each of these elements is dependent on the others, and provides important insight into the driving forces behind the success or failure of environmental labeling. Unfortunately, there is very little in-depth information in these areas.

The confusion and growing marketplace complexity surrounding environmental marketing terms has begun to create indifference and distrust among consumers toward advertised environmental attributes, and a reluctance among companies to advertise environmental attributes. Effectiveness of ECP relies on positive environmental certification programs. Marketers seem reluctant to provide clear, credible, and honest information in manufacturing and composition of a product. Some marketers are now adding complicated language to their product labels to avoid the liabilities associated with increasing numbers of governmental actions and regulations (Lawrence, 1992).

Although it is too early to discern their effectiveness, there has been widespread industry support for the U.S. Federal Trade Commission (FTC) guidelines. At the same time, industry has a number of reservations about third-party ECPs. A prominent concern is that companies are not willing to let an independent private party to dominate. Some of the criticisms are that: (1) companies will lose some control over their own production processes and marketing decisions; (2) an independent certification company may not be stable and credible, and the reputation of companies associated with it may be tarnished; (3) competing private labels will compromise the effectiveness of each; and (4) lack of a label or seal will be equated with being denied an award.

On the other hand, there have been some calls for governmental action to level the playing field for ECPs. Proposals for additional research and investigation include: public review of ECPs; development of uniform program operating principles; guidance on product evaluation methodologies; assistance in developing data sources; and harmonization with programs in other countries. Differences in product category definition and stringency of standards could potentially cause consumer confusion and act as trade barriers. International organizations such as the International Chamber of Commerce, the International Standards Organization, and the United Nations Environment Program, have recommended ways to standardize ECPs and increase the exchange of information and primary research conducted by various programs (USEPA, 1993a).

5.4.6
Labeling through life-cycle assessment

Life-cycle assessment (LCA) methodologies are being used in environmental labeling programs worldwide. Environmental labeling and life-cycle assessment have a common goal – the improvement of the environmental attributes of product systems. Environmental labeling can be viewed as an improvement analysis for a whole class of products. Most government-sanctioned third-party certification labeling programs are seal-of-approval programs that attempt to develop labeling criteria that address the overall environmental attributes of products.

LCA is being used in seal-of-approval environmental labeling programs to identify the most significant environmental impacts in the various stages of the life-cycle in order to guide the development of labeling criteria that address those impacts. Most labeling programs at least recognize the life-cycle concept in development of labeling criteria for classes of products, but only a few programs are using any part of the formal practice of LCA. LCA is also being used as a tool to inform the process of developing labeling criteria for labeling programs, but it has not taken the place of expert judgment and consensus-building in developing those criteria.

Most of the labeling programs are using simplified life-cycle inventories with elemental impact analysis and are not performing full quantitative LCA's. The time-consuming and expensive methodology is beyond the reach of most seal-of-approval labeling programs and may not be necessary for the purpose of developing labeling criteria. A streamlined LCA has been suggested for certifying most of the labeling programs. A streamlined LCA identifies the most significant environmental impacts throughout the product life-cycle for development of criteria. It does not necessarily require finding data for every input and output for every component of the product for every stage of the life-cycle. Unfortunately, little attention has been paid to the development of a consensus methodology for a "streamlined" LCA for environmental labeling, and the approaches being used by different labeling programs vary significantly (USEPA, 1993a).

In fiscal year 1991–1992 for projects related to pollution prevention, USEPA set aside 2% of its budgets to promote pollution prevention initiatives. One of the projects funded through that set-aside is the Consumer Products Comparative Risk project. The project seeks to simplify and standardize the methodology for studying product life-cycles and thereby create a feasible yet credible tool for comprehensively assessing and comparing the environmental impacts of consumer products. New techniques such as "multi-attribute analysis" are being developed by engineers to select the "best" materials for product design from among a growing number of alternatives. Moreover, for years environmental scientists have been refining techniques for preparing environmental impact statements. Some or all of these techniques may well serve as possible models for the assessment of the impacts of products on the environment. As of today, product life-cycle assessment has not become the decision-making documents due to four factors: Cost limitations, data limitations, lack of common unit, and inability to quantify environmental impacts (WWF, 1993).

5.4.7
Government procurement

Government procurement policies and practices have a profound impact on pollution prevention by using environmental labeling products, recycling and recovered materials. For example, an Executive Order (published in the US Federal Register, October 22, 1993) requires all agencies to acquire and use environmentally preferable products and services and implement cost-effective procurement preference programs favoring the purchase of these products and services (Bergeson, 1994).

The Executive Order requires the USEPA to issue guidance that recommends principles that executive agencies should use in making determinations for the preference and purchase of environmentally preferable products. It also requires each agency, in accordance with the Resource Conservation and Recovery Act to develop an affirmative procurement program for all designated EPA guideline items purchased by it. Currently designated EPA guideline items are: concrete and cement containing fly ash; recycled paper products; reused lubricating oil; retreated tires; and insulation containing recovered materials.

The Executive Order contains several provisions that address acquisition planning and affirmative procurement programs in general. For example, it requires agencies to consider the following factors in developing plans, drawings, work statements, specifications or other product descriptions in planning purchases for all procurements and in the evaluation and award of contracts:

- Elimination of virgin material mandates;
- Use of recovered materials;
- Reuse of product;
- Life-cycle cost;
- Recyclability;
- Use of environmentally preferable products;
- Waste prevention; and
- Ultimate disposal.

The Order requires the National Institute of Standards and Technology (NIST) to establish a program to test the performance of products containing recovered materials or deemed to be environmentally preferable. NIST must publish appropriate reports describing testing programs, their results and recommendations for testing methods and related specifications for use by executive agencies and other interested parties.

Each agency is required to establish goals for solid waste prevention and for recycling to be achieved by 1995. The Order requires agencies to strive to increase the procurement of products that are environmentally preferable or that are made with recovered materials, and to set annual goals to maximize the number of recycled products purchased, relative to non-recycled products purchased, relative to non-recycled alternatives. For example, to create incentives for reducing paper use, the Order contains several requirements:

- It requires agencies to purchase paper meeting minimum recycled content standards. All high-speed copier paper, offset paper, forms bond, computer printout paper, carbonless paper, file folders and white woven envelopes must contain no less than 20% post-consumer materials beginning Dec. 31, 1994. For other uncoated printing and writing paper, the minimum content standard should consist of 50% recovered materials, of which 20% is post-consumer materials.
- As an alternative to these two requirements, agencies may require that all printing and writing papers meet a minimum content standard of no less

than 50% recovered materials that are a waste material by-product of a finished product other than a paper or textile product which would otherwise be disposed of in a landfill.
- Related requirements in the Order address electronic means of communication and other ways to reduce paper use. It requires agency affirmative procurement programs to encourage that documents be transferred electronically, that all government documents printed internally be double-sided, and that contracts, grants and cooperative agreements include provisions requiring documents to be printed double-sided on recycled paper.
- It requires the executive branch to implement an electronic commerce system consistent with the recommendations adopted as a result of the National Performance Review, and to eliminate as many paper transactions in acquisitions as possible.

To ensure compliance with its requirements the Order provides that a Federal Environmental Executive (FEE) is designated by the President. The Order requires FEE to:

- Identify and recommend initiatives for government-wide implementation of the Order, including the development of Agency implementation and incentives to encourage the acquisition of preferred products; a federal implementation plan and guidance for instituting economically efficient federal waste prevention, energy and water efficiency programs, and recycling programs within each agency; and a plan for making maximum use of available funding assistance programs;
- Collect and disseminate information electronically on methods to reduce waste, materials that can be recycled, costs and savings from waste prevention and recycling, and current market sources of products that are environmentally preferable or produced with recovered materials;
- Guide and assist agencies in setting up and reporting on agency programs and monitoring their effectiveness; and
- Coordinate appropriate government-wide education and training programs. The FEE is also directed to form several interagency work groups and committees to recommend action to implement of the Order (Bergeson, 1994).

References

Abt Associates Inc (1990) Consumer Purchasing Behavior and the Environment: Results of an Event-Based Study, November
ADL (1983) "Benefits to Industry of Environmental Auditing". Report of the Center for Environmental Assurance, Arthur D. Little, Inc to USEPA
Allen TD et al. (1992) Pollution Prevention: Homework & Design Problems For Engineering Curricula. Dept. of Chemical Engineering, University of California at Los Angeles
Bailey PE (1991) "Life-Cycle Costing and Pollution Prevention," Pollution Prevention Review, Vol. 1, No. 1, p. 27

Bergeson LL (1994) "Federal Procurement Will Never Be The Same," Pollution Engineering, Vol. 26, No. 1, p. 87

Chemecology (1993) Design with the Environmental Mind. Chemecology Vol. 22, No. 3

Cude Brenda (1991) Comments prepared for FTC public hearings on environmental marketing and adverting claims on July 11, University of Illinois

Evers DP (1995) "Facility Pollution Prevention Planning," in: Industrial Pollution Prevention Handbook, ed. by HM Freeman. McGraw-Hill, Inc., pp. 155–170

Frankel CF, Coddington W (1993) "Environmental Marketing", in: Environmental Strategies Handbook, ed. by RV Kolluru., pp. 654–663

Fromm CR and MS Callahan (1986) "Waste Reduction Audit Procedures – A Methodology for Identification, Assessment and Screening of Waste Minimization Options." Waste Minimization Programs, Hazardous Materials Control Research Institute Conference Proceedings, Atlanta, Georgia. pp. 427–435

German Ministry for Environment (1990) Documentation of International Conference on Environmental Labeling – State of Affairs and Future Perspectives for Environment Related Product Labeling, July 5–6, 1990, Berlin, Germany

Guida JF (1982) "A Practical Look at Environmental Audits." J. of Air Pollution Control Association, Vol. 32, No. 5, pp. 568–573

Hashimoto M (1990) "The Ecomark System in Japan", in: German Federal Ministry for Environment, Nature Conservation, and Nuclear Safety. Documentation, International Conference on Environmental Labeling, July 5–6, 1990, Berlin, Germany

Hayes D (1990) "Harnessing Market Forces to Protect the Earth," Issues in Science and Technology, Winter 1990–1991

Henn CL and JA Fava (1993) "Life-Cycle Analysis and Resource Management," in: Environmental Strategies Handbook, ed. by RV Kolluru, pp. 550–533

ICC (1989) Environmental Auditing. International Chamber of Commerce, ISBN No. 92-842-1089-5

International Chamber of Commerce (1991). Environmental Labeling, Volume I, House of Commons Paper 474-I

Keoleian GA and D Meanery (1993) Life-Cycle Design Guidance Manual. Contract Report of the National Pollution Prevention Center, University of Michigan, Ann Arbor, MI., EPA/600/R-92/226, January

Keoleian GA, D Meanery and MA Curran (1993) "A Life-Cycle Approach to Product System Design." Pollution Prevention Review, pp. 293–306, Summer 1993

Keoleian GA and D Meanery (1994) "Sustainable Development: Review of Life-Cycle Design and Related Approaches," J. AWMA Vol. 44, p. 664

Kolluru RV and Rao (eds) (1994) Environmental Strategies Handbook: A Guide to Effective Policies and Practices, McGraw-Hill, New York

Kolluru RV and Rao (eds) (1996) Risk Assessment and Management Handbook for Environmental, Health, and Safety Professionals, McGraw-Hill, New York

Ling JT (1994) Personal communication

Molak V (1995) "Application of Risk Analysis to Set Pollution Prevention Priorities," in: Industrial Pollution Prevention Handbook, ed. by HM Freeman. McGraw-Hill Inc., New York, ISBN 0-07-022148-0, pp. 197–199

Neitzel H (1992) "Experience with the Blue Angel Scheme Working Methods, Problems, Balance, and Perspectives." Presented at the Annual Conference of the Environmental Law Network International, Athens, Greece, October 9–10, 1992

NYSDEC (1989) New York State Waste Reduction Manual. NYS Department of Environmental Conservation, Albany, NY

OTA (1992) Green Products by Design: Choice for a Cleaner Environment, Office of Technology Assessment, Congress of the United States, Washington, D.C.

Royston MG and TM McCarthy (1988) The Environmental Management Audit. Industry and Environment Review, Vol. 11, No. 4, p. 20

SCS (1993) Life-Cycle Assessment & the Environmental Report Card. Scientific Certification Systems, Oakland, Ca.

SETAC (1991). A Technical Framework for Life-Cycle Assessments, published by Society for Environmental Toxicology and Chemistry, Washington, D.C.

SETAC (1993) Guidelines for Life-Cycle Assessment: A Code of Practice. From the workshop held at Sesimbra, Portugal, March 31–April 3, 1993. Society for Environmental Toxicology and Chemistry, Washington, D.C.,

Shen TT (1996) "Beyond Pollution Prevention: Design for Environment and Sustainability". Presented at the 1996 Pollution Prevention Annual Conference" in Chungli, Taiwan, December 17–19

UK (1993) UK Ecolabeling Board Newsletter No. 5, London

UNEP (1991) Global Environmental Labeling: Invitational Expert Seminar, Lesbos, Greece, September 24–25, 1990. Working Group on Policies, Strategies and Instruments of the UNEP/IEO Cleaner Production Program

USEPA (1984) Risk Assessment and Management: Framework for Decision Making. EPA 600/9-85-002, pp. 13–22, October

UNEP (1988) Environmental Auditing. Published in "Industry and Environment Review" by the United Nations Environment Programme. Geneva, Switzerland

USEPA (1990) Environmental Labeling in the US: Background Research, Issues, and Recommendations. Draft prepared by Applied Decisions Analysis, Inc., Washington, D.C., February 1990

USEPA (1992) Chapter 7 Designing Environmentally Compatible Products of Facility Pollution Prevention Guide, Office of R&D, Washington D.C. EPA/600/R-92/088, May, 1992

USEPA (1993a) The Use of Life-Cycle Assessment in Environmental Labeling. University of Tennessee for Office of Pollution Prevention and Toxics, Washington, D.C., EPA/742-R-93-003, September

USEPA (1993b) Life-Cycle Assessment: Inventory Guidelines and Principles. Battle and Franklin Associates for Office of R&D, Risk Reduction Engineering Laboratory, Cincinnati, Oh., EPA/600/R-92/254, February 1993

USEPA (1993c) Life-Cycle Impact Assessment: Framework Issues. Research Triangle Institute, Center for Economics Research, Environmental Management Systems, RTP, NC, February 1993

USEPA (1993d) Status Report on Use of Environmental Labels Worldwide. Abt Associates Inc. for Office of Pollution Prevention and Toxics, Washington, D.C., EPA/742-R-9-93-001, September

Van Weenan JC (1990) Waste Prevention: Theory and Practice. CIP-Gegevans Koninklije Bibliostheek, The Hague

Vigon BW et al. (1993) Life-Cycle Assessment: Inventory Guidelines and Principles. Contract Report of Battelle Research Institute, Columbus, Oh., EPA/600/R-92/245

WWF (1993) Rethinking the Materials We Use: A New Focus for Pollution Policy. World Wildlife Fund Publication, TD 793.95.R48, Washington, D.C., ISBN 0-89164-140-8

6
Pollution prevention feasibility analyses

The level of required analysis depends on the complexity of the considered pollution prevention project. A simple, low-capital cost improvement, such as preventive maintenance, would not need much analysis to determine whether it is technically, environmentally and economically feasible. On the other hand, input material substitution could affect a product specification, while a major modification in process equipment could require large capital expenditures. Such changes could also alter process waste quantities and compositions, thus requiring more systematic evaluation (USEPA, 1992a).

Based on the use of potential pollution prevention techniques and options discussed previously in Chapter 4, and the knowledge and information discussed in Chapter 5, we can evaluate various options of pollution prevention projects, depending on the resources currently available. It may be necessary to postpone feasibility analyses for some options; however, all options should be evaluated eventually. This chapter describes how to screen the identified options to narrow the list to a few that will be evaluated in greater detail. Detailed analysis includes evaluation of technical, environmental, economical, and institutional feasibilities. It is important to note that many of the issues and concerns during pollution prevention feasibility analyses are interrelated as shown in Fig. 6-1.

6.1
Technical feasibility analysis

Technical feasibility analysis requires comprehensive knowledge of pollution prevention techniques, vendors, relevant manufacturing processes, and the resources and limitations of the facility. The analysis can involve inspection of similar installations, obtaining information from vendors and industry contacts, and using rented test units for bench-scale experiments when necessary. Some vendors will install equipment on a trial basis and payment after a prescribed time, if the user is satisfied.

Technical analysis should determine which technical alternative is the most appropriate for the specific pollution prevention project in question. Technical analysis requires considerations of a number of factors and asks very detailed questions to ensure that the pollution prevention technique will work as intended. Examples of facility-related questions to be considered, follow:

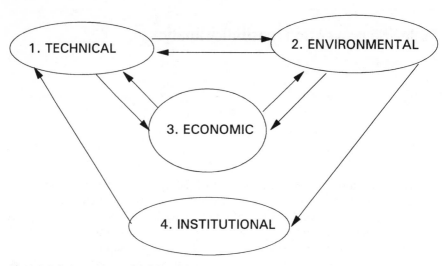

Fig. 6-1 Pollution prevention feasibility analyses

- Will it reduce waste?
- Is space available?
- Are utilities available or must these be installed?
- Is the new equipment or technique compatible with current operating procedures, work flow, and production rates?
- Will product quality be maintained?
- How soon can the system be installed?
- How long will production be stopped in order to install the system?
- Is special expertise required to operate or maintain the new system? Will the vendor provide acceptable service?
- Will the system create other environmental problems?
- Is the system safe?
- Are there any regulatory barriers?

Options that can affect production or quality need careful study. Although an inability to meet above constraints may not present insurmountable problems, they will likely add to the capital or operating costs (NYS, 1989).

All affected groups in the facility should contribute to and review the results of the technical analysis. Prior consultation and review with the affected groups (e.g., production, maintenance, purchasing) will ensure the viability and acceptance of an option. If a change in production methods is necessary, the project's effects on the quality of the final product must be determined.

For options that do not involve a significant capital expenditure, the team can use a "fast-track" approach. Material substitutions also can be accomplished relatively quickly if there are no major production rate, product quality, or equipment changes involved. Equipment-related options or process changes are more expensive and may affect production rate or product quality. Therefore, such options require more study. The assessment team will want to determine whether the option will perform in the field under conditions similar to the planned application. In some cases, they can arrange, through equipment vendors and industry contacts, visits to existing installations. A bench-scale or pilot-scale demonstration may be needed. It may also be possible to obtain scale-up data using a rental test unit for bench-scale or pilot-scale experiments. Some vendors will install equipment on a trial basis, with acceptance and payment after a prescribed time, if the user is satisfied (Evers, 1995).

6.2
Environmental feasibility analysis

The environmental feasibility analysis weighs the advantages and disadvantages of each option with regard to the environment. Most housekeeping and direct efficiency improvements have the obvious advantages. Some options require a through environmental evaluation, especially if they involve product or process changes or the substitution of raw materials. The environmental option of pollution prevention is rated relative to the technical and economical options with respect the criteria that are most important to the specific facility. The criteria may include:

- Reduction in waste quantity and toxicity
- Risk of transfer to other media
- Reduction in waste treatment or disposal requirements
- Reduction in raw material and energy consumption
- Impact of alternate input materials and processes
- Previous successful use within the company or in other industry
- Low operating and maintenance costs
- Short implementation period and ease of implementation
- Regulatory requirements

The criteria work well when there are many waste streams or options to consider, allowing a pollution prevention team to rate and assess the options quickly. A team must prioritize criteria and options with a number scale of 1 to 10 in terms of the company's goals and constraints. The "weighted sum method" is mostly used to analyze each options relative to the other options for each criteria. It calculates an overall weighted score for each option to reflect the company's goals and constraints. An illustrative example of the weighted sum method is given in Chapter 7.

The environmental evaluation is not always so clearcut. Some options require a thorough environmental evaluation, especially if they involve product or process changes or the substitution of raw materials. To make a sound evaluation, the team should gather information on the environmental aspects of the relevant product, raw material, or constituent part of the process. This information would consider the environmental effects not only of the production phase and product life cycle but also of extracting and transporting the alternative raw materials and of treating any unavoidable waste. Energy consumption should also be considered. To make a sound choice, the evaluation should consider the entire life cycle of both the product and the production process.

6.3
Economic feasibility analysis

Economic feasibility analysis is a relatively complex topic, which is only briefly discussed here. Economic analysis deals with the allocation of scarce, limited resources to various pollution prevention modifications; and compares various investments to help determine which investments will contribute most to the company.

A benefit is usually defined as anything that contributes to the objectives of the pollution prevention project; costs are defined as anything that detracts from the achievement of a project's objectives. Normally, benefits and costs are evaluated from the perspective of whether they contribute to (or detract from) the maximization of a company's income. Economic cost-benefit analysis uses a number of measures of profitability such as the:

- Net present value;
- Internal rate of return; and
- Benefit-cost ratio.

When measuring savings, it is important to look at not only the direct savings but also the indirect savings of pollution prevention. In addition, there are non-cost (or intangible) benefits of pollution prevention. In many cases, the indirect savings and non-cost benefits of pollution prevention are difficult to quantify in financial terms; nevertheless, they are an important aspect of any pollution prevention project, and should be factored into the decision-making process (USEPA, 1988).

The economic feasibility analysis of pollution prevention alternatives examines the incremental costs and savings that will result from each pollution prevention option. Typically, pollution prevention measures require some investment on the part of the operator, whether in capital or operating costs. The purpose of economic feasibility analysis is to compare those additional costs to the savings (or benefits) of pollution prevention.

6.3.1
Direct costs

For most capital investments, the direct cost factors are the only ones considered when project costs are being estimated. For pollution prevention projects, direct cost factors may only be a net cost, even though a number of the components of the calculation will represent savings. Therefore, confining the cost analysis to direct costs may lead to the incorrect conclusion that pollution prevention is not a sound business investment.

In performing the economic analysis, various costs must be considered. As with any project, the direct costs should be broken down into capital expenditures and operating costs:

(1) Capital expenditures – for purchasing process equipment, additional equipment, materials, site preparation, designing, purchasing, installation, utility connections, training costs, start-up cost, permitting costs, initial charge of catalysts and chemicals, working capital, and financing charges.

(2) Operating costs – typically associated with costs of raw materials; water and energy; maintenance; supplies; labor; waste treatment, transportation, handling, storage, and disposal; and other fees. Revenues may partially offset operating costs from increased production or from the sale or reuse of by-products or wastes.

6.3.2
Indirect costs

For pollution prevention projects, unlike more familiar capital investments, indirect costs are likely to represent a significant net savings. Indirect costs are hidden in the sense that they are either allocated to overhead rather than to their source (production process or product), or altogether omitted from the project financial analysis. A necessary first step in including indirect costs in an economic analysis is to estimate and allocate them to their source. Indirect costs may include:

(1) Administrative costs

(2) Regulatory compliance costs such as permitting, record keeping and reporting, monitoring, manifesting

(3) Insurance costs

(4) Workman's compensation

(5) On-site waste management and control equipment operation costs (USEPA, 1992a).

6.3.3
Liability costs

Estimating and allocating future liability costs involves mush uncertainty. It may be difficult to estimate liabilities from actions beyond our control, such as an accidental spill by a waste hauler. It is also difficult to estimate future penalties and fines for noncompliance with not yet exist regulatory standards. Similarly, it is difficult to estimate personal injury and property damage claims that result from consumer misuse, disposal of waste later classified as hazardous, or claims of accidental release of hazardous waste after disposal. Allocation of future liabilities to the products or production processes also presents practical difficulties in a cost assessment.

Some companies have found alternative ways to address liability costs in project analysis. For example, in the narrative accompanying a profitability calculation, a team could:

(1) include a calculated estimate of liability reduction;

(2) cite a penalty or settlement that could be avoid based on a claim against a similar company using a similar process; or

(3) qualitatively indicate without attaching a dollar value the reduced liability risk associated with the pollution prevention project.

Alternatively, some companies have chosen to loose the financial performance requirements of their projects to account for liability reductions. For example, the required payback period can be lengthened from three to four years, or the required internal rate of return can be lowered from 15 to 10% (USEPA, 1989).

6.3.4
Benefits

A pollution prevention project can benefit from water, energy, and material savings as well as from waste reduction, recycling, and reuse. It may also deliver substantial benefits from an improved product and company image or from improved employee health. These benefits remain largely unexamined in environmental investment decisions. Although they are often difficult to measure, they should be incorporated into the assessment whenever feasible. At the very least, they should be highlighted for managers after presenting costs that can be the more easily quantified and allocated.

Intangible benefits may include:

- Increased sales due to improved product quality, enhanced company image, and consumer trust in products;
- Improved supplier-customer relationship;

- Reduced health maintenance costs;
- Increased productivity due to improved employee relations; and
- Improved relationships with regulators (USEPA, 1992).

Since many of the liabilities and intangible benefits of pollution prevention will occur over a long period of time, it is important that an economic assessment look beyond the three- to five-year time frame typically used for other types of projects. When making pollution prevention decisions, select long-term financial indicators as criteria that account for: (1) all cash flows during the project; and (2) the time value of money. Three commonly used financial indicators meet the two criteria: Net Present Value (NPV) of an investment, Internal Rate of Return (IRR), and Profitability Index (PI).

The most common method for measuring economic feasibility is the discounted cash flow, or net present value (NPV) method. This method discounts projected cash flows to their present value, thus taking into account the time value of money. If the NPV is greater than zero at the company's required rate of return, the project is profitable. The net present value of a project can be calculated as follows:

$NVP = [Savings-Costs (year\ t)]/(1+MARR)^t$
for every year (t) of the project's life;

Where   Savings–Costs = the total estimated savings (anticipated revenues plus reduced future costs from the proposed project minus the total estimated cost of the proposed project for year t), and

MARR = the minimum attractive rate of return, defined as the average cost of capital for the company.

To illustrate how to apply the Net Present Value Method, we assume that a company will examine the economic feasibility of buying a solvent recovery system for pollution prevention:

- The total capital outlay for the system is $7500.
- The total ongoing average annual operating costs will be $500/year.
- The system will last for five years, then be discarded as scrap.
- The total estimated average annual savings will be $2800/year in avoided waste disposal and reduced raw material costs.
- The company's minimum attractive rate of return is 10%.

The net present value of the proposed solvent recovery project is then calculated as shown in Table 6-1.

As demonstrated in the Table, the total profit from this project begins the 5th year which is $1220. The initial cost is immediate and is not affected by discounting. Each of the following years savings must be discounted. The numbers in the

Table 6-1 An example of the NPV method (solvent recovery project)

Year	Savings (year t)	Costs (year t)	NPV year	NPV savings
	$	$	$	$
0	0	7500	−7500	−7500
1	2800	500	2300	2090
2	2800	500	2300	1900
3	2800	500	2300	1730
4	2800	500	2300	1570
5	2800	500	2300	1430
Total project NPV savings=sum of NPV column				$1220

last column are calculated by the above formula. For example, in the fourth year, the years savings minus the years costs are divided by one plus the MARR raised to the fourth power, or

$$[\text{Savings–Costs (year 4)}]/(1+\text{MARR})^4=2300/(1.1)^4=\$1570.$$

The total net present value of this 5-year project for year 0 through 5 are added together resulting in a total project NPV saving of $1220. Any other project's NPV would be calculated in the same way, and the result compared to the $1220 *which is the sum of saving column*. A larger NPV implies greater profitability (UOPARC, 1987).

Few companies allocate environmental costs to the products and processes that produce these costs. Without direct allocation of costs, businesses tend to merge these expenses into a single overhead account or simply add them to other budget line items where they cannot be disaggregated easily. The result is an accounting system that is incapable of: (1) identifying the products or processes most responsible for environmental costs; (2) targeting prevention opportunity assessments and prevention investments to the high environmental cost products and processes; and (3) tracking the financial savings of a chosen prevention investment. Total Cost Analysis (TCA) will help overcome each of these deficiencies (USEPA, 1992b).

6.4
Institutional feasibility analysis

Institutional analysis is concerned with evaluating the strengths and weaknesses of the company's involvement in the implementation and the operation of investment in pollution prevention projects. It includes, for example:

- staffing profiles;
- task analysis and definitions of responsibility;
- skill levels;

- processes and procedures;
- information systems and flows for decision-making; and
- policy positions on pollution prevention priorities.

The analysis should cover managerial practices; financial processes and procedures; personnel practices; staffing pattern; and training requirements. Issues of accountability need to be addressed. Proper incentives, in terms of money and career advancements will encourage employees to achieve pollution prevention goals (Shen, 1997).

References

Evers DP (1995) "Facility Pollution Prevention," in: Industrial Pollution Prevention Handbook, ed. by HM Freeman. McGraw-Hill, Inc., pp. 155–179

NYS (1989) New York State Waste Reduction Guidance Manual, Chapter 5. Division of Hazardous Substances Reduction, NYS Department of Environmental Conservation, Albany, N.Y.

Shen TT (1997) "Industrial Pollution Prevention," presented at the 1997 International Chinese Sustainable Development Conference in Los Angeles, July 4–5. Proceedings prepared by the ITRI of Taiwan

UOPARC (1987) Hazardous Waste Minimization Manual for Small Quantity Generators in Pennsylvania. Center for Hazardous Materials Research Center, Applied Research Center, University of Pittsburgh

USEPA (1988) Waste Minimization Benefits Manual. Phase I, Draft. Washington, D.C., Prepared by ICF Inc.

USEPA (1989) Pollution Prevention Benefits Manual, Phase II, Draft document from Pollution Prevention Clearance House (PPIC)

USEPA (1992a) Facility Pollution Prevention Guide, Chapter 6. Office of Solid Waste, Washington, D.C., EPA/600/R-92/083, May

USEPA (1992b) Pollution Prevention News, June issu, p. 4

7

Industrial facility pollution prevention plan

Any industrial facility is a complex entity which needs many different skills to operate properly (i.e., management, financing, marketing, purchasing, scheduling, labor). Planning and implementing a pollution prevention program is just as complex, because pollution prevention ethic must permeate every activity conducted at the facility. However, an industrial facility can benefit from a well-planned and executed pollution prevention effort, especially when it is integrated with existing corporate programs so as to avoid the effort and expense of establishing a new program.

This chapter describes how to develop a facility pollution prevention plan for a successful program. The plan consists of organizing a pollution prevention team, establishing goals, collecting and analyzing data, identifying pollution prevention opportunities, heightening employee awareness and involvement, promoting education and training, and presenting pollution prevention project proposals.

7.1
Organizing a pollution prevention team

A well-organized pollution prevention team is essential for a successful facility pollution prevention program. The team should be organized so that no one department should bear the pollution prevention responsibilities. The responsibility should lie with key representatives from maintenance, production, environmental, health and safety, purchasing, shipping and receiving, legal and engineering departments, and plant and executive managers. Not every industry will have all these departments, but any individuals knowledgeable about a particular process in the company should participate in the program from its beginning.

The specific composition of the facility pollution prevention team will depend on the size of the company and the nature of its production processes. In a very small facility, a team could consist of only one or two key persons. However, in a large facility, a team may consist of even several different sub-teams. A successful pollution prevention team will benefit from the contributions of not only company experts and technical consultants, but also personnel directly involved in the production processes. Production operators and line employees can help identify important waste streams and assess how changes would affect the manufacturing process. Personnel in customer service and supply can also contribute by reporting product usage to a pollution prevention team.

A pollution prevention team should be authorized to access and compile the necessary information about the facility and waste streams being analyzed. Such authority is necessary to complete a thorough feasibility assessment of potential waste pollution prevention options. The team should be responsible for developing a formal pollution prevention plan and priority projects. After the initial meeting, pollution prevention team members should informally discuss the essence of pollution prevention, the potential benefits for the company, and the company's processes, and operational procedures to carry out the plan of a program. A program may consist of several projects for planning.

The program leader should come from the highest possible level practical, because the leader must have the authority and the influence to maintain the program and ensure the integration of pollution prevention with the overall corporate plan. Because the leader must facilitate the flow of information among all levels in the company, a leader must be able to elicit broad-based support from the company's employees (USEPA, 1992).

7.2
Establishing goals

A successful plan depends on management commitment and employee input, which can be achieved when a pollution prevention team:

- provides a definition for pollution prevention
- states company policies and guidelines
- identifies company goals and objectives
- evaluates material use and handling
- identifies types and amounts of waste generated
- emphasizes the benefits of pollution prevention
- encourages employee participation
- encourages suggestions and new ideas for pollution prevention
- identifies and evaluates potential pollution prevention methods
- presents successful pollution prevention facts and figures
- equates savings from pollution prevention with company fiscal health.

When company goals for pollution prevention are consistent with individual departmental goals, the pollution prevention project will more likely succeed. Pollution prevention efforts are not necessarily a self-contained program. For example, a team may integrate pollution prevention into the existing corporate programs to save time and money of establishing a new program. In addition, a pollution prevention team does not immediately have to establish comprehensive pollution prevention practices within the facility; pollution prevention may be incorporated into the company's policies and practices in an evolving and incremental manner.

Goals of a pollution prevention plan should have the following attributes:

- Acceptable to those who will work to achieve them;
- Flexible and adaptable to changing requirements;
- Measurable over time;
- Consistent with overall corporate goals;
- Understandable; and
- Achievable.

Pollution prevention goals may be divided into waste specific goals and activity oriented goals as described below (NYS, 1989):

7.2.1
Waste specific goals

An example of a waste specific goal is source reduction: minimizing waste generation and replacing some or all toxic substances used with non-toxic substances to reduce risk to employees, the public, and the environment. Ideally, a facility should conduct waste audits on all its waste streams, but so many audits would cost too much money, time and expertise. If a facility has limited resources, it should focus on the top twenty percent of the waste streams by volume, since these usually account for the bulk of the generated waste. This focus would offer the greatest opportunity for significant gains in pollution prevention. Facilities with multiple waste generating operations of processes but limited resources may want to rank their processes and waste streams for priority analysis.

7.2.2
Activity oriented goals

Activity goals could include: (1) incorporating pollution prevention into performance evaluations of all management staff, (2) installing a revised accounting system that charges the cost back to the production line generating the waste, (3) training all employees in pollution prevention, or (4) holding monthly team meetings.

The concept of continuous quality improvement, an essential component of a pollution prevention, involves continually updated goals. Specific goals will vary over time and should be based on the size of the facility and the type of production processes undergoing change. It is essential to establish a number of measurable goals so that pollution prevention progress can be tracked within a given time.

7.3
Data collecting and analyzing

The first step toward understanding industrial processes and waste generation is to gather background information on the facility. Information is necessary for the accurate determination of the type and quantity of raw materials used, the type and quantity of wastes generated, the individual production mechanisms, and the interrelationships between the unit processes. A pollution prevention team should share the responsibility of obtaining this information. A time frame should also be established to assemble and present the data to the group.

A facility's waste streams data can come from a variety of sources, such as personnel involved with the process and waste streams, hazardous waste manifests, biennial reports, environmental audits, emission inventories, engineering studies, waste assays, and permits. This type of data can be requested and obtained from operating personnel of the facility (NYS, 1989).

7.3.1
Site reviews

Many data sources probably exist for a given site. The detailed data on target processes, operations, or waste streams may have to be obtained through site reviews and interviews with workers. Site reviews supplement and explain existing data. They should be planned well to save time, money and effort. This can be done by:

(1) Preparing an agenda and providing it to staff contacts in the area to be assessed several days before the visit.

(2) Scheduling the inspection to coincide with the particular operation that is of interest (e.g., makeup chemical addition, bath sampling, bath dumping, startup, shutdown, etc.).

(3) Monitoring the operation at different times during each shifts, and if needed, during all three shifts, especially when waste generation is highly dependent on human involvement.

(4) Interviewing the operators, shift supervisors, and work leaders in the assessed area; and discussing the waste generation aspects of the operation.

(5) Photographing or videotaping the area of interest, if warranted. Pictures are valuable in the absence of plant layout drawings, because they capture many details that otherwise may be later forgotten or inaccurately recalled.

(6) Observing the housekeeping aspects of the operation; checking for signs of spills or leaks; assessing the overall cleanliness of the site; and paying attention to odors and fumes.

(7) Assessing the organizational structure and level of coordination of environmental activities among various departments.

(8) Assessing administrative controls, such as cost accounting procedures, material purchasing procedures, and waste collection procedures (USEPA, 1992).

Some of the information that can be gathered through site reviews by asking the following typical questions:

– What is the composition of the waste streams and emissions generated in the process? What is their quantity?
– From which production processes or treatments do these waste streams and emissions originate?
– Which waste materials and emissions fall under environmental regulations?
– What raw materials and input materials in the company or production process generate these waste streams and emissions?
– How much of a specific raw or input material is found in each waste stream?
– What quantity of materials are lost in the form of volatile emissions?
– Are any unnecessary waste materials or emissions produced by mixing materials which could otherwise be reused with other waste materials?
– Which good housekeeping practices are already in force in the company to limit the generation of waste materials?
– What process controls are already in use to improve process efficiency?

7.3.2
Data analysis

It is worthwhile to predetermine the bases for calculating the material balances that will be worked out during data collection. A material balance for a given substance will reveal quantities lost to emission or to accumulation in equipment. Simplified mass balances should be developed for each of the important waste-generating operations to identify sources and to better understanding the origins of each waste stream. Essentially, as shown in Fig. 7-1, mass balance verifies what enters the process (i.e., the total mass of all raw materials, water, etc.), and leaves the process (i.e., total mass of the product, wastes, and by-products).

Input ----➤ Process -➤ Output   **Fig. 7-1** A simplified mass balance diagram

Raw Materials	+ Substance Generation	Products
Water	- Substance Consumption	By-products
Air		Wastes

Input = Output + Substance Generated - Substance Consumed

Mass balance calculations are particularly useful to quantify fugitive emissions, such as evaporative losses. Waste stream data and mass balances will enable a pollution prevention team to track the flow and characteristics of waste streams over time. The first step in preparing a mass balance is to draw a process diagram, which is a visual means of organizing the data on the material flows and on the composition of the streams entering and leaving the process. The amount of material input should be equal to its output, corrected for generation and consumption during the process.

The limitations of material balance should be understood. They are useful for organizing and extending pollution prevention data and should be used whenever possible. Below are such limitations:

- Most processes have numerous process streams, many of which affect various environmental media.
- The exact composition of many streams is unknown and cannot be easily analyzed.
- Phase changes occur within the process, requiring multi-media analysis and correlation.
- Plant operations or product mix may change frequently, so the material flows cannot be accurately characterized by a single balance diagram.
- Many sites lack sufficient historical data to characterize all streams.

Despite the limitations, material balances are essential to organize data, identify gaps, and permit estimation of missing information. They can help calculate concentrations of waste constituents where quantitative composition data are limited. They are particularly useful if there are points in the production process where it is difficult or uneconomical to collect or analyze samples. Data collection problems, such as an inaccurate reading or an unmeasured release, can be revealed when "mass in" fails to equal "mass out". Such an imbalance can also indicate fugitive emissions. For example, solvent evaporation from a parts cleaning tank can be estimated as the difference between solvent put into the tank and solvent removed by disposal, recycling, or dragout (USEPA, 1992).

The result of those data collecting and analyzing activities is a catalog of waste streams that provides a description of each waste, including quantities, frequency of discharge, composition, cost of management, and other important information. The pollution prevention team should also collect data on the facility itself, including process design, raw materials, production data, operating costs, environmental reports and permits, as well as company policies and organizational information:

(1) Design data
   - Process flow diagrams
   - Material and heat balances for production processes and pollution control
   - Operating manuals and process descriptions

- Equipment lists
- Equipment specifications and data sheets
- Piping and instrument diagrams
- Plot and elevation plans
- General arrangement diagrams and work flow diagrams

(2) Environmental data
- Toxic release inventory
- Hazardous waste manifests
- Emission inventories
- Water discharge inventories
- Annual hazardous waste reports
- Waste assays
- Environmental audit reports
- Permits and/or permit applications

(3) Raw material and production data
- Purchasing records
- Raw materials inventory records
- Material application diagrams
- Material safety data sheets
- Operator data logs
- Production schedules
- Product composition and records
- Operating costs

(4) Other data
- Company environmental policy statements
- Organization charts
- Results of previous audits (NYS, 1989).

The above data and other useful information can be obtained from line workers, maintenance staff, process engineers; purchasing, inventory, shipping and receiving records; and accounting personnel. These employees can be interviewed to determine how the processes are run; what types of raw materials, cleaning agents, lubricants, and other materials are used; what types of waste are generated and how they are handled; what other types of records are kept; and what information is not recorded on a regular basis. When gathering this information, the team will be able to track wastes to determine if there are seasonal or shift variations in wastes generated. Once such information is assembled and data are collected and evaluated, a pollution prevention team can analyze the general process for potential pollution prevention.

A pollution prevention questionnaire is essential to collect and coordinate all necessary data. It also guides a pollution prevention team through the analysis. Typical items in a questionnaire include:

- A list with full characterization of all waste streams and generation points (verification of data collected from the review of site records);
- Operating data, such as the number of workers, schedule of operations, materials and equipment specifications, and utilities requirements;
- A list of all pollution prevention practices, including housekeeping which the pre-inspection review identifies as already exiting (to be verified and evaluated);
- A list of proposed pollution prevention options identified during the pre-inspection review (to be verified); and
- A list of other general pollution prevention opportunities to look for: (a) administrative control of materials; (b) housekeeping, handling, and storage; (c) raw material substitution; (d) recycle or reuse of waste streams; (e) modification of process, equipment, or operation; and (f) potential for redesign of process.

The questionnaire is especially useful when a substantial number of off-site personnel are involved in the analysis process; however, it may not be necessary for smaller companies or operations. A careful discussion led by plant personnel provides an alternative to questionnaires for smaller companies. It accomplish two purposes: (1) compiling all pertinent information about the facility and processes of concern into one source; and (2) familiarizing pollution prevention teams (especially outside members) with both the processes to be examined and the types of data required. The questionnaires need to be prepared for each specific industry, and they need to specify what operations will be reviewed as well as help guide a pollution prevention team through a plant or process area visit.

Questions should target the specific data required to analyze potential pollution prevention options. Ideally, a separate questionnaire should be prepared for each process to be examined. Although common data, such as material costs and usage rate, are required for many of the processes, each process will require unique information that might not be included in a more generic form. A process specific-form, although more time consuming to develop, will make the actual audit inspection and feasibility analysis easier and more efficient. It will minimize the need for follow-up visits and phone calls to collect information missed in the original inspection (USEPA, 1992).

7.4
Identifying pollution prevention opportunities

This section provides guidelines for prioritizing streams and/or unit processes for beginning pollution prevention assessments and identifying pollution prevention opportunities in the facility (USEPA, 1991).

7.4.1
Prioritize streams

Waste streams and unit processes should be prioritized to determine which should be examined first. A good starting point for prioritization is the use of process flow diagrams which show all of the input and output streams for each unit process.

When establishing priorities for pollution prevention, all of the input and output streams should be ranked – beginning with those which require immediate attention, followed by those which are less urgent. Factors should be considered when ranking the streams are as follows:

- US EPA's 17 target chemicals from the 33/50 program
- Toxic Release Inventory waste
- High purchase and/or disposal cost
- Highly toxic and carcinogens
- Resource Conservation and Recovery Act (RCRA) waste
- Particular regulatory and citizens' concerns
- High use and/or release rate
- Potential for removing bottlenecks in production or waste treatment
- Potential liability due to endangerment of employees or the public
- Potential for successful implementation
- High volume waste (may include tonnage)
- Hazardous Air Pollutants (HAPs)
- CFCs and other future banned materials

Once the streams are ranked, candidate input and output streams, especially waste streams, can be identified, keeping in mind the goals set at the beginning of the program, for the initial pollution prevention assessment. As the assessment proceeds, these priorities may change.

7.4.2
Assessing and identifying opportunities

It involves first looking at the processes associated with the candidate streams and then expanding the assessment to the entire facility so that all potential opportunities are addressed. The pollution prevention team should discuss the potential waste streams and the staffing of the overall facility to determine who should conduct the initial process assessment. Typically a team of two to three people is effective.

The team should first become familiar with the targeted processes. The flow diagrams developed provide an understanding of the process but may not explain why wastes are generated. For this information, the team must go into the facility and study the processes in detail. This study should be conducted while the process is in operation (ideally during all shifts) and, if possible, during a shut-down, clean-out, or start-up period to identify what materials are used and wastes are generated by this procedure. When studying the process, the team should note any potential pollution prevention opportunities and should pay particular attention to the following:

- Observe procedures of operation by workers
- Quantities and concentrations of materials (especially wastes)
- Collection and handling of waste (note if wastes are mixed)
- Any recordkeeping – and obtain copies of these if not already done
- Flow diagram – follow through actual process
- Leaking lines and/or poorly operating equipment
- Any spill residues
- Damaged containers
- Physical and chemical characteristics of the waste or release.

It may also be helpful to photograph the process to recall specific details later. Often, details can be better captured visually than with words. However, this should be cleared with the appropriate personnel first. Picture taking within a facility may be prohibited under certain policies.

The team should talk with the line personnel, including operators, supervisors, and foremen, as much as possible. In doing so, they should determine the required operating conditions, product specifications, and equipment specifications for the process. They should discuss the points previously listed as well as the daily routine the workers follow. Specifically, the team should try to identify when waste is generated – not just by the regular process but by upsets, off-spec products, spills, etc. The team should also talk with the maintenance and housekeeping personnel who service the process to determine when, why, and how the process is serviced. It is important to talk with these individuals as they generally have the best working knowledge of the processes.

The team should look at the other processes in a similar manner, if possible, after examining the targeted processes. The team should also conduct an overall survey of the facility. This survey consists of investigating supplemental operations such as shipping/receiving, purchasing, inventory, vehicle maintenance, waste handling/storage, laboratories, powerhouses/boilers, cooling towers, and maintenance.

The team should note potential opportunities for pollution prevention by ask questions as listed below:

- Shipping/receiving
  Packaging materials – what is done with waste?
  How are materials shiped/received – drums, bulk?
  Can containers be returned/recycled?
  Are you required to return empty containers to vendor?
  What happens to pallets?

- Purchasing
  Who orders materials?
  How for in advance are materials ordered?
  Can materials be ordered as needed (just-in-time)?
  Is the minimum amount ordered?

- Inventory
  What is the shelf-life of all materials?
  Is there an inventory control system? Bar-coding?
  Is there a central stockroom (no individual orders)?
  Do you operate by "just-in-time" philosophy?
  Do you operate by "last in, first out" principle?

- Vehicle maintenance
  Are solvents used for parts clearing?
  Are solvents recycled?
  Have solvent alternatives been tested?
  Do you recycle batteries, used oil, or antifreeze?
  How are used oil filters/carburetor cleaners handled?

- Waste handling and storage
  Are waste streams segregated?
  Do you know the sources of all waste?
  Do you have a "waste inventory" control system?
  How often is waste shipped off-site? Treated on-site?
  How is waste handled once shipped off-site?

- Laboratories
  How are chemicals ordered? In what quantities?
  What is the shelf-life of all chemicals?
  How are expired chemicals handled?
  Are solvents recycled/reused (e.g., first rinse)?
  How are gases stored?
  How are laboratory wastes handled?
  Are laboratory wastes segregated?

- Powerhouse/boiler
  How is fly ash/slag handled?
  How is tube clean-out material handled?
  What type of fuel is used? Are alternatives used?
  What type of boiler water treatment chemicals are used?
  How is boiler blow-down handled?

- Cooling towers
  What type of chemical additives are used?

How is bottom sediment handled?
What is your water source? Is water recycled?

- Maintenance
  What types of cleaners are used?
  Are solvent used? Are they recycled/reused?
  Have solvent/cleaner alternatives been tested?
  How are waste oil/greases handled?
  How are other wastes generated and handled?

Once the process assessments and plant survey are completed, the data obtained should be reviewed for thoroughness by all members of the team. This review will also initiate the brainstorming process for ideas to reduce waste at the source.

7.4.3
Identifying reduction options

The team members are encouraged to generate ideas and discuss options. They should also solicit ideas from other personnel at all levels of the entire facility. Some of the options may be simple to identify and implement such as:

- Ship/receive materials in bulk to eliminate drum disposal if large quantities are used
- Reuse containers where possible
- Order materials "just in time" to avoid expiration
- Establish a central stockroom/inventory control system
- Investigate solvent/cleaner alternatives or reducing the total number of different solvents used
- Reuse solvents where possible
- Segregate waste streams

Other options that may not be as easily identified but must definitely be considered involve source reduction and in-process recycling. A priority approach in selecting options may be developed. Ranking options on a high, moderate, or low continuum helps to ensure that pollution prevention is not a "one-shot" approach. Moderate and low priority options should still be considered since circumstances such as a change in raw materials, regulations, or technology could occur.

Once these options have been applied to specific streams/processes, further investigation or change in product composition may be required. For example, it may be necessary to implement new or existing techniques/technologies or to identify raw material alternatives. It may be helpful to contact other facilities, vendors, trade associations, state and local environmental assistance agencies, and publications for ideas. These groups may be aware of material alternatives or similar pollution prevention technologies that have been successfully implemented. Further pollution prevention opportunities may be identified through "upstream" suppliers and "downstream" consumers. These individuals should also be allowed input into the company's program (USEPA, 1992).

7.4.4
Determining costs

For economic feasibility study (see Section 6.3), the full cost of waste generation must determined. The economics of pollution prevention technologies must be developed, including calculating the cost savings and payback periods. Methods for true cost determination and economic analysis are presented. A cost accounting system for all wastes generated in the facility are also described.

The full cost of waste generation includes more than just treatment or disposal costs; it includes all the costs incurred by producing and handling waste. all of the expenditures associated with the waste stream, both direct and indirect, should be identified. These include, but are not limited to the following: purchasing, storage and inventory, and inprocess use of materials; flue gas and waste water releases, solid waste collection, waste storage, on-site treatment or recycling; waste disposal; waste transportation; lost raw materials; and labor costs.

Often, wasted raw material costs are three-fourths of the full cost of generating waste. Waste disposal costs are typically less than half the total costs. Some examples of waste associated costs to consider in the U.S. are listed below:

(A) Hazardous substance use
  - Purchasing includes taxes on hazardous products, safety training, MSDS filing, safety equipment, extra insurance premiums, and labor.
  - Storage and inventory include special storage facilities, safety equipment, storage area inspection and monitoring, storage container labeling, safety training, emergency response planning, spill containment equipment, lost product from spills and evaporation, labor, SARA Title III (TRI) reporting.
  - In-process use includes safety training, safety equipment, containment facilities and equipment, clean-up supplies, and labor.
  - Lost raw materials include labor for handling, equipment for clean-up, and reporting.

(B) Waste generation
  - Flue gas and waste water releases include air emission permits and controls, TRI estimates, TRI reporting, TRI fees, worker health monitoring, sewer discharge fees, NPDES permits, water quality monitoring, sampling training, pretreatment equipment, pretreatment system operation and maintenance.
  - Solid waste collection cost includes safety training, safety equipment, collection supplies, container labels, container labeling, recordkeeping, truck maintenance.

- Waste storage cost includes storage permits, special storage facilities, spill containment equipment, emergency response planning, safety training, storage area inspection and monitoring.
- On-site treatment or recycling cost includes capital and operating costs, depreciation, utilities, operator training, safety equipment, emergency response planning, permits, inspection and monitoring, insurance.
- Disposal cost includes sewer fees, container manifesting, disposal vendor fees, preparation for transportation, transportation, insurance and liability, disposal site monitoring.

Once the full costs of the waste streams are determined, an economic analysis of each specific pollution prevention project can be conducted. This analysis will provide management information on the costs and benefits associated with the technologies so they can decide whether it is economically feasible to proceed with implementation. Certain benefits, such as reduced long-term liability, reduced worker exposure to toxic chemicals, and improved community relations, will be difficult to quantify. Economic feasibility analysis of pollution prevention in terms of direct costs, indirect costs, liability costs, and benefits can be found in Section 3 of Chapter 6.

In the United States, EPA has identified and selected 17 industries that have greater potential opportunities to prevent pollution. These industries are shown in Table 7-1.

Table 7-1 Industries selected for pollution prevention opportunities for the 1990s

Industry	SIC Group*	Processes that could reduce wastes
Textile dyes and dyeing	226	Dye and scouring agent recovery
Wood preserving	2491	Development and use of less toxic preserving agents
Pulp and paper	26	Improved ability to recycle coated stock; fiber strength/restoration opportunities
Printing	271–275	Improved solvent recovery; use of low and non-VOC inks
Chemical manufacturing	281	Solvent recycling and substitution; improved catalysts
Plastics	2821	Scrap segregation and increased ability to mix different materials
Pharmaceuticals	283	Solvent recycling and substitution
Paint manufacturing	285	Use of low and non-VOC*_ paints; improved application
Ink manufacturing	2893	Development of low and non-VOC inks; elimination of metallic pigments
Petroleum processing and distribution	291	Spill prevention and materials recovery

Table 7-1 (continued)

Industry	SIC Group*	Processes that could reduce wastes
Steel production	331	Recycling of tars, electric arc furnace dust, calcium fluoride
Non-ferrous metals	333–334	Arsenic isolation; sulfur oxide emission reduction
Metal fishing	3471	Improved bath constituent recovery; alternate corrosion protection approaches
Electronics/semiconductors	3674	"Clean" fabrication
Automobile manufacturing/assembly	371	Life extension of oils and coolants; paint substitution and improved paint application
Laundries/dry cleaning	721	Improved solvent recovery; solvent substitution
Automobile refinishing/repair	753	Improved solvent recovery; solvent substitution

*Standard Industrial Classification Code
**Volatile organic compound

7.5
Employee awareness and involvement

A corporate/facility awareness program could effectively increase the pollution prevention knowledge of employees. Supervisors should discuss the status of the pollution prevention plan and project at weekly meetings. They should encourage the employees to bring pollution prevention ideas to the facility pollution prevention team meetings. A team should work with all employees to develop ideas for pollution prevention initiatives. Volunteers should be encouraged to participate in pollution prevention programs. All volunteers should be commended in some way for their interest in helping the company, their co-workers, and the environment. For example, employee suggestions should continually be heard and recognized, because they generate innovative ideas (IHWRIC, 1993).

7.6
Education and training

An education and training plan for industrial multi-media pollution prevention may be divided into technical and non-technical areas. Technical areas include products, processes, recycling and reuse. Non-technical areas include educational programs and dissemination of information, incentives and disincentives, economic costs and benefits, sociological and human behavioral trends, and management strategies. Cross-disciplinary training must be available for professionals to understand the importance of multimedia pollution prevention principles and strategies (Shen, 1990).

Pollution prevention education and training is still developing but has made rapid progress. It should be tailored for management, line and maintenance staff, and the education and training activities should be incorporated into company procedures. A pollution prevention orientation program would be helpful for all industry employees, regardless of their job function. Industrial employees will need training for any new technologies or techniques added to unit processes. Depending upon the size of the facility, training may be required in more than one shift. Another option that would give industrial employees pollution prevention initiative, is to have performance evaluation systems reflect pollution prevention responsibilities. As a pollution prevention team identifies strategies, they must consider training before implementation (Foecke, 1991).

In plant design, there is a growing emphasis on pollution prevention rather than end-of-pipe treatment and disposal. But pollution prevention needs a fundamental conceptual base to facilitate research and teaching, this conceptual base is still being developed. Innovative approaches to consumer product design are required, especially for items widely dispersed throughout society. Engineering education should incorporate environmental constraints into the routine design procedures of existing engineering disciplines. The environmental consequences of technology and the basis of the regulatory standards should be part of the engineering curriculum (Friedlander, 1989).

7.7
Proposing a pollution prevention plan

A pollution prevention team is responsible for developing a proposed plan for pollution prevention implementation. The plan should include all the ideas developed by the team such as:

- the statement of support from management;
- the pollution prevention team's organizational structure;
- the organizational guidelines and statement of purpose;
- the methods for fostering participation by all employees;
- the process flow diagram and materials balances;
- the material tracking system and possible technology transfer;
- the specific pollution prevention goals and projects;
- the procedures, criteria and schedule for implementation;
- the support of research and development if necessary;
- the structure of an incentive/reward program; and
- the provisions for employee training (Pojasek, 1991; USEPA, 1992).

The pollution prevention plan should be presented to and agreed upon by management so that they understand how the pollution prevention team will proceed and what resources/support will be required from them. The plan should be modified annually as pollution prevention experience increases and goals are reached, thereby allowing a company to continually improve the entire program.

References

Allys GH (1993) "Developing A Plan To Prevent Pollution." Environmental Protection, Vol 4, No. 12, pp. 19–23

Foecke T (1991) "Training for Pollution Prevention," in: Pollution Prevention Review, Spring 1991

Friedlander SK (1989) "Pollution Prevention: Homework and Design Problems for Engineering Curricula," Rollelot Department of Chemical Engineering, University of California at Los Angeles

IHWRIC (1993) Pollution Prevention: A Guide to Program Implementation. Illinois Hazardous Waste Research and Information Center, Champaign, Ill., TR-009, February

NYS (1989) Waste Reduction Guidance Manual. Prepared by ICF Technology Inc. for New York State Department of Environmental Conservation, pp. 3–3 to 3–11, March

Shen TT (1990) "Educational Aspects of Multimedia Pollution Prevention," in: Environmental Challenge of the 1990's, Proceedings of the International Conference on Pollution Prevention: Clean Technologies and Clean Products, USEPA/600/9-90/039, September

USEPA (1991) Industrial Pollution Prevention Opportunities for the 1990's. EPA/00/S8-91/052. Risk Reduction Engineering Laboratory, Cincinnati, Oh.

USEPA (1992) Facility Pollution Prevention Guide. Office of R&D, Washington, D.C., EPA/600/R-92/088, May

8
Implementation of pollution prevention plan

After we establish a pollution prevention project or plan of program and analyze its technical, environmental, economic, and institutional feasibilities, we will be able to more easily encourage management to implement chosen projects. All members of the company may not embrace a pollution prevention project immediately, especially if they do not fully understand the benefits and the cost savings of pollution prevention. To implement a pollution prevention plan or program most effectively, we must constantly emphasize the true cost of waste generation and management. The true cost includes all environmental compliance costs such as manifesting, training, reporting, accident preparedness; future liability costs; and intangible costs such as product acceptance, labor relations, and public image. A detailed description of true cost is discussed previously in Section5.5.3 "Product Life-Cycle Costing Analysis" and Section 6.3 "Economic Feasibility Analysis."

A pollution prevention plan is a written guide used to chart the progress of the program. It reiterates management support, lists reasons for the program, identifies the pollution prevention team, describes how waste will be characterized, provides a strategy and schedule for pollution prevention assessments, institutes a cost allocation system, indicates how technology transfer will be evaluated and implemented. The plan needs to be periodically updated to reflect the continuous nature of a pollution prevention program. Projects are the specific activities undertaken to reduce or eliminate waste.

This chapter describes the essential elements and methods of (1) understanding processes and wastes, (2) selecting a pollution prevention project, (3) obtaining funding, (4) implementing projects through various engineering steps, and (5) reviewing and revising projects.

8.1
Understanding processes and wastes

To effectively implement a pollution prevention plan, it is important to understand the various unit processes and where in these processes waste is being produced as discussed previously in Chapter 3. This section will explain how to determine the various unit process steps in materials extensive amount of data gathering may be necessary in this step in order to achieve a complete process characterization.

Generally two methods are used to understand processes and waste generation. One method begins with gathering information on total multi-media (air, land, and water) waste releases at the end of each process, and then backtracks to determine waste sources. Another method tracks materials from the point at which they enter the plant until they exit as wastes or products. Both methods provide a baseline for understanding where and why wastes are generated and a basis to measure waste reduced after implementation of pollution prevention projects. The steps involved in these characterizations include gathering background information, defining a production unit, general process characterization, understanding unit processes, and completing a material balance.

8.1.1
Gathering background information

The first step toward understanding processes and waste generation is gathering background information on the facility. This allows for the accurate determination of the type and quantity of raw materials used, the type and quantity of wastes generated, the individual production mechanisms, and the interrelationships between the unit processes. The pollution prevention team should divide up the responsibilities for obtaining this information. A time frame should be established for assembling the data and presenting it to the group. Figure 8-1 indicates possible sources of background information. It provides suggestions on data that should be assembled and where this information might be found.

In addition to these data, useful information can be obtained from line workers, maintenance staff, process engineers; purchasing, inventory, shipping and receiving; and, accounting personnel. These employees can be interviewed to determine how the processes are run; what types of raw materials, cleaning agents, lubricants, etc. are used; what types of waste are generated and how it is handled; what other types of records are kept; and what information is not recorded on a regular basis. When gathering this information, begin to track wastes to determine if there are seasonal or shift variations in wastes generated. Once this information is assembled, the general process can be characterized.

8.1.2
Define production units

To compare the amounts of waste generated during different time periods, and subsequently measure relative waste reductions, a production unit should be defined for each process either the unit process or the overall process depending on the nature of the facility. A production unit is simply a set quantity of product characteristic of the process-unit of plastic, gallons of acid, number of copies, etc.

Once the production unit is defined, wastes generated can be quantified as waste per production unit. Since total production can vary, comparing the

Information On:	From:
Raw Materials Use	Purchasing Records Production logs Inventory Records Packaging Material Discarded MSDSs Shipping and Receiving Logs Vendor Information Annual Report
Waste Generated	Waste Manifests Environmental Reporting TRI data Waste collection and storage Sewer Records (POTWs) Production Logs Permits/applications Environmental violations Flow diagrams Laboratory analyses Annual Report Obsolete expired stock Rejected Product Spill & leak reports
Production Mechanisms	Operations manuals (SOPs) Production logs Vendor information Flow diagrams Control diagrams Product specifications Quality control guidebook
Process Interrelationships	Product-to-raw material data Production logs Flow diagrams Product specifications Quality control data Facility layout Dependencies on preceding processes (e.g., how change in one affects another)
Economic Information	Cost accounting reports Pollution control costs Operating costs for waste handling and disposal Costs for products, utilities, raw materials, labor

Source: Illinois Hazardous Waste Research and Information Center
    TR-009. February 1993

Fig. 8-1 Possible sources of background information

amounts of waste generated for different time periods will not reflect the reductions achieved due to pollution prevention activities (i.e., waste will increase or decrease with production changes). For example, a printing press may use 1000 copies for a production unit and might then define wastes as "waste per 1000 copies".

By assembling background information, process flow diagrams for both the general process and individual processes can be developed. These diagrams, along with the material balances, help provide an understanding of the processes and the wastes generated. The production unit can be used for waste reduction comparisons throughout the pollution prevention program.

8.1.3
Characterize general process

A typical process has raw material inputs, product outputs, and waste generation. It can be represented by a general process flow diagram. This diagram may not physically resemble the process but will show the movement of raw material through the process as well as the generation of final product and waste. Fig. 8-2 presents a simple diagram of industrial processes.

In addition to the raw material, final product, and waste flows, other inputs can be represented on the general flow diagram such as lubrication fluids, cleaning agents, cooling water, etc. This will provide an understanding of the overall process and the associated wastes. The general process can then be separated into individual or unit processes.

8.1.4
Understand unit processes

Most production operations can be subdivided into a series of unit processes. For example, the general process of metal parts fabrication can be represented by at least seven individual processes.

1. Receiving and storing bulk metal
2. Cutting, bending, or shaping metal
3. Cleaning metal
4. Painting or coating metal
5. Assembling parts
6. Packaging
7. Shipping of assembled parts

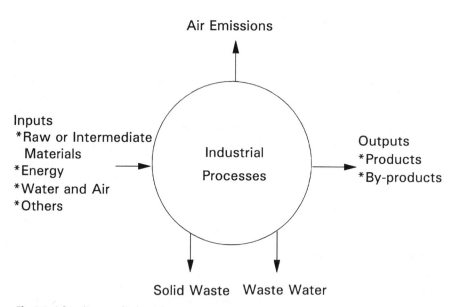

Fig. 8-2 A flow diagram of industrial processes

Each unit process has its own inputs and outputs; the product from one step becomes the input material for the following step. The raw materials, products, and wastes for each unit process can be shown on a more detailed flow diagram. This diagram should contain the type/composition and quantity of raw materials, products, and wastes to all media. The diagram should also include other inputs (lubrication fluids, cooling water, cleaning agents, etc.) along with the quantities used. The background information obtained previously will be helpful to determine the types/compositions and quantities of the general process. For example, a simplified general process flow diagram of metal parts fabrication is illustrated in Fig. 8-3 (no recycle of any material).

The flow diagrams for the unit processes can be completed using either of the two approaches: (1) start with the waste and products generated and then determine the sources of the waste by going backwards through each of the processes, or (2) start with the raw materials and track them through each of the unit processes until products and waste material are generated. For cases where waste streams are not separated but rather are combined prior to handling, the second method may be the preferred initial approach. The two methods may also be combined to complete the unit process flow diagrams and thus a detailed overall process diagram.

It is critical to determine the types/compositions and quantities of raw materials consumed, product yield, and wastes generated as accurately as possible for each unit process. All wastes released to the environment (gas, liquid, solid) should be characterized. These wastes can include: emissions from stacks; vent emissions from process areas; fugitive emissions from pipes, tanks, or vessels and leaking equipment; spent wash waters/cleaning solvents; cooling water; overspray from painting operations; cleaning rags; material scrap (e.g., metal, packaging, etc.); and other wastes. By subdividing the process into individual components, these types of wastes become more evident. With the information,

Metal Parts Fabrication

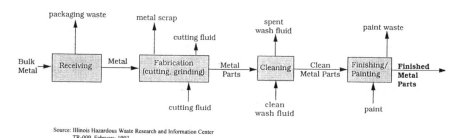

Source: Illinois Hazardous Waste Research and Information Center
TR-009. February 1993

Fig. 8-3 A process flow diagram of metal parts fabrication

a materials balance can be performed for the unit processes and then for the overall facility.

8.1.5
Perform materials balance

A materials balance accounts for all inputs and outputs into a process; in other words, what goes in must come out. A material balance should be performed for each unit process and for the overall production line. Although this typically is a very involved procedure, and while it is usually possible to identify sources of waste without having completed a materials balance, there are long term benefits to having done a materials balance. This material balance can help determine if fugitive loses are occurring in the process (e.g., fugitive loss from a solvent tank = difference between solvent in and solvent out). In a physical process, one in which there is no chemical change of materials, the raw materials that are not converted to product generally end up as waste. For example, a materials balance of metal parts fabrication process is shown in Fig. 8-4.

Key elements of a materials balance may include:

1. Quantity of raw material brought on-site;
2. Quantity produced on-site including amounts produced as production by-product;
3. Quantity consumed on-site;
4. Quantity shipped off-site as, or in, product;
5. Total waste generation (before recycling and treatment);
6. Amount of raw material in beginning and ending inventory;
7. An indicator of production levels involving the chemical; and
8. Release and transfer rate.

Metal Parts Fabrication

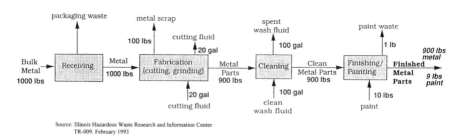

Source: Illinois Hazardous Waste Research and Information Center
TR-009. February 1993

Fig. 8-4 A material balance for metal parts fabrication

For a chemical process, the materials balance becomes more complicated as raw material inputs are converted to products through one or more chemical reactions. Some un-reacted raw materials may also end up as waste along with reaction by-products. For these processes, a standard material balance may already be available as part of the daily production log or cycle. Where possible, however, actual measurements of the amounts of materials used and generated should be used to produce the mass balance. The reason for this is that manufacturing processes can change over a period of time to a point where the actual materials balance would differ from that derived from the standard operating procedures.

Once the material balance has been performed, the actual amount of each waste generated by a process and the source becomes apparent if not already known. These numbers are the baseline amounts of total waste generated at the start of the pollution prevention assessment and can be used for comparison throughout the implementation of the program.

8.2
Selecting projects

Final selection of a project among the various proposed projects for implementation depends primarily upon the pollution prevention feasibility analyses. The selection should generally rely on the hierarchy for waste reduction which emphasizes more source reduction; results of the waste reduction assessment; availability of specific clean technologies or procedural applications; qualitative assessment of technical and economic feasibility; institutional feasibility and other considerations as described in Chapter 6. The next step is to make a schedule for implementation. The selected pollution prevention projects should be flexible enough to accommodate possible alternatives or modifications. The pollution prevention team should be willing to do background and support work, and anticipate potential problems in implementing projects.

Two basic procedures for screening waste reduction options are:

(1) an informal evaluation involving a consensus of team members to pursue selected options. This option works well for a facility with only a few waste streams or for a team with only a few members; and

(2) a more structured evaluation in which each of the options is compared against the criteria most important to the company. This more structured method works well when there are many waste streams or options to consider, allowing a team to rate and assess the options quickly. A team must determine what the important criteria are in terms of the company's goals and constraints (NYS, 1989).

One of the popular structured evaluations is the weighted sum method which calculates an overall weighted "score" for each option to reflect the company's goals and constraints. The essential steps of this method (USEPA, 1988) are to:

- Express the goals and constraints as criteria;
- Assign a different numeric value to each criterion in order of importance, giving the most important criterion the highest number;
- Calculate the weighted rating (R×W) for each criterion and each option, and then sum these weighted ratings for each option, and compare each option to other options, indicating the performance of that option relative to other options;
- Rank each option in order of highest to lowest weighted sum. Select among those options with the highest weighted sums for further technical and economic evaluation.

For example, in applying the Weighted Sum Method, a pollution prevention team member working on the project selection at ABC Company decided that the most important criteria to the company are:

- Reduction of treatment/disposal costs;
- Reduction of safety hazards;
- Improvement of on product quality;
- Short implementation period;
- Ease of implementation;
- Reduction of liability; and
- Reduction in waste quantity.

Based on the company's priority and constraints, this individual weighted the relative importance of these criteria and then entered the criteria and their weights onto a worksheet. ABC Company was considering three pollution prevention options (X, Y and Z), as shown in Table. 8-1. Each option was rated relative to the other options for each criterion.

The weighted rating for each option was calculated (W×R) for each criterion, and these weighted ratings were summed for each option. Option Y had the highest rating (370), followed by Option X (360). Option Z (282) had the lowest rating which can be calculated by the reader. The calculation of W×R for Options X and Y are illustrated in Table. 8-2.

Option Z will not be considered any further because its rating is so much lower than the others. Options X and Y will be further evaluated to determine which one is most preferred. Ranking options on a high, moderate, or low in priority help ensure that pollution prevention is not a one-shot approach. Moderate and low priority options should still be considered since raw materials, technologies or regulations change.

Once a project has been selected, further detailed investigation or change in product composition may be required. For example, it may be necessary to im-

Table 8-1 Weighted rating of pollution prevention options

Criteria	Weight W	Option R–X	R–Y	R–Z
Reduction of treatment/disposal costs	10	8	6	3
Reduction of safety hazard	8	6	9	8
Efficient on product quality	10	10	9	10
Short implementation period	2	2	10	1
Ease of implementation	5	2	2	3
Reduction of liability	7	4	4	5
Reduction in waste quantity	9	10	10	4

Table 8-2 Weighted rating of pollution prevention decision

Criteria	Weight W	Option X R	R×W	Option Y R	R×W
Reduction of treatment/disposal costs	10	8	80	6	60
Reduction of safety hazard	8	6	48	9	72
Efficient on product quality	10	10	100	9	90
Short implementation period	2	2	4	10	20
Ease of implementation	5	2	10	2	10
Reduction of liability	7	4	28	4	28
Reduction in waste quantity	9	10	90	10	90
Final evaluation: Sum of R×W			360		370

plement new or existing technologies or to identify raw material alternatives. At this point, it may be helpful to contact other facilities, vendors, trade associations, government agencies, and publications for ideas. They may have knowledge of or experiences in material alternatives or similar pollution prevention technologies that have been successfully implemented. Further pollution prevention opportunities may be discovered through material or equipment suppliers and product consumers (USEPA, 1988).

The total cost of possible pollution prevention projects may be assessed by breaking down into the following categories: process equipment costs, construction material costs, utility connections/systems costs, site preparation cost, installation costs, training costs, engineering costs, star-up costs, permitting costs, input material costs, salvage value, waste disposal costs, R&D costs, other O&M costs, and by-product recovery. The cost categories of assessment can be ranked by the average contribution of each category in terms of a project's net present value (NPV). This helps ensure that the most significant cost categories are always analyzed first. Experiences indicate that among the above 15 cost categories, a model was developed to prioritize the pollution prevention projects. The model is intended to correctly identify P2 investment options as either a negative or positive net present value. The model illustrates the four out of 15 cost categories were critical for priority selection. The top four cost categories are:

- Process equipment costs: Equipment costs required for implementation. Information on substitute process equipment should be readily available from vendors.
- Input material costs: Changes in input material costs brought about by the pollution prevention project. The cost of substitute material should be readily available from the vendor. Conversely, the current cost of material should be a matter of company record.
- Waste disposal costs: Handling, transporting, treating, and disposal of wastes; subcategories were TSDF fees, on-site treatment, on-site storage/handling costs, permitting, and reporting costs. Information on existing waste disposal costs should already be part of the company's records. In many cases, hazardous waste disposal will have been targeted specifically as an area for reduction or as a major expense to be examined.
- Other O&M costs: Costs associated with the day-to-day operation and maintenance of the pollution prevention system; subcategories were labor, supplies, analytical, and miscellaneous. Information on the labor hours and duties associated with new processes or equipment should be readily available from the vendor or contractor.

The screening model for cost analysis by eliminating projects with negative net present values was demonstrated at a major Air Force aircraft maintenance facility. It was proved to be more than double the efficiency of analysis if used in the screening stage of an opportunity assessment (Aldrich and Williams, 1988).

8.3
Obtaining funding

The pollution prevention team must seek funding for those selected projects that will require expenditures. Within a company, there are probably other projects, such as expanding production capacity or moving into new product lines, that compete with the pollution prevention projects for funding. If the team is part of the overall budget decision-making process, it can make an informed decision whether the selected pollution project should be implemented right away or whether it can await the next capital budgeting period. The team needs to ensure that the pollution prevention projects will be reconsidered at that time.

Some companies will have difficulty raising funds internally for capital investment, especially companies in developing countries. External funds are available to implement pollution prevention projects. Private sector financing includes bank loans and other conventional sources of financing. Financial institutions are becoming more cognizant of the sound business aspects of pollution prevention.

Government financing is also available in some cases. Small- and Medium-sized plants may be worthwhile to contact local, state, national, and international agencies, both public and private, for loans or financial assistance. In the United States, it may be worthwhile to contact the local or state's department of commerce or U.S. Small Business Administration for information regarding loans for pollution prevention and control. Some states provide financial assistance for pollution prevention projects. For developing countries, the United Nations' Environment Program (UNEP), United Nations' Development Program (UNDP), and the World Environment Center (WEC) provide not only financial assistance (grants or loans), but also technical assistance to environmental projects. Since 1993, WEC has been providing technical assistance to various industries of developing countries for waste minimization projects (WEC, 1997).

- International institutions provide loans, grants and technical assistance for developing countries, including:
- The World Bank (WB): It invests In projects with primarily environmental objectives, in addition to its regular investment portfolio. All proposed investments are screened for potential environmental impacts and categorized accordingly.
- The International Bank for Reconstruction and Development (IBRD): It is the oldest and the largest institution owned by the governments of 180 countries that have subscribed to its capital, and makes loans only to creditworthy borrowers. Assistance is provided only to those projects that promise high real rates of economic return to the country(WB, 1997).
- The International Development Association (IDA): It has provided assistance to poorer developing countries on terms that bear less heavily on their balance of payments than IBRD loans. IDA's assistance is concentrated on the very poor countries(WB, 1997).
- The International Finance Corporation (IFC): Its function is to assist the economic development of developing countries by promoting growth in the private sector of their economies and helping to mobilize domestic and foreign capital for this purpose(WB, 1997).
- The Multilateral Investment Guarantee Agency (MIGA): Its principal responsibility is promotion of investment for economic development in member countries through guarantees to foreign investors against losses caused by non-commercial risks and through advisory and consultative services to member countries to assist in creating a responsive climate and information base to guide and encourage the flow of capital (WB, 1997).
- The World Environmental Center (WEC) is a not-for-profit environmental organization. In 1982, WEC launched an industry and government supported environmental technology transfer program under the International Environment Development Services (IEDS). Using volunteers and experts from

industries, government, academia and non-government organizations, IEDS provides assistance to developing countries throughout the world. WEC implements its industrial waste minimization programs in Eastern Europe and Asia. The objective is to demonstrate measurable environmental and economic benefits from no-cost or low-cost waste minimization activities and subsequently, to introduce waste minimization practices throughout the selected industry. WEC organizes "train-the-trainers" workshops in waste minimization and supports Pollution Prevention Centers in developing countries with WEC projects (WEC, 1997).

– The Global Environment Facility (GEF) of United Nations Development Programme (UNDP): It assists developing countries to deal with four main global environmental problems: Global warming, pollution of international waters, destruction of biological diversity, and depletion of the stratospheric ozone layer. It is a three-year experiment that provides $1.3 billion grants (1992–1995) for investment projects, technical assistance and research (GEF, 1996).

8.4
Engineering implementation

A strong engineering approach helps ensure proper implementation of the selected projects. Outside process engineering support may be required if company personnel do not have the time to implement tasks. Many pollution prevention projects may require changes in operating procedures, purchasing methods, materials inventory control, equipment modification or new equipment. Such changes may affect a company's policies and procedures. But the phases of implementation resemble those of most other company projects.

Personnel who will be directly affected by the project (line workers and engineers), should participate from the start. Those personnel indirectly affected (e.g., controllers, purchasing agents) should also participate as project implementation proceeds. Any additional training requirements should be identified and arrangements made for instruction. All employees should be periodically informed of the project status and should be educated as to the benefits of the projects to them and to the company. Encouraging employee feedback and ideas may ease the natural resistance to change (IHWRIC, 1993).

Implementation of a pollution prevention project will generally follow the procedures established by the company for implementing any new procedure, process modification, or equipment change. Implementing a major pollution prevention project typically involves several steps:

(1) Preparing a detailed design. It is helpful to discuss the project with representatives from production, maintenance, safety, and other departments who may be affected by the change or who may have suggestions regarding equipment manufacturers, layout, scheduling, or other aspects of implementation.

(2) Preparing a construction bid package and equipment specifications. If construction is required, details of the necessary construction will need to be assembled into a construction bid package. Depending on the established procedures in the company, specifications for new equipment or particular manufacturers and models may be necessary.

(3) Selecting construction staff and purchasing materials. Construction may be performed by an in-house or outside company, depending on cost and availability.

(4) Installing new equipment. Construction will generally involve installation of the necessary equipment. Timing and scheduling of installation may be critical in some operations.

(5) Training personnel. Personnel from maintenance as well as production may need training. Proper operation and maintenance are critical for effective, safe, and trouble-free operation. Training is sometimes available from equipment vendors.

(6) Starting operation. Extra care is necessary during the initial stages of operation to ensure that all proceeds smoothly; often first impressions of the effectiveness of the system are often lasting.

(7) Monitoring and evaluating performance. Monitoring typically is an integral part of operations, and is performed by the production staff. Depending on the complexity of the chosen pollution prevention project, only a few of these steps may be necessary (NYS, 1989; IHWRIC, 1993).

8.5
Reviewing and revising projects

The pollution prevention process does not end with implementation. After the pollution prevention plan is implemented, we need to track its effectiveness and compare it to previous technical and economic assessments. Options that do not meet the original performance expectations may require re-work or modifications. This can be done through the knowledge gained by continuing to evaluate and fine-tune the pollution prevention projects. Chapter 11 will provide details on measuring pollution prevention progress after implementing and evaluating it against the established goals.

The success of the initial pollution prevention project may be only the first step along the road to establishing pollution prevention as either a stand-alone program or as an important criterion for consideration in other ongoing plant programs within the company. Developing the pollution prevention ethic is usually a step-by-step evolutionary process. It starts slowly and gradually builds momentum. Each successful pollution prevention experience provides even more incentive for management to support and diversify pollution prevention within the company.

To ensure that the pollution prevention momentum is preserved from the end of one pollution prevention project to the beginning of the next, it is important to provide an opportunity for feedback and evaluation. This opportunity should be used not only to critique the previous effort, but also to promote it and raise its visibility; that is, to "wave the banner" for pollution prevention and demonstrate that pollution prevention has even broader applications in the company. In this way we can use each pollution prevention experience as a promotional device for the next project, so that the role of pollution prevention within the company will continue to grow. With time, pollution prevention will become a natural part of the company's infrastructure and operating practices (NYS, 1989; USEPA, 1992).

References

Aldrich JR and WM Williams (1998) "Screening P2 Investment Opportunities," in: Pollution Prevention Review, pp. 1–9, Spring 1988

GLF (1996) Global Environmental Facility: Small Grants Programme. Project Document, GEF Office, UNDP, New York

IHWRIC (1993) Pollution Prevention: A Guide to Program Implementation. Illinois Hazardous Waste Research and Information Center, TR-009

NYS (1989) Waste Reduction Guidance Manual. New York State Department of Environmental Conservation, March

USEPA (1988) The EPA Manual for Waste Minimization Opportunity Assessments, EPA 600/2-88/025, April

USEPA (1992) Facility Pollution Prevention Guide. Office of Research and Development, Washington, D.C., EPA/600/R-92/088, May

World Bank (1993) Environmental Bulletin of the World Bank, Vol. 5, No. 4, p. 2, Fall 1993. Environmental Department, Washington, D.C.

World Bank (1997) World Bank Environmental Projects: Environment Matters at the World Bank. WB Environment Department Publication, Washington, D.C.

WEC (1997) Personal communication with Mr. Antony Marcil, President of the World Environmental Center, New York

9
U.S. P2 laws, regulations, strategies and programs

The mandates of the federal media-specific laws and regulations (air, water and land) are implemented by environmental regulatory agencies dominated by separate air, water, and waste offices. Many of the professionals in government and industry were trained to apply end-of-pipe controls appropriate for the specific media. Most professionals do not have the opportunity to see environmental problems from more than one perspective, and even today meeting the goals of individual acts continues to keep each office occupied. The more broadly focused laws such as National Environmental Pollution Act (NEPA), Toxic Substance Control Act (TSCA), Clean Air Act Amendment (CAAA), and Pollution Prevention Act (PPA) remain outside the scope of most pollution control work and are often implemented in limited ways of integration or multimedia approach.

The following six options were suggested for USEPA for moving forward integration in the United States:

1. Add multi-media provisions to existing laws and regulations as they are reauthorized.
2. Amend laws in other policy sectors with environmental measures.
3. Transform existing NEPA into a stronger integrative statute.
4. Build on pollution prevention and regional ecosystem approaches.
5. Transform the Toxic Substances Control Act into a law that can drive EPA programs to assess, reduce, and control toxic substances.
6. Build an Organic Act for USEPA on a function-by-function basis (Irwin, 1991).

The growing concern over the possible links between exposure to toxic chemicals and human cancers provided the impetus for regulatory control over the chemical industry that produces toxic chemicals and hazardous wastes. The ultimate goal is to assure that all toxic chemicals and hazardous wastes are handled, treated, and disposed of properly. The legal structures established for the control toxic chemicals and hazardous wastes in air, water, and land, usually represent a compromise between the public health and welfare on the one hand, and technical, economic, and political factors on the other. The manner in which this compromise is achieved largely depends on the situation existing in each jurisdiction or country. In the United States, regulation of toxic chemicals and hazardous wastes became a major undertaking of the U.S. Environmental Protection Agency (EPA) beginning in 1976 with the enactment of the Toxic Substance Control Act (TSCA) and the Resource Conservation and Recovery Act (RCRA). Until then, only pesticides were regulated under the 1972 Federal Insecticide, Fungicide, and Rodenticide Act (FIFRA).

To control adverse impacts of toxic chemicals and hazardous wastes on human health and the environment, environmental laws and regulations have grown exponentially in the past 20 years, not only in pages and cost of compliance, but in influence over the kind of world in which we live. During the 1980 s, the Conservation Research and Foundation undertook a series of research projects to examine the nature and extent of the problems created by fragmented laws and institutions and develop solutions for the problems. The researchers revealed that the components of our natural environment are interrelated in many complex ways, and pollutants tend to travel from one part of environment to another. The basic underlying rationale for paying attention to the cross-media problems is the disparity between the multiplicity of environments defined by laws, regulations, and jurisdictions and the unity of the natural phenomena with which those policies and institutions try to deal (Irwin, 1991).

Much environmental laws and regulations in the United States have been initiated at the federal government level and they are continually amending and evolving. Some states have enacted laws and regulations to protect unique environments within their jurisdiction. With the emphasis on giving states the responsibilities for enforcing such regulations, states are increasingly responsible for enforcing many of the environmental regulations. This chapter provides major pollution prevention laws, regulations and executive orders with brief discussing their objectives and provisions and specifying pollution prevention authority and requirements relating to implement pollution prevention programs. State environmental laws and regulations in general have been patterned after the federal programs and will not be discussed.

9.1
The U.S. pollution prevention related laws and regulations

The public policy emphasis on environmental protection over the past 25 years has spawned numerous laws. The major environmental statutes to implement pollution prevention are as below:

- The Pollution Prevention Act (PPA) of 1990
- The Resource Conservation and Recovery Act (RCRA) of 1976, 1984 (HSWA Amendments of 1984)
- The Clean Air Act (CAA) of 1967, 1990
- The Clean Water Act (CWA) of 1948, 1956, 1972, 1977, 1987
- The Toxic Substances Control Act (TSCA) of 1976
- The Federal Insecticide, Fungicide, and Rodenticide Act (FIFRA) of 1975, 1978, 1980, 1988

The shift in focus from pollution control (treatment and disposal) to pollution prevention (waste reduction and recycling) began in 1976 with the Resource Conservation and Recovery Act (RCRA). RCRA identified reducing waste at the source as the most desirable waste management option. With the passage of the Hazardous and Solid Waste Amendments (HSWA) in 1984, regulations required the generators of hazardous waste to submit a biennial report summarizing their efforts to reduce the volume and toxicity of waste generated. HSWA also requires generators who ship their waste off-site to certify that they have a program in place to reduce the volume and quantity of these wastes to the degree determined by the generator to be economically and technically practical.

The EPA is committed to making pollution prevention the guiding principle of all its environmental efforts. The new policy has five key parts:

1. Incorporate multi-media prevention as the principle of first choice in all the mainstream activities, including regulatory development, permitting and enforcement.
2. Help build a national network of prevention programs among state, local and tribal governments.
3. Expand those environmental programs that emphasize cross-media prevention, reinforce the mutual goals of economic and environmental well-being, and represent new models for government/private sector interaction.
4. Increase its efforts to generate and share information to promote prevention and track progress through measurement systems such as the Toxic Release Inventory.
5. Develop partnerships to increase technological innovation.

Since many environmental statutes focus on the control, management, and disposal of pollutants, rather than on their prevention, this section describes only those pollution prevention related major environmental statutes. However, some environmental statutes do indirectly promote pollution prevention through establishing regulatory programs that increase the cost, potential liability, and public scrutiny associated with managing hazardous materials. This section also presents the major significance on the federal pollution prevention regulations or policies promulgated pursuant to the statutory provisions.

9.1.1
Pollution Prevention Act (PPA)

In October 1990, the Pollution Prevention Act (PAA) was established. It is based on the premise that "wherever feasible, the generation of hazardous waste is to be reduced or eliminated as expeditiously as possible" and that "source reduction is fundamentally different and more desirable than waste management and pollution control." This legislation greatly expands the Environmental Protection Agency's (EPA) role in encouraging industrial source reduction and recycling in its regulatory and nonregulatory programs, as well as in collecting and assessing data about such activities. The law was enacted in part because Congress concluded that even where feasible, opportunities for source reduction are often not realized because of institutional barriers, such as governmental regulations and industrial management which focus upon treatment and disposal and regulations which focus on a single type of pollution rather than using a multi-media (air, water, and land) approach (PPA, 1990).

The EPA issued the Pollution Prevention Strategy (56 Fed. Reg. 7649, February 1991) to help clarify its pollution prevention mission and objectives to be accomplished. The strategy is designed to accomplish two main goals: (1) to provide guidance and focus for current and future efforts to incorporate pollution prevention principles and programs in existing EPA regulatory and nonregulatory programs, and (2) to set forth a program that will achieve specific pollution prevention objectives within a set, reasonable timeframe (USEPA 1991).

Most government programs have emphasized control of wastes after they are produced. The policy shift is reflected in the excerpt from the CAA which mandates a national policy creating a hierarchy of preferred waste management approaches: source reduction, recycling, treatment, and disposal; directs the EPA to create an Office of Pollution Prevention at EPA Headquarters in Washington, DC to implement this statute and develop a comprehensive pollution prevention strategy; provides states matching grants for programs designed to promote the use of source reduction by businesses with appropriate means of measuring the effectiveness of these state grant programs; requires EPA to establish the Source Reduction Clearinghouse to compile information on management, technical, and operational approaches to source reduction in a computerized database format; and requires EPA to submit a biennial report to Congress. More CAA regulations and programs will be discussed in Section 9.1.3.

The Pollution Prevention Act also states that: "Each owner or operator of a facility required to file an annual toxic chemical release form shall include with each filing a toxic chemical source reduction and recycling report for the preceding calendar year." This requirement became effective in 1992. The facility reporting provisions include eight requirements:

1. The quantity of the chemical entering any waste stream or otherwise released to the environment;
2. The amount of the chemical which is recycled in a calendar year, including the percentage change from the previous year;
3. Source reduction practices used with respect to that chemical during the year (this includes a variety of technologies and techniques such as improvement in management, training, inventory control, materials handling, or other general operational phases of industrial facilities);
4. The amount expected to be reported under (1) and (2) for the two years immediately following the reported year, expressed as a percentage change;
5. A ratio of production in the reporting year to production in the previous year;

6. The techniques which were used to identify source reduction opportunities (such as employee recommendations, external and internal audits, participative team management, and material balance audits);
7. The amount of any toxic chemical released into the environment; and
8. The amount of the chemical from the facility that is treated during such calendar year and the percentage change from the previous year.

Before the passage of the Pollution Prevention Act of 1990, the American Institute for Pollution Prevention (AIPP) was jointly established by EPA and the University of Cincinnati in 1989, to develop analyses, techniques, and programs to assist government and the private sector in further pollution prevention efforts. The AIPP is comprised of experts in association with professional societies and trade associations (USEPA, 1991).

9.1.2
Resource Conservation and Recovery Act (RCRA)

The Resource Conservation and Recovery Act (RCRA) addresses the management of solid waste, hazardous waste, and underground storage tanks that contain petroleum or hazardous substances. The objective of RCRA is to minimize the generation of hazardous waste and the land disposal of hazardous waste by encouraging process substitution, materials recovery, properly conducted recycling and reuse, and treatment. The RCRA establishes a comprehensive, or "cradle-to-grave," policy and regulatory scheme applicable to hazardous wastes. Its policy states that, to the extent feasible, the reduction or elimination of hazardous waste generation should be achieved as expeditiously as possible. However, its hazardous waste provisions regulate wastes after they are generated.

Hazardous waste generators are required to certify that they have a program in place to reduce the volume or quantity and toxicity of the materials they manage. Hazardous waste treatment, storage, and disposal facilities are required to certify that they have a program in place to reduce the volume or quantity and toxicity of the materials they manage. Those programs must exist to the extent that they are economically practical. The RCRA does not provide extensive authority to mandate pollution prevention. However, in 1984, the Hazardous and Solid Waste Amendments (HSWA) added several new provisions to the RCRA, some of which require pollution prevention. These provisions make it clear that pollution prevention is a fundamental element of United States hazardous waste management policy.

Major RCRA statutory restrictions or prohibitions include:

- The placement of any bulk hazardous waste in salt domes, salt bed formations, underground mines or caves is prohibited until the facility receives a permit.
- The landfilling of bulk or noncontained liquid hazardous waste or free liquids contained in hazardous waste is prohibited.

- The placement of any non-hazardous liquid in a landfill operating under interim status or a permit is prohibited unless the operation will not endanger groundwater drinking sources.
- The land disposal of solvent and dioxins is prohibited unless human health and the environment will not be endangered.
- Land disposal of wastes listed in Section 3004(d)(2) is prohibited unless human health and the environment will not be endangered.
- New units, lateral expansions and replacement of existing units at interim status landfills and impoundments are to have double liners and leachate collection systems; waste piles, a single liner.

EPA comprised a hazardous list of over 200 commercial chemical products and chemical intermediates by generic name. Substances became hazardous wastes when discarded. If a commercial substance is on the list, its off-spec species and spill residues are also considered hazardous. The overall listing further includes acutely toxic commercial chemical wastes. Essential requirements under RCRA (40CFR Part 262) for waste generators are:

- Identification – Hazardous wastes must be identified by list, testing or experience, and assigned waste identification numbers.
- Notification – No later than 90 days after a hazardous waste is identified or listed, a notification is to be filed with EPA or an authorized State. An EPA identification number must be received.
- Manifest system – Implement the manifest system and follow procedures for tracking and reporting shipments.
- Packing – Implement packaging, labeling, marking and placarding requirements prescribed by Department of Transportation (DOT) regulations.
- Annual report – Waste generators are required to submit an annual report by March 1.
- Exception reports – When a generator does not receive signed copy of manifest from designated TSDF within 45 days, the generator sends an Exception Report to EPA including a copy of manifest and a letter describing efforts made to locate waste and findings.
- Accumulation – When waste is accumulated for less than 90 days, generators shall comply with special requirements including contingency plan, prevention plan and staff training.
- Permit for storage more than 90 days – If hazardous wastes are retained onsite more than 90 days, a generator is subject to all requirement of TSDFs including the need for RCRA permits.
- Essential requirements under RCRA for transporters include:
- Notification – same as for generators.
- Manifest system – The transporter must fully implement the manifest system. The transporter signs and dates the manifest, returns one copy to the generator, assures that the manifest accompanies the waste, obtains the date and signature of the TSDF or next receiver, and retains one copy of the manifest for himself.

- Delivery to TSDF – The waste is delivered only to designated TSDF or alternate.
- Record retention – The transporter retains copies of the manifest signed by the generator, himself and the accepting TSDF or receiver, and keeps these records for a minimum of 3 years.
- Discharges – If discharges occur, notice shall be given to National Response Center. Appropriate immediate action shall be taken to protect health and the environment and a written report shall be make to the Department of Transportation (DOT).

Essential requirements for treatment, storage or disposal facilities (TSDFs) are:

- Notification – same as for generators.
- Interim status – These facilities include TSDFs; onsite hazardous waste disposal; onsite storage for more than 90 days; in transit storage for greater than 10 days and the storage of hazardous sludges, listed wastes, or mixtures containing listed wastes intended for reuse.
- Interim status facility standards – The standards and requirements shall be met including: general information, waste analysis plan, security, inspection plan, personnel training, handling requirements, preparedness and prevention, contingency planning and emergency procedures, records and reports, manifest system, operating logs, annual and other reports, groundwater monitoring, closure and post-closure plans, financial requirements, containers, tanks, surface impoundments, piles, all treatment processes, landfills and underground infection.
- Permits – Facilities with interim status must obtain a Part B RCRA permit.

Although RCRA, quite detailed in coverage, is considered one of the best legislations in the U.S., inadequacy and loopholes of the regulations have been identified including:

- Regulatory structure is unnecessary complex and no incentives for compliance or better waste management.
- Regulations have too many cross-references.
- Regulations complicate the permitting process. Voluminous details on the means of carrying out the regulations result in a very slow and costly process.
- Regulations insufficiently weigh important features. Little flexibility exits for the important characteristics of a particular facility, population, groundwater aquifers, atmospheric conditions, waste properties and the facility's compliance history.
- Priority setting has not been realistic. There is a widespread perception of inadequate progress and outputs.
- Problems exist between EPA and the States such as differences in priorities and stringencies.
- Public education has been ineffective such as hearings and communicating the risks posed from exposure.

- Vital data and analytical techniques are still lacking. Major technical and scientific uncertainties still exist.
- Data management systems need improvement.

9.1.3
Clean Air Act (CAA)

The Clean Air Act (CAA), originally passed in 1967 and amended in 1990, seeks to protect our nation's air quality through imposing emission standards on stationary and mobile sources of air pollution. The CAA provides the EPA's authority to require pollution prevention measures, including the installation of control equipment as well as process changes, the substitution of materials, changes to work practices, and operator training and certification. Under the amendment Sec. 112, the EPA is required to regulate 189 substances presumed to merit regulation as air toxics.

For mobile source, the CAA Amendment requires increasing percentage of fleet vehicles to the alternative fuels and establishes alternatively fueled vehicle pilot program in California. It also requires the development of reformulated gasoline and oxygenated fuels to reduce air pollutants. For new source review, it requires that new sources located in nonattainment areas use most stringent controls and provide offsets. Such offsets may be achieved through pollution prevention.

The CAA Amendment imposes stringent sulfur dioxide emissions limits for reducing acid rain and creates system of tradable emissions allowances. It forces power plants to either reduce their emissions to the specified rates or develop or buy emissions allowances, thereby creating market-driven incentives to reduce emissions. It also requires the phaseout of production and sale of chlorofluorocarbons (CFCs), and CFC-containing products that have been shown to contribute to the destruction of the stratospheric ozone layer (Sec 42 USA. Sec 7671).

9.1.4
Clean Water Act (CWA)

The Clean Water Act (CWA) seeks to restore and maintain the chemical, physical, and biological integrity of the nation's waters. The EPA has the authority to develop technology-based, industry-specific national limits (implemented through NPDES permits) on the amounts of regulated pollutants that a facility is allowed to discharge into the nation's waters. The CWA subjects point-source discharges of contaminated stormwater to regulation, including pollution prevention plans.

9.1.5
Toxic Substances Control Act (TSCA)

The Toxic Substances Control Act (TSCA) of 1976 provides the EPA with the authority to test chemicals for their potential health and environmental effects prior to their introduction into commerce and to regulate these substances where they pose an unreasonable threat to health or environment. The TSCA provides the authority to regulate the manufacturing, processing, distribution in commerce, use, and disposal of chemical substances and mixtures.

TSCA Sec 5 addresses chemical manufacturing and processing notices. It prohibits any person from manufacturing or importing a new chemical substance without notifying the EPA. Notification is accomplished through the submittal of a Premanufacture Notice (PMN). Submittal of a PMN enables the EPA to screen new chemicals before commercial production or importation begins. Under Sec. 5f order, the EPA may prohibit the manufacture, processing, or distribution of a chemical substance or may seek an injunction to prohibit the manufacture, processing, or distribution of the chemical.

TSCA Sec 6 addresses the regulation of hazardous chemical substances and mixtures. When the manufacturing, processing, distribution, use, or disposal of an existing chemical substance or mixture presents an unreasonable risk of injury to health or the environment, the EPA has broad authority to promulgate rules to prohibit or limit production or to impose labeling or other requirements. Such other requirements include imposing record keeping and testing requirements, imposing restrictions on the commercial use and disposal of the chemical, and requiring that notice of potential hazards be provided to distributors, users, and the public.

TSCA Sec 8 consists of five subsections that establish reporting and record-keeping requirements. Chemical manufacturers, importers, processors, and, in certain cases, distributors, may be required to submit reports and/or maintain records These requirements provide the EPA with considerable leverage for obtaining detailed data about chemical substances, which supports actions to restrict or prohibit chemical use. Section 8e requires any person who manufactures, processes, or distributes in commerce a chemical substance or mixture and who obtains information which reasonably supports the conclusion that such substance or mixture presents a substantial risk of injury to human health or the environment shall immediately inform the EPA of such information unless such person has actual knowledge that the EPA has been adequately informed of such information.

TSCA can also be divided into five parts as follows:

(1) Testing – Under TSCA, Section 4, EPA can require product testing of any substance which may present an unreasonable risk of injury to health or the environment. Some testing standards are proposed.

(2) Inventory and Pre-manufacture Notification (PMN) – EPA has published an inventory of existing chemicals. A substance that is not on this list is considered "new" and requires PMN to EPA at least 90 days before the chemical can be manufactured, shipped or sold (TSCA, Section 5). EPA may reject PMN for insufficient data, negotiate for suitable data, prohibit manufacture or distribution until risk data are available. EPA may completely ban the product

from the market, or review the product data for an additional 90 days (TSCA, Section6).

(3) Regulation under Section 6 – EPA can impose a Section 6 rule if there is reason to believe that the manufacture, processing, distribution, or use or disposal of a chemical substance or mixture causes, or may cause, an unreasonable risk of injury to health or the environment. Regulatory action can range from labeling requirements to complete prohibition of the product. A Section6 rule requires informal rulemaking, a hearing, and a cost-benefit analysis.

(4) Imminent hazard – This is defined as a chemical substance or mixture causing an imminent and unreasonable risk of serious or widespread injury to health or the environment. When such a condition prevails, EPA is authorized by TSCA, Section7, to bring action in U.S. District Court. Remedies include seizure of the chemical or other relief including notice of risk to the affected population or recall, replacement or repurchase of the substance.

(5) Reporting and recordkeeping – TSCA, Section 8(a) deals with general reporting. Section8(c) calls for records of significant adverse effects of toxic substance on human health and the environment. It requires that records of alleged adverse reaction be kept for a minimum of 5years. Section8(d) allows EPA to require that manufacturers, processors and distributors of certain listed chemicals submit to the EPA lists of health and safety studies include individual files, medical records, daily monitoring reports, etc. Section (e) requires action upon discovery of certain data.

9.1.6
Federal Insecticide, Fungicide, and Rodenticide Act (FIFRA)

Federal Insecticide, Fungicide, and Rodenticide Act (FIFRA) and its amendments of 1975, 1978, 1980, and 1988 establish the federal authority to regulate the distribution, sale, and use of pesticides in the United States. FIFRA's major provisions include product registration, pesticide use, and removal of pesticides from the market. All pesticides that are distributed or sold must be registered with the EPA, unless they are the subject of experimental use permits of an exemption.

As of 1991, there were approximately 1200 registered active ingredients in over 20,000 different pesticide products (USEPA 1991). Registration allows a pesticide product to be sold and distributed for specified uses in accordance with specified use instructions, precautions, and other conditions.

For pesticides that are not in compliance, or if their use causes unreasonable adverse effects on the environment, the EPA may issue a notice of intent to cancel the pesticide's registration, change its classification, or hold a hearing on these issues. The EPA has the authority to supersede the classification process and immediately suspend the registration if necessary to prevent an imminent hazard.

The FIFRA authorizes the EPA to cancel a pesticide if it is causing unreasonable effects on human health and the environment, or suspend its use immediately in order to prevent an imminent hazard. The EPA has the authority to seize pesticides, or to issue "stop sale, use, or removal" orders if the pesticide is in violation of any FIFRA provision, or if the registration has been canceled.

9.1.7
Basic principles of laws and regulations

Ideally, laws and regulations should follow four basic principles: simple and understandable, reasonable and fair, implementable, and affordable. However, as a law abiding country, the U.S. government tries to be very specific and explicit on what is required and that makes for a very thick regulatory package of documents which is basically a lawyer's document and tough to read, understand, and interpret. As a result, such regulations can defeat their intended purpose. The fourteen thousand pages of laws and regulations that convey the intents to specify just how the protection of air, land and water will be achieved. To further complicate matters, the above mentioned laws and regulations typically have counterparts in each of the states and each state law usually has its own corresponding set of regulations to fit their local social and cultural conditions. As if this weren't enough, outcomes of court proceedings often affect either the statutes or the regulations. The legal and administrative overhead in hazardous waste work is extremely high. A large percentage of Superfund moneys, for example, have been spent to legal and administrative effort, rather than to technical research, cleanup, remediation or reclamation activities.

9.2
Executive orders promoting federal pollution prevention

In addition to federal statutory laws, there are several executive orders that also require or promote pollution prevention. Generally, these executive orders are binding on the federal government and affiliated entities and they are briefly described below.

9.2.1
Energy efficiency and water conservation at federal facilities
EO 12902 (3/8/94)

EO 12902 requires federal agencies to (1) develop and implement programs to reduce energy consumption and increase energy efficiency at their facilities and buildings by using prioritization studies, facility audits, and technologies for energy efficiency, water conservation, and renewable energy; (2) increase the sue of solar power and other renewable energy sources; and (3) use cleaner, less-polluting fuels and energy sources instead of petroleum-based products, and reduce the use of petroleum where such alternatives are not practical or cost efficient.

9.2.2
Federal compliance with right-to know laws and pollution prevention requirements
EO 12856 (8/3/93)

EO 12856 ensures that all federal facilities operate in a manner that (1) reduces the amount of toxic chemicals entering any waste-stream through source reduction and recycling activities, ensuring compliance with the provisions of the Pollution Prevention Act; (2) complies with toxic chemical and hazardous substance reporting requirements, to ensure public knowledge and awareness; and (3) revises existing procurement process to reflect source reduction principles, and use and test innovative pollution prevention technologies on site to encourage market development and wider use of more effective pollution prevention methodologies.

9.2.3
Requiring agencies to purchase energy-efficient computer equipment
EO 12845 (4/21/93)

EO 12845 encourages energy efficiency and reduced emissions by power plants such as coal-fired utilities by purchasing and using power-conserving computer equipment that meet the criteria of the EPA's "Energy Star" computer purchasing program; and educates federal users about the energy-saving and environmental benefits of using less energy and generating fewer pollutants, thereby encouraging wider use and future benefits.

9.2.4
Federal use of alternative-fueled vehicles
EO 12844 (4/21/93)

EO 12844 mandates use of alternatively fueled vehicles by federal agencies to substantially reduce toxic and hazardous air pollutants; and directs all federal agencies to increase their purchases of alternatively fueled vehicles by 50% over levels specified by Energy Policy Act for fiscal year 1993 through 1995.

9.2.5
Procurement requirements and policies for federal agencies for ozone-depleting
substances EO 12843 (4/21/93)

EO 12843 directs all federal agencies to (1) revise procurement regulations and policies to conform with the provisions of the Clean Air Act Amendments addressing stratospheric ozone layer protection; (2) maximize their use of alternatives to ozone-depleting substances, evaluate present and future needs for ozone-depleting substances, and develop recycling initiatives to reduce and prevent further ozone-layer degradation; and (3) modifies procurement specifications and practices to substitute non-ozone-depleting substances for ozone-depleting substances currently used.

9.2.6
Federal agency recycling and the Council on Federal Recycling and Procurement Policy
EO 12780 (10/31/91)

EO 12780 directs federal agencies to (1) promote cost-effective waste reduction and recycling activities, providing a positive forum for the development and study of policy options and procurement practices that enhance environmentally sound and protective waste reduction and recycling practices; (2) develop an alternative procurement program designed to purchase products with recycled content; and (3) create the Council on Federal Recycling and Procurement Policy whose broad mission is to encourage federal agencies to purchase products that reduce waste generation, assist in the development of waste reduction and recycling programs, and collect and disseminate federal agencies information concerning waste reduction methodologies, recycling program costs and savings, and current market prices of recycled content products as well as those that reduce wastes.

9.2.7
Federal energy management
EO 12759 (4/17/91)

EO 12759 encourages wise energy management practices by all federal agencies to a variety of ways including using alternative, less-polluting fuel, minimizing use of petroleum products, and encouraging employee outreach programs.

9.3
EPA strategy and programs

To carry out the provisions of the Pollution Prevention Act, EPA adopted a strategy to implement the policy of pollution prevention throughout all its offices, programs and activities. An Office of Pollution Prevention was established to coordinate all agency pollution prevention activities, and a Pollution Prevention Advisory Committee of top-level agency staff to ensure that a pollution prevention strategy is the preferred approach emphasized in agency activities affecting all environmental media. In the pollution prevention policy statement of 1993, EPA summarized its pollution prevention program into seven different themes, characterizing EPA's pollution prevention activities (USEPA 1997).

(1) Incorporating prevention into the mainstream work of EPA.
(2) Building a national network of prevention programs.

(3) Promoting cross-media prevention programs.
(4) Establishing new federal partnerships.
(5) Generating environmental information on pollution prevention.
(6) Developing partnerships for technological innovation in pollution prevention.
(7) Changing existing federal laws to encourage pollution prevention.

9.3.1
Prevention as EPA's the first choice

EPA has undertaken a concerted effort since 1993 to find the best ways to incorporate prevention into regulations and permitting, through such efforts as the
Source Reduction Review Project (SRRP) and EPA's Common Sense Initiative.
EPA has also looked inward, at activities in its own facilities, and committed to
taking advantage of prevention opportunities. In February 1996, EPA published
an assessment of the SRRP experience. The assessment identifies the successes
and obstacles encountered in SRRP and makes a number of recommendations.
EPA also found ways to increase flexibility in some instances by adding innovative incentives in the rules to encourage businesses to choose multimedia pollution prevention approaches. These incentives allow companies to find the least-
cost ways of meeting standards, and will enable many facilities to achieve better
environmental results (USEPA, 1996).

The assessment provides case studies for seven SRRP rule-makings, consisting of NESHAPs under the Clean Air Act, Effluent Guidelines under the Clean
Water Act, and a RCRA hazardous waste listing. The seven individual assessment cases are: pulp and paper manufacturing; halogenated solvent cleaners;
carbonates hazardous waste listing; pesticide formulating, packaging, and
repackaging effluent guideline; metal products and machinery effluent guideline; pharmaceuticals effluent guideline; and wood furniture manufacturing.

The positive lessons that can be drawn from their experience include: (1) focus on cross-media data collection and cross-media analysis of regulatory options; (2) write the standard to be flexible; (3) test new territory with stakeholders; (4) coordinate agenda-setting; (5) use preambles and development documents to explain P2 multi-media approaches; (6) use statute-specific approaches
developed in SRRP; and use program-tailored P2 training.

The fundamental obstacles to more effectively integrating multi-media and
P2 approaches into regulatory and other mainstream activities are: (1) the lack
of incentives for inter-office coordination in planning and budgeting; (2) the
piecemeal nature of the statutory framework; (3) challenges in promoting P2
process changes and innovative technologies; (4) the lack of understanding
about cross-media impacts; (5) the lack of resolution about collecting source reduction data through industry surveys; and (6) unclear roles for sharing P2 leadership among all parts of the EPA on various aspects of P2. These obstacles have
had the following effects: (1) resource allocations are not conducive to the coordination and cross-media analysis and information sharing among offices is

limited; (2) deadlines for rules affecting the same industry are generally not synchronized, and regulatory strategies are not developed on an industry-sector basis; (3) P2 process changes and innovative technologies can be difficult to promote; (4) potential cross-media impacts sometimes remain unknown; (5) missing P2 data from surveys sometimes impedes development of source reduction options; and (6) P2 and cross-media policy issues are not explored as creatively as they might be.

Based on EPA's rule-making experience, implementing source reduction in permitting will raise issues regarding flexibility in timetables, resource demands on permit writers, cross-media coordination in permitting, and risk sharing for innovative source reduction approaches. To make multi-media and P2 perspectives more central in the development and implementation of EPA rules, the assessment report recommends to:

(1) Emphasize the key link between cross-media solutions and source reduction;
(2) Continue placing special attention on targeted rules, especially during their implementation through the permitting and compliance phases;
(3) Apply some of the positive SRRP lessons to more rule-makings;
(4) Take steps to start systematically addressing the obstacles to fostering prevention;
(5) Reinvent the planning and budgeting processes to enhance cross-media and P2 outcomes;
(6) Develop a cross-media legislative strategy;
(7) Deepen Agency understanding of cross-media impacts;
(8) Address Paperwork Reduction Act concerns about collecting source reduction data from industry;
(9) Clarify P2 roles within the Agency; and
(10) Link efforts to address these obstacles to full implementation of Common Sense approach and reinvention of EPA (USEPA, 1997).

9.3.2
A national network of pollution prevention

EPA is providing funding support, technical assistance, information dissemination, and forming federal/state/local government partnerships to focus efforts on pollution prevention as the national goal for environmental management. The cornerstones of this support are: (1) EPA funding of state and local pollution prevention projects, (2) dissemination of pollution prevention related information and technical assistance, and (3) combined participation of federal, state, and local environmental leaders in supporting prevention as a main focus of achieving environmental protection.

Through the Pollution Prevention Incentives for States (PPIS) grant program, EPA provides $5.8 million each year for state and tribal, local, and community pollution prevention programs and initiatives. The goal of the PPIS grant program is to assist businesses and industries in identifying pollution prevention strategies and solutions for complying with environmental regulations. EPA's Environmental Justice through Pollution Prevention grants assist community-based groups in developing collaborative approaches to achieving environmental justice through pollution prevention.

9.3.3
Cross-media prevention programs

EPA is pioneering voluntary programs such as Partners for the Environment which are EPA/industry interactions aimed at educating industry, citizens, state and local governments, and other stakeholders to participate in pollution prevention. These partnerships include the 33/50 Program, Climate Wise, Green Lights, National Industrial Competitiveness Through Efficiency (NICE), Agriculture in Concert with the Environment (ACE) Program, Pollution Prevention Incentives for States (PPIS) Program, Design for the Environment (DfE), Waste Wise, Project XL, Environmental Accounting, Water Alliances for Voluntary Efficiency (WAVE), and Pesticide Environmental Stewardship (USEPA 1993, 1997).

(A) EPA 33/50 PROGRAM

The EPA 33/50 Program of 1990 is one of the first major programs developed specifically in response to the Pollution Prevention Act. The program involves the voluntary cooperation of industry in reducing environmental releases of specific chemicals. a voluntary pollution prevention initiative that derives its name from its overall goals, i.e., a 33% reduction of emissions nationwide of 17 high-priority toxic chemicals by 1992 and a 50% reduction by 1995. USEPA is asking companies to examine their own industrial processes to identify and implement cost-effective pollution prevention practices for these chemicals. As required by the Pollution Prevention Act, industrial reporting requirements have been expanded, beginning in 1991, to include information on pollution prevention. The 17 chemicals are:

- benzene
- cadmium and compounds
- carbon tetrachloride
- chloroform
- chromium and compounds
- cyanides
- dichloromethane
- lead and compounds
- mercury and compounds
- methyl ethyl ketone
- methyl isobutyl ketone
- nickel and compounds

- 1,1,1-trichloroethane
- tetrachloroethylene
- trichloroetylene
- toluene
- xylenes.

The selection of 17 toxic chemicals was based upon criteria including health and environmental effects, possibility of exposure, volume of production and release, and potential for pollution prevention. Because air emissions accounted for over 70% of the releases and transfers of the 17 selected chemicals, the program overlaps with the Clean Air Amendments of 1990. The top seven 33/50 chemicals accounted for approximately 84% of the releases and transfers of the total 17.

EPA includes the 33/50 Program within its broader pollution prevention strategy. Crucial to the success of the program is that the toxic release inventory (TRI) provides data permitting the evaluation of the program's progress and the status of its target chemicals relative to overall trends. Reduction is measured against 1988 TRI data. The three significant groups of 33/50 chemicals are metals and metal compounds, non-halogenated organic chemicals, and halogenated organic chemicals.

(1) Metals and metal compounds – Of the 33/50 chemicals, five are in the metals and metal compound's class. They are cadmium, cadmium compounds, chromium, chromium compounds, lead, lead compounds, mercury, mercury compounds, nickel, and nickel compounds. The metals all exist naturally in trace amounts; however, mining, purification, use, and incineration concentrate their presence in the environment. Their relative mobility and tendency to bioaccumulate heighten their environmental impact. Metals are used commercially as alloys, catalysts, and in batteries and in metal plating processes.

(2) Non-halogenated organic chemicals – The chemical class of non-halogenated organics includes the 33/50 chemical's benzene, toluene, the xylenes, methyl ethyl ketone and methyl isobutyl ketone. The term "non-halogenated" refers to the absence of elements known as the halogens, including fluorine, chlorine, bromine, iodine, and astatine. This category may be broken into two subgroups for descriptive purposes: the aromatic chemicals and the ketones. Benzene, toluene, and the xylenes belong to a group of compounds known as aromatic compounds (so named for their odors). They are produced as a major byproduct of petroleum refining. They are used in gasoline, solvents, and as raw materials for plastics and other chemicals industry products. These chemicals contribute to smog formation, and benzene is a known human carcinogen. Methyl ethyl ketone and methyl isobutyl ketone belong to a group of compounds known as ketones. They are common chemical solvents and intermediates noted for their contribution to smog forma-

tion. Because of their pleasant scents, ketones have been used in the flavoring and perfume industries.

(3) Halogenated organic chemicals – The six chlorinated organic chemicals on the 33/50 list, carbon tetrachloride, chloroform, methylene chloride, tetrachloroethylene, 1,1,1-trichloroethane, and trichloroethylene, belong to the larger TRI class of halogenated organic chemicals. The halogenated organic chemicals all contain varying amounts of halogen(s), which in this case is chlorine (other halogens include fluorine, bromine, iodine, and astatine). Several of these chemicals deplete ozone in the upper atmosphere, making them apt targets for the 33/50 pollution prevention initiative. 1,1,1-trichloroethane and carbon tetrachloride, both 33/50 chemicals, are ozone-depleters. These six chemicals see widespread use as industrial chemicals employed to manufacture other chemicals, as common components of many consumer products, and as solvents. Because of their photochemically reactive nature, some of these chemicals contribute to smog formation. Trichloroethylene and tetrachloroethylene have short half-lives in the lower atmosphere, which results in smog formation but which prevents them from depleting the ozone layer.

Industry participating in the 33/50 Program has proven remarkably successful and is responsible for an accelerated reduction in the 17 chemicals targeted by the program. The 1,300 corporate participants in the 33/50 Program own more than a quarter of the total number of TRI facilities, and were able to meet the program's 1995 goal of 50% reduction a year ahead of schedule (USEPA, 1997).

(B) Climate wise recognition program

It is a joint EPA/Department of Energy voluntary pledge program that encourages private industry and others to adopt flexible, comprehensive approaches to reducing greenhouse gas emissions. The program provides technical assistance and puts companies in touch with financial services to "jump start" energy efficiency and pollution prevention actions. With 13 charter companies, Climate Wise companies already represent almost 4% of U.S. industrial energy use. Climate Wise participants expect to save more than $80 million annually by the year 2000.

(C) The green lights program

The Green Lights Program was officially launched on January 16, 1991. The goal is to prevent pollution by encouraging major institutions – business, governments, and other organizations – to use energy efficient lighting. Because lighting consumes electricity (about 25% of the U.S. total energy) and more than half the electricity used for lighting is wasted, the program offers a substantial opportunity to prevent pollution, and to do so at a profit. Lighting upgrades reduce electric bills and maintenance costs and increase lighting quality; typically, investments in energy efficient lighting yield 20% to 30% rates of return per year. It is also a voluntary, non-regulatory program sponsored by the EPA to encourage companies to convert to more energy-efficient lighting, and thus reduce pollution produced from energy generation (e.g., carbon monoxide, sulfur dioxide, nitrogen oxide emissions; scrubber waste; boiler ash, etc.). By switching to new lighting technologies, companies can also lower electric bills and improve lighting quality.

EPA promotes energy efficient lighting by asking major institutions to sign a Memorandum of Understanding with the Agency to install energy-efficient lighting in 90% of their space nationwide over a 5-year period, but only where it is profitable and where lighting quality is maintained or improved. EPA, in turn, offers program participants a portfolio of technical support services to assist them in upgrading their buildings. A computerized decision supports system provides Green Lights corporations and governments a rapid way to survey the lighting systems in their facilities, assess their retrofit options, and select the best energy-efficient lighting upgrades. EPA has established a national lighting product information program in conjunction with utilities and other organizations. This program provides brand name information so that purchasers will be able to choose products with confidence. In addition, it will allow innovative products to be qualified rapidly, removing a significant barrier for new technologies.

As part of the support program, EPA helps Partners identify financing resources for energy-efficient lighting. Green Lights Partners receive a computerized directory of financing and incentive programs offered by electric utilities, lighting management companies, banks, and financing companies. The database is updated and distributed on a regular basis. EPA has also developed Green Lights Ally programs for lighting manufactures, equipment vendors, service providers, electric utilities, and lighting management companies to promote the environmental, economic, and quality benefits of energy efficient lighting. These groups are encouraged to give rebates to those customers who use energy-efficient lighting. Allies commit to undertake the same level of retrofits as Green Lights Partners, but also will assist in developing the technical support programs (USEPA, 1993). Inspired by the success of Green Lights, RPA introduced Energy Star Building, a program that takes pollution prevention to new heights. Enjoying the same rapid growth as Green Lights, Energy Star Buildings allow participants to maximize profitability, increase productivity, and improve occupancy comfort through increased energy efficiency. To date, Green Lights and Energy Star Buildings participants have distinguished themselves by prevention more than 4.5 billion pounds of greenhouse gas emission per year and saving more than $250 million per year (USEPA, 1997).

(D) The National Industrial Competitiveness Through Efficiency (NICE) Program

This is a joint program of the Department of Energy (DOE) and EPA's Office of Pollution Prevention and Toxics (OPPT). The NICE grant program strives to improve energy efficiency, advance industrial competitiveness, and reduce environmental emissions of industry. Large-scale research and demonstration projects are targeted at industries with the highest energy consumption and greatest levels of toxics and chemicals released.

Projects are established to use the one-time grant funds as seed money to overcome start-up risks. Industry needs to finance continuation of projects past the initial grant funding period. As part of the grant funded phase, awardees must design, test, demonstrate, and assess the feasibility of new processes and/or equipment which can significantly reduce generation of high-risk pollution. DOE Regional Support Offices and USEPA Regional Offices work through state energy and environment offices to actively seek out interested state developmental, energy, and industry organizations(USEPA, 1993).

(E) The Agriculture in Concert with the Environment (ACE) Program

This is a joint program of the USEPA and the Department of Agriculture (USDA) to undertake a grant program. The purpose is to promote the adoption of sustainable agriculture practices and reduce the use of highly toxic herbicides and other pesticides. Establishing a harmonious relationship between agriculture and the environment offers the opportunity for multiple gains on all sides – for farm owners, farm workers, consumers, and communities as a whole.

ACE grants have been distributed from a joint pool by USEPA and USDA Cooperative State Research Service. Host institutions in four regions (northeast, south, north central, and west) manage the evaluation, project selection, and distribution of funds for their regions. Evaluation panels in each of the four regions include representatives from government, academic and other research institutions, the farming industry, the environmental community, and other private or public organizations (USEPA, 1993)

(F) The Pollution Prevention Incentives for States (PPIS) Program

PPIS Program is intended to build and support state pollution prevention capabilities and to test, at the state level, innovative pollution prevention approaches and methodologies. The program types and activities include:

- Institutionalizing multi-media pollution prevention as an environmental management priority, establishing prevention goals, developing strategies to meet those goals, and integrating the pollution prevention ethic into state or regional institutions.
- Other multi-media pollution prevention activities, such as providing direct technical assistance to businesses; collecting and analyzing data; conducting outreach activities; developing measures to determine progress in pollution prevention; and identifying regulatory and non-regulatory barriers and incentives to pollution prevention.

- Initiating demonstration projects that test and support innovative pollution prevention approaches and methodologies.
- Eligible applicants are states and federally recognized Indian tribes. Awards are made through USEPA regional offices.
- Organizations selected for an award must support half of the total cost of the project in order to receive the 50% match required by the Pollution Prevention Act (USEPA, 1993).

(G) The Design for the Environment (DfE) Program

Established in October 1992, it is a voluntary cooperative program which promotes the incorporation of environmental considerations, and especially risk reduction, at the earliest stages of product design. The program has initiated a number of wide-ranging projects which operate through two levels of involvement.

(1) Industry cooperative projects work with specific industry segments to apply substitute assessment methodology, share regulatory and comparative risk information, and invoke behavioral change.

(2) Infrastructure projects are aimed at changing aspects of the general business environment which affect all industries in order to remove barriers to behavior change and provide models which encourage businesses to adopt green design strategies.

Through its DfE program, EPA created voluntary partnerships with industry, professional organizations, state and local governments, other federal agencies, and the public. EPA's efforts are directed at giving businesses the information needed to design for the environment and at helping businesses use this information to make environmentally informed choices. Within each business, the DfE program works to make sure that the information reaches the people who make the choice – from buyers to industrial design engineers (USEPA, 1994).

The program has been working closely with trade associations and individuals in three specific industry segments. These cooperative projects develop substitutes with assessments, which compare risk and environmental trade-offs associated with alternative chemicals, processes, and technologies; and which provide models for other businesses to follow when including environmental objectives in their designs. The DfE program has awarded grants to universities which fund research into alternate synthesis of important industrial chemical pathways. Results of the research provide the chemical industry with tools for production which reduce risk and prevent pollution. The grants are providing a model for further National Science Foundation grants (USEPA, 1993).

(H) The Waste Wise Program

Waste Wise is a voluntary partnership between EPA and U.S. companies aimed at reducing inefficient materials use and thus reducing municipal solid waste, and conserving energy and natural resources. Through this program, firms establish cost-effective goals of their choice to reduce their municipal solid waste through prevention, recycling, and by buying or manufacturing recycled products. This voluntary partnership program clearly demonstrates the cost benefits of incorporating prevention programs into American businesses. A related program, the Waste Minimization National Plan, targets prevention of hazardous waste. The plan's goal is to reduce persistent, bioaccumulative, and toxic chemicals in hazardous waste by 50% by the year 2005 (USEPA, 1997).

(I) The Environmental Accounting Project

A collaborative effort with business, academia and others, the project promotes sound management accounting and capital budgeting practices which better address environmental costs. It encourages and motivates business to understand the full spectrum of environmental costs, and integrate these costs into decision making. As of 1993, the project has over 650 members who are actively participating or interested in environmental accounting. The project has produced numerous tools, such as P2/Finance software, that help companies incorporate environmental costs in their capital budgeting decisions (USEPA, 1997).

(J)The Pesticide Environmental Stewardship Program (PRSP)

The PESP program is a voluntary public-private partnership of EPA, the Department of Agriculture, the Food and Drug Administration, and groups that use or influence the use of pesticides. The program's goals are to develop specific use/risk reduction strategies that include reliance on biological pesticides, integrated pest management, and other safer approaches to pest control. Voluntary partnerships with PESP are expected to have 75% of U.S. agricultural acreage adopt integrated pest management practices by the year 2000.

(K) The Water Alliances for Voluntary Efficiency (WAVE)

The WAVE is another voluntary program similar to Green Lights, that is dedicated to achieving water use efficiency. Hotel and motel chains are targeted for participating in the program, but other groups such as hospitals and schools will be involved as well. The primary goals of WAVE are to reduce water and energy consumption through the installation of water-efficient equipment, linking water-use efficiency to reduce costs, and educating business staff and the public on the benefits of reduced water use.

(L)The Project XL

In 1995, President Clinton launched Project XL to encourage development of alternative strategies to achieve greater environmental benefits over current regulatory programs. As part of Project XL, EPA is creating partnerships with states to provide a limited number of companies with the opportunity to demonstrate their environmental excellence and leadership. These companies will be given the opportunity to modify or replace current regulatory system requirements at specific facilities with company-developed alternative, flexible strategies. Each alternate strategy must meet a number of conditions, such as environmental performance, transparent to public, worker safety, community support, and enforceable.

(M) The New Consumer Labeling Initiative (CLI) Project

The CLI pilot project launched in March 1996 to provide better environmental information on products to consumers, including improved product labels. The project invites ideas from consumers, industry, and health and safety professionals to improve the environmental, health, and safety information appearing on household product labels, specifically targeting home and garden pesticides and household hard surface cleaners. The primary goal is to ensure that consumers have and understand the information they need to make responsible product choice based on their own needs and values.

(N) The Environmental Technology Initiative (ETI) Program

The ETI program is to encourage and support private sector development of pollution prevention technological innovations. The goal is to identify and reduce barriers to the development, introduction, and use of safer chemicals and technologies. The project offers industry in opportunity to explore new, non-traditional ways to manage risk, as opposed to the traditional regulatory approach (USEPA, 1996)

(O) The Green Chemistry Program

The Green Chemistry Program aims at spurring development of innovative technology to reduce or eliminate the use or generation of toxic substances in the design, manufacture, and use of chemicals. The program supports research in environmentally benign chemistry and promotes partnerships with industry in developing green chemistry technologies.

In March 1995, the Green Chemistry Challenge program was announced. The program promotes innovative uses of green chemistry for pollution prevention by grants and awards in three areas: (1) use of alternative synthetic pathways, (2) use of alternative reaction conditions, and (3) the design of chemicals that are less toxic than current alternatives or that are inherently safer with regard to accident potential.

(P) The Environmental Leadership Program (ELP)

The ELP program supports facilities that have volunteered to demonstrate their innovative approaches to environmental management and compliance, including pollution prevention. The purpose of ELP is two-fold: (1) to recognize facilities that develop and implement innovative environmental management systems and "beyond compliance," and (2) to work with these facilities and understand their systems and programs and to share that information with the regulated community to improve environmental management and increase compliance. In exchange for volunteering to demonstrate their innovative approaches, EPA offers facilities several incentives, including public recognition by RPA as an environmental leader, a limited grace period to correct any violations discovered during the facilities participated in the pilot program.

9.4
EPA's Pollution Prevention Research Program

The Pollution Prevention Research Branch at EPA's Risk Reduction Engineering Laboratory is responsible to manage pollution prevention (P2) research programs. The Research Branch supports projects and provides technical assistance to encourage the development and adoption of technologies, products, and pollution prevention techniques to reduce environmental pollution. The P2 research activities may be described by dividing them into two time-periods: prior to 1994 and after 1994.

9.4.1
Research programs prior to 1994

EPA's main emphases of the P2 research were:
- assisting small business sectors (such as printing, metal finishing, dry cleaning) in achieving pollution reduction;
- developing tools (such as life cycle assessment) to analyze and measure pollution prevention potentials;
- partnering with industry (such as pulp and paper) in evaluating cleaner technology demonstration;
- evaluating innovative waste reduction technologies developed in universities, research institutes, and industry;
- working to partnership with other government departments in developing specific technologies or assessing opportunities for pollution prevention.

There was also a program to examine the socioeconomic aspects of pollution prevention. Most of the P2 research programs have been and are still being supported by EPA.

(A) Clean Technology Projects

Projects aim to develop, demonstrate and evaluate innovative processes for reducing pollution through source reduction. Included are projects to develop techniques for measuring the benefits of utilizing cleaner technologies.

(B) Pollution Prevention Assessments

The purpose is to identify pollution prevention opportunities in a variety of industries and to transfer that technical information to assist others in implementing pollution prevention.

(C) Clean Products Projects

Projects aims to support the design and development of products whose manufacture, use, recycle and disposal represent reduced impacts on the environment. Included are projects to develop and demonstrate life cycle methodologies for determining the overall environmental impact of products.

(D) Long Term Pollution Prevention Research

The research provides the technical foundation for breakthrough advancements that will lead to increased pollution prevention.

(E) Cooperative Researches

The purpose is to promote pollution prevention in other Federal Agencies through cooperative research and demonstration projects.

9.4.2
Research programs after 1994

EPA's current pollution prevention research programs can be described in four categories:

1. Development of analytical tools and methods needed to assess or measure pollution, and quantify improvements;
2. Development of generic technologies that have simultaneous appeals to many industry sectors and to agricultural practices;
3. Development of, in collaboration with industry, sector-specific technologies requiring systems approaches; and
4. Demonstration and verification of cleaner technologies on large scales.

There has been a strategic shift in research direction since 1994. The main strategic shift was the use of risk as a driver and motivator for doing prevention research. Thus, only the high priority problems need to be addressed first. An expanded grants program at the Office of Research and Development (ORD) supports high-risk issues of pollution prevention.

EPA emphasizes tools and methods to meet design for the environment or exceed regulatory requirements such as life cycle assessment, simulation, guidance and design tools, assessment tools, impact assessment tools, and control algorithms. A compilation of summaries of current Branch projects is available from the Risk Reduction Engineering Laboratory, Mail Stop 466, Cincinnati, Ohio, 45268, USA.

9.5
Pollution prevention education and training

Education is the key to any successful programs. Almost all corporate executive officers and industrial managers receive some business education. Unfortunately, prior to 1989, there was virtually no graduate or undergraduate business school program or coursework focused on environmental problems. Some business school faculty discussed environmental problems as a topic in "business, government, and society" or "business ethics" courses, but such treatment was inevitably cursory. Change began to occur in the late 1980 s when the National Wildlife Federation and Corporate Conservation Council (NWF/CCC) focused on business education as one of its outreach program projects.

The first environmental business management course was offered at Boston University in the fall 1989 semester; the Loyola and Minnesota courses followed in the spring 1990 term. In 1990, Tufts University President Jean Mayer convened 22 university presidents from 13 countries in Tallorires, France, to discussed the role of universities in working toward an environmentally sustainable future. The conference resulted in a declaration of actions to make environmental education and research a principal goal of universities around the world (Cortese, 1992). By the end of 1992, the three schools had offered the environmental management courses to several hundred business school students and shared course information, teaching materials, and environmental management case studies with hundreds of faculty from 15 other countries. The NWF/CCC curriculum project has proved to be the right idea at the right time in bringing environmental management into business education (Post, 1994).

The Tallorires Declaration (excerpt) of University Presidents' Environmental Action Agreement states that: university heads must provide the leadership and support to mobilize internal and external resources so that their institutions respond to this urgent challenge. The 22 university presidents agreed to take the following actions:

(1) Use every opportunity to raise public, government, industry, foundation, and university awareness by publicly addressing the urgent need to move toward an environmental sustainable future.

(2) Encourage all universities to engage in education, research, policy formation, and information exchange on population, environment, and development to move toward a sustainable future.

(3) Establish programs to produce expertise in environmental management, sustainable economic development, population, and related fields to ensure that all university graduates are environmentally literate and responsible citizens.

(4) Create programs to develop the capability of university faculty to teach environmental literacy to all undergraduate, graduate, and professional school students.

(5) Set an example of environmental responsibility by establishing programs of resource conservation, recycling, and waste reduction at the universities.

(6) Encourage the involvement of government (at all levels), foundations, and industry in supporting university research, education, policy formation, and information exchange in environmentally sustainable development. Expand work with nongovernmental organizations to assist in finding solutions to environmental problems.

(7) Convene school deans and environmental practitioners to develop research, policy, information exchange programs, and curricula for an environmentally sustainable future.

(8) Establish partnerships with primary and secondary schools to help develop the capability of their faculty to teach about population, environment, and sustainable development issues.

(9) Work with the U.N. Conference on Environment and Development, the U.N. Environment Programme, and other national and international organizations to promote a worldwide university effort toward a sustainable future.

(10) Establish a steering committee and a secretariat to continue this momentum and inform and support each other in carrying out this declaration (Post, 1994).

The EPA's educational programs emphasize four specific themes: (1) wise use of natural resources, (2) prevention of environmental problems, (3) the importance of environmentally sensitive personal behavior, and (4) the need for additional action at the community level to address environmental problems. The EPA has been and is working in partnership with State and local governments, industry, educational institutions, textbook publishers, and other entities to a project which would ultimately produce pollution prevention education materials for students and teachers. The project established an environmental ethic and work toward improved environmental quality. The materials produced concentrating on kindergarten through grade 12 or a specific segment of this broad group, and emphasizing preventing pollution at the source. The goal is to provide youth with an appreciation and an understanding of the potential benefits of pollution prevention, including conservation and recycling (USEPA, 1993; 1997).

The growth of pollution prevention programs in educational institutions has been helped by two laws passed in 1990: the National Environmental Education Act and the Pollution Prevention Act. Both these acts helped build a framework for integrating pollution prevention into educational programs across the United States. Spurred by the legislation at the federal, state, and local levels, the field of environmental education is rapidly expanding both in the United States and world-wide. Organizations such as the EPA-funded Environmental Education and Training Partnership Project are accelerating the pace of environmental education through improved information transfer, basic training materials, and training for educators. Many colleges and universities go beyond engineering research and development and provide forums for regulators, businesses, and local communities to come together to resolve environmental issues through pollution prevention. Colleges and universities have internalized pollution prevention and are making broad sweeping institutional changes to reduce environmental impacts and consumption of natural resources on their campuses.

Faculty from the science and engineering departments of colleges and universities have prepared problem sets and new courses devoted exclusively to preventing pollution and have woven prevention concepts into existing courses. Engineering faculty teach students how to incorporate pollution prevention in process design (applying the concept of Design for Environment), and also how to spot opportunities for waste reduction in unit operations. Life cycle analysis is another active area of research in which universities are engaged, often as a cooperative research effort with EPA.

EPA supports schools, universities, and nonprofit organizations in developing innovative ways to incorporate pollution prevention ideas into educational initiatives. As universities have recognized the need for an interdisciplinary approach to environmental studies, they realize that faculty must be prepared to teach these new courses. The Tufts Environmental Literacy Institute (TELI) offers interdisciplinary professional development on environmental issues for university and secondary school faculty. TELI training equips faculty to teach environmental issues both from an interdisciplinary perspective and with specific reference to their own fields. Carnegie Mellon University developed a university-wide pollution prevention research effort, the Green Design Initiative (GDI). The GDI consists of interdisciplinary teams whose goal is to prepare new environmental management and pollution prevention tools for product and process design, policy, and environmental management.

The National Pollution Prevention Center (NPPC) at the University of Michigan established since 1991 offers many tools and strategies to incorporate pollution prevention concepts into the curricula of universities and colleges for faculty, students, and professionals. Course compendia and other educational materials being developed by the NPPC are based on a systems approach to pollution prevention. In addition, the NPPC offers a unique national internship program that provides practical experience to undergraduate and graduate students in waste prevention process assessments. NPPC publishes Pollution Prevention Educational Research Compendia in a variety of disciplines, including business law, chemical engineering, chemistry, accounting, industrial engineering/operation management, agriculture, architecture, and strategic management. Each compendium offers a discipline-specific resource list, as annotated bibliography, selected readings, syllabi, and assignments. The information is available on the World Wide Web at http://www.Umich.edu/nppcpubResLists/.

9.6
Summary of Pollution Prevention Act of 1990

Section 1 – Short title: Pollution Prevention Act
Section 2 – Findings and policy: This section establishes a pollution prevention hierarchy as a national policy, declaring that:

1. pollution should be prevented or reduced at the source wherever feasible;
2. pollution that cannot be prevented should be recycled in an environmentally safe manner whenever feasible;
3. pollution that cannot be prevented or recycled should be treated in an environmentally safe manner whenever feasible; and
4. disposal or other release into the environment should be employed only as a last resort and should be conducted in an environmentally safe manner.

Section 3 – Definitions: Although the statute does not provide any definition of pollution prevention, it is clear that Pollution Prevention is intended to include source reduction and to exclude most forms of recycling. Source reduction is defined to mean any practice which reduces the amount of any hazardous substance, pollutant or contaminant entering any waste stream or otherwise released into the environment (including fugitive emissions) prior to recycling, treatment or disposal; and reduces the hazards to public health and the environment associated with the release of such substances, pollutants or contaminants. The term includes equipment of technology modifications, process or procedure modifications, reformulation or redesign of products, substitution of raw materials and improvements in housekeeping, maintenance, training, or inventory control.

Section 4 – EPA activities: EPA is to establish an office, independent of the single medium programs, to carry out its functions under this Act, and develop and implement a strategy to promote source reduction. As part of the strategy, the Administrator will:

1. establish standard methods of measurement for source reduction;
2. review regulations before and after proposal to determine their effect on source reduction;
3. coordinate source reduction activities in each Agency office and promote source reduction practice in other Federal agencies;

4. develop improved methods for providing public access to data collected under Federal environmental statutes;
5. facilitate the adoption of source reduction techniques by businesses;
6. identify measurable goals that reflect the policy of the Act, tasks to meet these goals, dates, required resources, organizational responsibilities, and means for measuring progress;
7. have a senior level liaison group to provide guidance to the Administrator, to provide outreach to the industrial community, and to provide liaison with the educational community;
8. establish an advisory panel of technical experts to advise the Administrator on ways to improve the collection and dissemination of data;
9. prepare recommendations to Congress to eliminate barriers to source reduction;
10. provide opportunities to use Federal procurement to encourage source reduction;
11. establish model source reduction auditing procedures;
12. conduct an annual award program.

Section 5 – Grants: Grants are to States for programs to promote source reduction by businesses. Federal funds are to be no more than 50% of the funds made available to a State in each year of that State's participation in the program.

Section 6 – Clearinghouse: The Administrator shall establish a Source Reduction clearinghouse containing information on management, technical and operational approaches to source reduction. Information on source reduction gathered pursuant to this Act shall be made available to the public.

Section 7 – Source reduction and recycling data collection: Facilities required to report under the Toxic Release Inventory (TRI) provisions of Section 313 of SARA must provide information on pollution prevention and recycling activities with each annual filing. These requirements take effect with the annual report for the first calendar year beginning after the date of enactment. The report shall set forth the following information on a facility-by-facility basis for each toxic chemical:

- quantities entering waste streams, and the percentage change from the previous year;
- quantities recycled, the percentage change from the previous year, and the recycling process used;
- source reduction practices;
- a ratio of production in the reporting year to production in the previous year;
- techniques used to identify source reduction opportunities;
- amounts of chemicals released through one-time, catastrophic events; and
- the amount of the chemical that is treated, and the percentage change from the previous year.

Section 8 – EPA report: A report to Congress is required with 18 months of enactment, and biennially afterwards. The report will contain a detailed description of the actions needed to implement the strategy to promote source reduction, and an assessment of the Clearinghouse and grant program. After the first report, each subsequent report will also contain additional analyses of data, barriers to source reduction, recommendations concerning incentives, priorities for research, and an evaluation of source reduction opportunities.

Section 9 – Savings: The Act does not modify or interfere with implementation of Title III of SARA, or responsibilities or liabilities under other State or Federal law.

Section 10 – Authorization: 8 million dollars is authorized for State grants, and 8 million for other functions in fiscal years 1991, 1992, and 1993.

Section 11 – Implementation: Provides rulemaking authority.

References

Bower BT (1990) "Economic, engineering and policy options for waste reduction," in: National Research Council, Committee on Opportunities in Applied Environmental Research and Development of the Waste Reduction Workshop Report. National Research Council, p. 3

Irwin, FH (1991) Environmental Law. World Wildlife Fund, Publications Vol. 22:1, Washington, D.C.

Koenigsberger MD (1986) "For Pollution Prevention and Industrial Efficiency," presented at Governor's Conference on Pollution Prevention Pays, Nashville, Tenn., March

NRC (1966) Waste Management and Control. National Academy of Sciences, National Research Council. National Academy Press, p. 3

NYSDEC (1989) Waste Reduction Guidance Manual. ICF Technology Incorporation, Fairfax, Va., March

Post JE (1994) "Regulation, Markets, and Management Education," in: Environmental Strategies Handbook, ed. by RV Kolluru. McGraw-Hill, Inc., pp. 26–30

PPA (1990) Pollution Prevention Act, Public Law 101–508, Section 6604(4)

Purcell AH (1993) "Pollution Prevention in the 21st Century: Lessons From Geneva". Proceedings of the International Conference on Pollution Prevention, June 10–13, 1993, Washington, D.C., p. 529

Shen TT and GH Sewell (1986) "Control of Toxic Pollutants Cycling," Proceeding of the International Symposium on Environmental Pollution and Toxicology, Hong Kong, September 9–11, 1986. Hong Kong Baptist College, Hong Kong

Shen TT, CE Schmidt and TR Card (1993) Chapter 1 of Assessment and Control of VOC Emissions From Waste Treatment and Disposal Facilities. Van Norstrand Reinhold, New York

USEPA (1989) "Draft Pollution Prevention Strategy", Washington, D.C.

USEPA (1991) The National Report on Toxics in the Community – National and Local Perspectives. Chapter 5: Pollution Prevention. EPA 560/4-91-014, September

USEPA (1991) Pollution Control 1991: Progress on Reducing Industrial Pollutants. Office of Pollution Prevention, EPA 21P-3003, October, p. 130

USEPA (1992) Facility Pollution Prevention Guide. Office of R&D, Washington, D.C., EPA/600/R-92/088, May

USEPA (1993) 1993 Reference Guide to Pollution Prevention Resources, prepared by Labat-Anderson Incorporated for the Office of Pollution Prevention and Toxics, Washington, D.C., EPA/742/B-93-001, February

USEPA (1993) Shareholder's Action Agenda: A Report of the Workshop on Accounting and
    Capital Budgeting for Environmental Costs. EPA 742-R-94-003, December 5–7
USEPA (1994) EPA's Design for the Environment Program. Office of Pollution Prevention
    and Toxics, EPA 744-F-94-003, February
USEPA (1996) Environmental Technology Initiative for Chemicals, EPA 743-K-96-001, May
USEPA (1996) Prevention Pollution Through Regulations: The Source Reduction Review
    Project – An Assessment. EPA-742-R-96-001
USEPA (1997) Pollution Prevention 1997 – A National Progress Report, EPA-742-R-97-00

10
State, city and local pollution prevention programs

In the United States, state-based environmental programs have made a unique contribution to pollution prevention (P2) through their direct contact with industry and awareness of local needs. Most state programs are focusing on teaching businesses about pollution prevention through outreach and technical assistance. In doing so, the states sought to instill the pollution prevention ethic throughout the business community. Many states have increased efforts to integrate pollution prevention into the state environmental protection programs and implementations. Initiatives have included training state and county regulators in pollution prevention, reviewing state regulations to identify barriers to pollution prevention, increasing referrals from the regulatory program to the technical assistance program, and incorporating pollution prevention considerations into permits, notices of violation, and settlement agreements.

All state pollution prevention programs are established independently according to each of its State laws. However, each State must implement Federal laws and regulations. States may establish their rules and regulations more stricter but not less stricter than that of Federal rules and regulations. In 1996, EPA targeted its Pollution Prevention Incentives for States (PPIS), Small Business Development Centers (SBDCs), local governments, nonprofit organizations, and state regulators. The EPA 1997 grants cycle further supported this effort to develop networks and create partnerships. To receive funding under PPIS, states are required to assess local needs and design a program to meet those needs. The grant program also encourages the states to combine forces with other state organizations actively promoting pollution prevention.

Many states have enacted legislation to establish pollution prevention programs or to institutionalize state waste reduction policies. State programs may undertake a variety of activities to achieve their pollution prevention goals. In general, four approaches are used by the states to implement their programs: technical assistance/outreach, mandatory facility planning, regulatory integration or coordination, and voluntary partnerships. States often use a combination of all three of these approaches. Most environmental protection is implemented through state media programs. In order for pollution prevention to take hold, state media programs need to see how prevention can help achieve their goals. Prevention is important for regulatory programs because single media programs may have the effect of shifting waste across environmental media. The single media regulatory stricture is not conducive to understanding these cross-media issues, or acting on them. Due to the difficulty in changing organizational biases and the time required to develop a pollution prevention mentality among state regulatory and compliance staff, states will continue to struggle with this issue over the next several years.

Technical assistance activities include opportunity assessments, information clearing houses, facility planning, hotlines, computer searches, and research projects. Outreach and education activities include workshops, seminars, training, publications, and grants and loans. Several states employ retired engineers and graduate students to conduct assessment. The retired engineers enhance the credibility of state programs with industry. Involving graduate student in the process helps the students to learn the pollution prevention approaches and encourages them to employ it in their careers. Information clearinghouses compile pollution prevention documents for regulatory staff, targeted audiences, and the general public. They also provide technical information on request. Facility programs are designed to encourage facilities that generate pollution to evaluate their processes with an eye toward eliminating waste and pollution. Pollution prevention technical assistance programs face a major challenge in piecing together a stable level of funding from a variety of sources, and maintaining political support for these programs (USEPA, 1997).

EPA's Office of Pollution Prevention and Toxics (OPPT) convenes the Media Association P2 Forum, which consists of program directors that sit on state waste, water, and air associations and members of the National Pollution Prevention Roundtable. Pollution prevention can be a common thread for single-media state programs, and the quarterly forum meetings provide a rare opportunity for these organizations to discuss pollution prevention. One of the importance discussions was how to integrate the pollution prevention ethic into all areas of their environmental regulations. In fact, some states have already begun to integrate pollution prevention into their regulatory activities; in other states, regulatory integration is only in the planning stages.

This Chapter briefly addresses state pollution prevention programs and selecting the New York State Multimedia Pollution Prevention (M2P2) Program and City of Cincinnati P2 Program as case studies. It also discusses local and community pollution prevention programs and their implementation approaches.

10.1
Overview of state and local P2 programs

State pollution prevention programs vary widely in scope to meet different needs of the states. States are required to assess local needs and design a program to meet those needs for federal grants. The EPA grant program also encourages the states to combine forces with other state organizations actively promoting pollution prevention. State program activities include:

- Technical assistance, outreach, and education – opportunity assessment, information clearinghouses, facility planning assistance, hotlines, computer searches, research and collaborative projects, workshops, seminars, training, publications, grants and loans.
- Regulatory integration – enforcement settlements, permutation, compliance inspections, waste management.

State programs have made a unique contribution to pollution prevention through their direct contact with industry and awareness of local needs. Most state programs were focused on teaching businesses about pollution prevention through outreach and technical assistance. Many states have increased efforts to integrate pollution prevention into the state environmental programs. Initiatives have included training state and county regulators in pollution prevention, reviewing state regulations to identify barriers to pollution prevention, increasing referrals from the regulatory program to the technical assistance program, and incorporating pollution prevention considerations into permits, notices of violation, and settlement agreements. The development of methods to measure pollution prevention progress and to evaluate state program effectiveness has emerged as an important new trend. Both the states and EPA are struggling with selection of the best approach (USEPA, 1997).

State and local outreach program forms partnerships with state and local governments to build capacity to understand impact of climate change and reduce their greenhouse gas emissions. State and local authorities are critical players in the effort to reduce these emissions because they have jurisdiction over activities that contribute greenhouse gases, including land use, transportation, building codes, and waste management. Moreover, states and localities account for a significant percentage of national emissions of greenhouse gases. The mission of the program is to empower decision-makers by providing them with appropriate products and services, including pollution prevention management concept and practice. Cities and counties become partners in the State and Local Outreach Program through initiatives coordinated by the International Council for Local Environmental Initiatives (ICLEI). One of these initiatives, the Green Fleets Project, provides incentives for energy savings in the transportation sector, while the most recent initiative, the Cities for Climate Protection Campaign," addresses energy consumption in other sectors, including buildings and transportation (USEPA, 1998).

10.2
NYS Multimedia Pollution Prevention (M2P2) Program

10.2.1
Background

The New York State Department of Environmental Conservation (DEC) has actively participated in various multi-media pollution prevention initiatives since

March 1992. This approach reflects that the DEC is beginning to change its attitude and approach in dealing with environmental management as it moves towards the 21st century.

This effort was initiated when the Commissioner of DEC set up a small working group representing senior level management from various programs within the Department. The programs in this group called the Multi-Media Pollution Prevention Planning Group include Air, Water, Solid Waste, Hazardous Waste, Hazardous Site Remediation, Spills Management, Fish and Wildlife, Regulatory Affairs, Legal, Information Management, Natural Resources and Public Affairs. This group includes both central and regional office representation. The group examines what the Department does, how it implements programs, levels of efficiency, where changes needed to be made and what kind of adjustments should occur. The group's responsibilities range from minor adjustments within the Department to total DEC restructuring. The planning group has recommended changes in the following areas:

– structure
– culture
– organization

In addition, the group has recommended the creation of a Pollution Prevention Unit (PPU) to oversee the DEC's implementation of these activities. In July 1992, a DEC policy directive was issued to create a PPU which accepts the planning Group's report and recommendations.

10.2.2
Themes of the M2P2 program

The overall theme of the Multimedia Pollution Prevention initiative (M2P2) is to integrate the various DEC efforts related to policies, programs and implementation. Since the DEC's creation in 1970, it has established various single media programs within the Agency as well as in its regional offices. Now for the first time, it is beginning to adopt the multi-media pollution prevention approach.

The M2P2 program involves an integrated facility management approach that consists of integrating the various types of permitting, compliance, enforcement, inspection and pollution prevention initiatives into a cohesive set of actions. The DEC's initiatives will help achieve a 50% reduction of toxic pollutants to the environment by the year 2000, focusing their efforts on a manageable number of facilities. Within the DEC program, approximately 400 facilities account for 95% of all the toxic pollutants generated, discharged and emitted to the environment as of 1990. DEC's pollution prevention initiative includes the following approaches in order of preference:

– source reduction or elimination stressed first and foremost;
– recovery, recycling and reuse activities as the next preferred;

– detoxification or destruction technologies for materials which cannot be reduced, recovered, reused or recycled;
– disposal as the least preferred method of waste management.

A wealthy economy and healthy environment needs environmental and economic partnerships. DEC recognizes that it must work with industries to ensure
that environmental protection strategies are compatible with the business climate within New York State. The regulated community as well as the DEC must
also trust each other for the partnership to succeed.

In addition, the Pollution prevention program in New York State must continue to expand toward the number of pollution prevention strategies to include:

– the implementation of the state's Hazardous Waste Reduction Act of 1990;
– the state's Anti-degradation Policy for Water;
– the development of regulations for chemical and petroleum bulk storage facilities;
– the implementation of the state's solid waste policy including not only municipal refuse but also ash and industrial sludges and an approach that is consistent with the level of toxicity;
– a reduction of persistent toxins in the Great Lakes;
– implementation of the Clean Air Act requirements such as low sulfur fuel, carbon monoxide reductions, and VOC reductions.

10.2.3
Key program elements

Below describes the major elements of DEC's pollution prevention initiatives.

Permitting process: The existing DEC environmental programs are based on
traditional single media statutes. The DEC plans to streamline its regulations
and to evaluate whether an in-depth line-by-line review of a permit application
is necessary. The DEC also reviews its environmental permits in programs that
involve construction and operation.

Some programs currently have two permitting actions for the same facility.
When these requirements were first put into regulations a long time ago, there
may have been a legitimate need to have such actions. However, as progression
occurs into the 1990's, the DEC has considered whether to continue such an approach or to issue one permit for a particular action with a permit condition that
must be satisfied before the facility can fully operate. These dual permitting activities occur in the operation of solid waste facilities as well as air pollution control and other facilities. In the fall of 1993, the DEC's solid waste regulations were
revised to consolidate the two permitting processes into one for a specific facility
application.

At present, DEC is re-assessing its activities such as rollover permits and environmental benefit strategies because resources are finite and should be devoted to high priority permits. One DEC activity in its water quality program has

been to automatically roll over minor permit renewals until the DEC reviews them. This approach allows the DEC to focus its efforts on more major permitting activities.

The Pollution Prevention Unit is reviewing together with other programs, the various regulations that have been promulgated by the DEC for inclusion of various pollution prevention components. Various single media programs are being factored into pollution prevention. For example, DEC is setting water quality standards based on health risks and toxicity, as well as source reduction.

DEC is reviewing its requirements to streamline the hazardous waste regulations. When those regulations were first enacted many years ago, there were a number of reasons why DEC was more stringent than the federal EPA requirements, one reason being that New York State lists polychlorinated biphenyls (PCBs) as a hazardous waste in its regulations. However, now the DEC is reviewing the differences between its hazardous waste regulations and EPA requirements so as to make its regulation more consistent and compatible with the federal program. Proposed regulatory changes in hazardous waste management are substantially reducing a number of differences between the state's program requirements and the federal program requirements. DEC will positively and significantly benefit the regulated community because the community will be able to better understand the state's regulations if they are similar to the federal programs.

DEC is currently pursuing integrated facility permitting. An integrated approach to permitting, compliance, enforcement, inspection and pollution prevention activities at a facility will more efficient deliver DEC's services. An integrated facility permit considers the following factors:

- the corporate attitude and spirit is proactive and cooperative;
- integrated facility management is applicable and desirable;
- the risk of violations is low;
- the facility needs operational flexibility;
- the corporation is small with environmental staff that are knowledgeable in the overall policies, operations, administration and maintenance of the facility;
- major expansions and frequent operational modifications are anticipated at the facility;
- the facility is not the target of a major enforcement effort ;
- the facility is not burdened by any activities or issues that would impede or hamper the program ;
- the facility is on the DEC's "400/95" list (see Section 16.1.6).

DEC believes that the permit content of an integrated facility should consist of a single document with common expiration dates. It authorizes discharges to all media and includes specific permit limits to all media. In addition, control technology, monitoring requirements and reporting requirements should cover all media and other conditions such as on-site monitors, safeguards, pollution prevention programs. The benefits of such an approach will allow the facility owner and the DEC staff to easily look at the facility in a holistic fashion.

10.2.4
Environmental and economic partnership

Because good environmental practice makes good business sense, the DEC is working with the Department of Economic Development to create a prevention strategy that not unfair to New York State businesses. The DEC is thus implementing its multi-media pollution prevention (M2P2) initiative with two of its major components that will positively effect on both the environment and industrial economy.

(a) Integrated facility approach
Larger industries in New York State seem to be more responsible for the generation and release of toxic pollutants to the environment. Environmental regulations have been created through an integrated facility management (IMF) approach. This approach provides a regulatory program that is more efficient and more effective because of multi-media inspections and other coordinated regulatory activities at major facilities. It addresses programs such as implementing significant risk reduction and pollution prevention measures as well as permitting and enforcement. The IMF approach will save time and resources of both industry and government.

(b) Approach to small business
New York State's small business will benefit from the M2P2 initiative through an environmental self-audit manual for small businesses developed by the DEC and the Department of Economic Development (DED). The manual provides a quick and easy reference guide to help ensure environmental compliance. In addition, other activities are being made available both within the DED, DEC and the Environmental Facilities Corporation (a non-profit public benefit corporation), such as non-regulatory technical assistance, workshops and other forms of guidance. The Pollution Prevention Clearinghouse provides free access to information on technical, policy, program and legislative aspects of pollution prevention. It currently has over 7,000 entries.

The Governor's Award recognizes successful efforts by industry to increase efficiency of operation and to provide long-term environmental protection through reduction of hazardous waste and toxic emissions and discharges. This award is an important part of establishing a strong partnership between government and industry which increases DEC's efficiencies while optimizing industry's competitiveness and survival. The first annual Governor's Pollution Prevention Award was presented at the 7th Annual Pollution Prevention Conference in Albany in June 1994.

10.2.5
Regulatory reform initiatives

Sensitive to economic needs and impacts, DEC is assessing its regulatory programs with modifications in areas related to:

- Reducing transactional costs by making the environmental regulatory, permitting, and compliance processes and procedures more efficient and speedier without sacrificing the underlying environmental and resource protection objectives established in major federal and state statutes.
- Assessing more specific environmental compliance requirements for air and wastewater discharges, for solid and hazardous waste management and other regulatory programs. They can be achieved in a more cost effective manner.
- Targeting the environmental protection objectives to meet the highest priority needs so as to reduce excessive burdens and costs for the regulated community. The M2P2 initiative's focus on the "400/95" list that comprises 95% of the toxins, is an example of this effort.

(a) Inspections

Multi-media inspection is one of the key components of the DEC's M2P2 initiatives. Several years ago DEC piloted a multi-media inspection program with one of the DEC's local regional offices. This pilot effort had two primary purposes:

(1) It allowed a media-specific inspector to gain knowledge and to observe media-specific inspections. For example, if a RCRA inspector is conducting a routine inspection, he needs to observe any visible impacts to air or water, and natural resources as well as record and relay information to program-specific people. This process allowed DEC to prioritize routine reconnaissance inspections.

(2) It trained people to function as a team. As the DEC increasingly moved into the M2P2 arena, it undertook team inspections. A Regional Office inspection model and inspection process were established which tested a few local industries. The process worked, but not very well. Each program lumped everything under a generic umbrella of items such as solvent usage and storage. DEC has revised its inspection forms based on these observations and has circulated them to other regional offices for feedback. The forms have subsequently been revised and now work well. Ultimately, this approach render DEC more efficient.

(b) Compliance and enforcement

Compliance and enforcement are also key components of the DEC's M2P2 initiative. As a part of the environmental and economic partnership, DEC is also seeking new ways to encourage compliance and enforcement through hybrid agreements instead of traditional orders on consent. DEC believes that in certain and appropriate circumstances, enforceable memorandum of understanding and agreements could allow DEC far more flexibility. In this regard, the DEC reviews the activities beyond normal compliance, to those even in settlements: third party compliance audits, management audits, accident prevention, community awareness, best management practices, tank assessment and replacement, toxic reduction measures, risk assessment and reduction, permitting and other pollution prevention measures. These activities are important because as they become integrated into the various programs, DEC will be able to regulate a facility holistically. An action plan or management plan of a facility becomes positive and beneficial not only in achieving compliance, but also in securing measures that surpass compliance for both industry and regulatory agency.

10.2.6
Information management and confidentiality

Information management is an important DEC element as this initiative related to pollution prevention. The DEC has prepared a "400/95" list which includes approximately 400 industries that represent 95% of all the toxic air emissions, water discharges and hazardous waste generation (based on 1990 data). DEC focuses only these 400 facilities. The list represents a manageable number of facilities that can be handled in this integrated facility management approach. When the Pollution Prevention Unit prepared the original 400/95 list, it was discusses that the various databases within the Department of Environmental Conservation were difficult to integrate. The initial list included facilities that did not belong, such as certain one-time remedial activities and cleanups that have subsequently been removed from the list. As to the quality of the 400/95 list, DEC is assessing whether changes in the current reporting formats for air, water, solid waste, hazardous waste and others are necessary. As this process evolves, it will be imperative that industry works with DEC to improve the quality of information received about specific processes.

DEC recognizes that confidentiality must be preserved to the extent by law. The DEC has begun to communicate with several industries in New York State to improve reported information on industrial processes. Currently, much of the information that is submitted by industries within New York State is transmitted throughout the Agency to various divisions and regional offices. Confidentiality is protected under state law. Electronic filing of information will further tighten the controls and help DEC to deal with sensitive information. Industries must feel that they can rely on the assurances of DEC to maintain the confidentiality of the information that they send to DEC.

Within DEC, a Regulatory Data Management Workgroup composed of representatives from all the major program elements has been assembled. The function of the Workgroup is to assess information needed to develop a workable model and to integrate the data. DEC is currently developing this long-term goal while improving information management strategies.

10.2.7
Training and outreaching

Multi-media pollution prevention requires the following three types of training:

- team building training
- technical training
- supervisory training

The Department of Environmental Conservation (DEC) has begun to assess the process of quality management and train its staff so that DEC professionals in its central and regional offices can work more efficiently and effectively. Team building and team approaches are difficult for individuals accustomed to single media approaches. Some professionals within the DEC told their managers and superiors that they could work more quickly alone than as a team. Training will help avoid the misconception.

DEC has built team concepts into performance programs for its employees. As of June 1994, two-thirds of the staff within the DEC accepted this M2P2 approach. About 20% embraced this approach less enthusiastically; 5 or 10% of the staff showed resistance to the change. To operate as a team, the ultimate goal is to move all elements forward so that the whole process does not take a long time.

For the past years, DEC has conducted a number of pollution prevention workshops for small and large businesses. These workshops have included automotive repair shops, the printing industry, the electronics industry, the electroplating industry, and others. The DEC has worked closely with industries, local government and trade associations in conducting these workshops. The feedback was positive and encouraging. The Pollution Prevention Unit has been contacting additional industries, trade associations and local governments to ask them to continue to participate in various forums and workshops and to continue the dialogue and outreach. DEC recognizes that it must first encourage small businesses to comply before they actually prevent pollution.

The DEC has conducted annual conferences for the past ten years. These conferences have provided a forum and a focal point for discussion regarding successes, opportunities for improvement, and program directions. Case studies of people sharing their ideas and expertise are main topics of the conference. DEC also mails Pollution Prevention Bulletin to over 7,000 businesses, local governments, chambers of commerce, and universities. The bulletin conveys the message of DEC's activities so that industries know what the DEC is doing, how it is doing it and where it is going. Finally, issuance of fact sheets and success stories is important as DEC recognizes industry pollution prevention efforts.

10.2.8
USEPA and DEC partnership

The United States Environmental Protection Agency (EPA) has played an important role in the DEC's M2P2 initiatives by providing many millions of dollars in grant to the DEC to implement federally mandated programs. EPA sets a national strategy which states implement. However, to implement M2P2 program, DEC requires EPA to provide necessary funding for delegated and non-delegated programs for pollution prevention work plans and outputs along with quality assurance. For this new state and federal partnership, DEC expects that:

- EPA should agree with DEC's integrated multi-media pollution prevention approach to ensure cost effective initiatives as opposed to traditional bean-counting programs. EPA provides sufficient federal grant flexibility to allow implementation of multi-media pollution prevention program.
- The success of this partnership has to be gauged both administratively and technically on true and meaningful environmental measurement capabilities including appropriate indicators and methodologies.
- EPA must clearly invest additional funds in the research and development programs as well as enhance the development of proven pollution prevention technologies.
- EPA must provide additional resources on a continuing basis to support state multi-media pollution prevention programs.

New York State can learn from EPA's organizational structure and approach to implementing programs. EPA is set up with a headquarters in Washington, DC and 10 regional offices. EPA headquarters issues policy guidance and direction while the 10 regional offices implement the various program elements by working closely with the states. In New York State, DEC has its headquarters offices in Albany and 9 regional offices throughout the State. DEC programs are decentralized, but not all. Some programs like the Bulk Storage Program, the Spill Management Program and the Hazardous Waste Remediation programs are generally decentralized programs managed in the regions. However, other programs such as hazardous waste management are fragmented, where certain activities are undertaken by regional offices, and other activities are handled by Albany headquarters personnel. The current approach needs to foster a more integrated management style with a team concept.

The federal auditors oversee state expenditures of federal grants. They must be familiar with the process so that they make more efficient use of resources and streamline the administrative approach. DEC is working closely with EPA regions to ensure that expenditures of various grant moneys are appropriate.

10.2.9
DEC's Pollution Prevention Unit

The Pollution Prevention Unit (PPU) was established in July 1992 based on recommendations by the multi-media planning group and by an earlier directive to implement the DEC's M2P2 initiative. The PPU's mission is to plan, manage, monitor and coordinate information management, technical assistance and support activities related to DEC's M2P2 initiative. The PPU currently has only 20 staff members who are involved with influencing an Agency of approximately 4,000 employees.

The PPU is assisted by the commitment of one man each from the Information Management and Systems DEC program, Public Affairs Office and General Counsel. In addition, each of the major program areas in the Department (Water, Air Resources, Solid Waste, Hazardous Waste, Regulatory Affairs, Spills Management, Remediation and Natural Resources) has designated a liaison to work with the PPU on various M2P2 activities. Furthermore, an M2P2 coordinator has been assigned in each of the nine regional offices to integrate and coordinate program activities in the regions as well as work closely with the PPU to develop and implement the M2P2 initiatives. Finally, a Multi-Media Pollution Prevention Advisory Group has been established to advise the DEC on the development and implementation of its M2P2 initiative. The Group also provides a forum for the regulated community, environmental groups, other state agencies and the Legislature. It also participates in issues that relate to this initiative. The Advisory Group currently consists of 30 members representing DEC, the state departments of Health and Economic Development, large and small businesses, local governments, environmental groups, legislative staff and trade groups (NYS-DEC, 1993a).

(a) Goals
The goals of the Pollution Prevention Unit within the DEC are:

- To prioritize and coordinate the various program activities to ensure that the M2P2 initiatives are implemented.
- To develop and communicate policy in an effective and timely manner.
- To ensure that the New York M2P2 approach becomes a model for the federal government, states and associations.
- To direct the M2P2 educational and technical initiatives to the industrial sector, general public, state and federal policy makers and department staff.
- To ensure progress and development of the M2P2 initiative to achieve a 50% reduction of toxic chemicals by the year 2000.
- To manage and coordinate the M2P2 initiatives so that an integrated facility management approach is achieved.
- To ensure that a training curriculum for multi-media pollution prevention is institutionalized.

- To ensure the development and implementation of an integrated information management system that allows measurement and tracking.
- To ensure that all state agencies comply with the state agency environmental audit provisions and the Environmental Conservation Law.
- To implement a small business environmental audit and outreach program.

- The major tasks of the PPU include:
- providing technical support for and coordination of major M2P2 enforcement cases
- coordinating information management activities
- developing and implementing internal and external M2P2 technical assistance and outreach activities
- planning and coordinating DEC's overall M2P2 activities
- coordinating and reviewing interprogram toxicity and risk assessments
- developing and disseminating information on Environmental Self Audits for Small Businesses
- implementing the State Agency Environmental Audit Program (NYSDEC, 1993a).

(b) Culture change

Many of the DEC's single media program activities over the last twenty years has resulted in the creation of parochial views and fiefdoms. This occurred because these programs were set up based on single media state statutes and corresponding regulatory development. However, in December of 1992, certain senior level program managers were rotated into other programs within the agency. The initial reaction among the agency was one of cautious optimism. Management jobs require management skills and expertise in one specific program. Some felt that the change from media-specific to M2P2 foster sharing of knowledge and experiences across other program areas. Others felt that the specific knowledge gained within a specific media by some of these managers, would result in a loss of certain technical skills and service delivery. However, in the succeeding months, the evaluation of this process has shown otherwise. Managers had to constantly learn the best way to manage environmental programs and develop new knowledge and skills to meet the future needs. They must develop the trust and relationship with their professional staffs to make informed sound professional decisions.

(c) Empowerment

Empowerment is a key operative term of DEC's in 1990 s management. In the initial years effort of the DEC's integrated multi-media pollution prevention program, there were approximately 50 facilities that were targeted for this effort. The regional offices carried out the bulk of the work. As a result, the regions have set up a team as well as facility managers (i.e. those who coordinate team efforts for a given facility). Departmental facility managers need to be empowered to make professional judgments to deal with the routine situations that occur daily in a facility. In the past, managers did not always respond quickly and effectively, so when a facility requested information or decisions from the agency it was necessary to involve a number of people in the process often responses or decisions took several weeks. Professionals must be empowered. Early in this process, senior managers had some concern with because of the traditional structured hierarchical management approach within the Agency. But DEC must deliver services more efficiently to flatten out the organizational structure and let those who manage the projects make decisions. Although DEC is still in the early stages of M2P2 initiative, the team approach and the facility manager concept are beginning to effectuate some of these changes.

Program managers have not only been rotated at the senior level, but also moved down to middle management bureau directors within DEC. When staff members are first hired, especially professional staff members, they generally seek to experience different programs, gain additional knowledge, share their expertise, and move forward in their careers. These goals need to be reinforced. However, when professionals reach middle and upper level management, it becomes difficult for them to move because opportunities for advancement are limited. To challenge these people within the Agency, DEC needs to rejuvenate them and to challenge their professional and management skills.

A resource allocation plan is being developed to move the necessary resources to the regions where they are more needed. While the central office in Albany makes policies, guidance and direction, the regions deal with facilities on a daily basis.

(d) Work plan

Within DEC, a work plan is necessary to allow DEC to manage its technical, professional and support resources as well as to meet the federal grant requirements. DEC's work plans are designed to incorporate actual time and activity distribution to better allocate resources. In DEC, the Office of Environmental Quality (consisting of the Divisions of Air Resources, Water, Hazardous Substances Regulation, Solid Waste and the Pollution Prevention Unit) has developed an integrated work plan and an integrated multi-media pollution prevention work plan. The Office of Environmental Remediation is in the process of developing an integrated work plan for the Divisions of Hazardous Waste Remediation and Spill Management. The Office of General Counsel has already begun to review work plans, needs and their priorities in upcoming planning cycles. This process will certainly benefit professional and support staffs in other offices who will then be able to reallocate and direct their resources to support legal efforts.

10.2.10
Pollution prevention regulatory approach

There has been a substantial amount of debate over the last several years over how to most effectively implement pollution prevention within programs and as part of a state's environmental policy. At issue is whether the pollution prevention requirements imposed on industry should be voluntary nor regulatory. New York State strongly believes in encouraging voluntary programs. USEPA's "33/50" program of voluntary efforts to reduce emissions of a select number of chemicals is certainly an important initial step towards an overall pollution prevention strategy. In New York State, industry itself or as a result of pending regulatory requirements, has reduced its generation, discharges, and emissions of toxic pollutants. However, voluntary efforts are insufficient because they lack the necessary tracking, reporting, and feedback which ensure true progress on source reduction activities.

In New York State, a regulatory approach has been adopted to complement the existing voluntary efforts. Not every facility that generates hazardous waste, emits, or discharges toxic chemicals will be subject to this proposed regulation, which will generally affect larger companies. This regulation will require that emitters, discharges, and generators of toxic pollutants should plan, evaluate and implement source reduction, and other appropriate pollution prevention measures to help the state ensure a 50% reduction in toxic pollutants by the year 2000. This proposed regulation that is expected to affect approximately 700 regulated entities in New York State has several key elements, one of which is the thresholds described below.

(1) For air, the threshold is 40,000 lbs. or greater of combined stack and fugitive emissions based on toxic release inventory reporting.

(2) For water, industries that discharge directly to the waters or indirectly into a publicly owned treatment facility, more than 12,000 lbs. per year of toxic pollutants, will be required to submit and implement a toxic chemical reduction plan.

(3) For hazardous waste, the threshold of 25 tons is derived directly from New York State's 1990 Hazardous Waste Reduction Statute and requires a submission and implementation of a Hazardous Waste Reduction Plan. Publicly owned treatment works (POTWs) are part of this proposed regulatory process because some industries discharge toxic pollutants to a POTW. The POTW needs to treat and ensure reductions in its discharges as well as work with those industries to reduce discharges before reaching the POTW.

This proposed regulation seeks to stress source reduction as the preferred waste management scheme. The source reduction preference is consistent with the Federal Pollution Prevention Act as well as New York State's program. This proposed regulation requires industries to plan and implement those measures that are technically feasible and economically practicable. It will allow those that are regulated, a flexible, rather than a prescriptive course, to pursue a pollution prevention. This regulation will further require an integrated approach so that no cross-media transfer will be allowed.

This regulation recognizes the inevitability of a planning process. Industries must develop pollution prevention plans which will evolve over time as the economy, technology, and other management and technical changes occur. Past achievements of true source reduction, pollution prevention measures will be recognized. Strong public participation is also a component of the program so that the public can participate in plan preparation both before and after submission to the Department. Also, this regulation will be compatible with existing regulatory requirements such as the Clean Air Act and pretreatment to the extent possible. Finally, this proposed regulation will protect the confidentiality of certain processes and trade secret information to the extent allowed by New York State law.

New York State is pursuing an M2P2 regulatory approach to challenge industries to undertake those source reduction measures that are technically feasible and economically practicable. However, the DEC must have assurances of understands the assessments that were conducted, the projects that are being implemented, and how they relates to true source reduction measures. DEC has developed a series of indices for the facility to follow so that the DEC can measure how well a facility's implementation is progressing.

This proposed regulation will challenge industry to meet certain percentage reductions over time and to ease DEC's administrative burden. DEC feels that reduction in reported numbers of toxic release inventory data is not attributable just to pollution prevention measures. Economic slow-downs, discontinued processes and other factors have also contributed to the decrease in numbers. This proposed regulation, among other things, seeks to help industry improve efficiency in its processes and quality of its products. Industry must strive to produce less waste per pound of product or less waste over a period of time for true source reduction and pollution prevention (NYSDEC, 1993b).

10.2.11
Facility management planning

Another area of regulatory impact will be facility management planning (FMP), as part of the Department's overall integrated facility management approach. Facility management planning is a set of planning and implementation actions that encompass a matrix approach over time, based on a number of factors. The FMP approach considers environmental and health risks, the cost and benefit of such action, and the cost effectiveness of implementation. Considerations of such actions are appropriate in planning or rolling cycles of about five years or so. This time frame allows for unforeseen or unplanned high priority initiatives that may occur during this time period. Such an approach allows operational flexibility to meet industrial concerns, while providing strong environmental protection and economic benefits. Facility management planning activities may be implemented by specific permitting conditions, enforcement or compliance related conditions, or other regulatory mechanisms. The elements of facility management planning include but are not limited to:

- environmental, compliance, and management audits
- air permitting and best management practices including fugitive reductions and compliance
- water permitting and best management practices and compliance
- hazardous waste reduction planning and permitting if applicable
- toxic chemical reduction planning, and pollution prevention and risk reduction activities
- tank assessment and replacement
- accident prevention
- emergency response planning
- community awareness programs including public participation
- remediation and corrective action including groundwater cleanup
- natural resource impacts such as fish and wildlife habitats, marine resources and wetlands
- solid waste hierarchical approaches involving paper, ash, plastic, metal, glass, and sludge
- stormwater run-off management
- other facility-specific actions

10.2.12
Conclusions

The New York State Department of Environmental Conservation has undertaken the M2P2 initiative over the last couple of years. The evaluation of M2P2 programs shows the following:

- Team concept has worked; the facility manager is a good focal point for industry;
- DEC's regional offices have implemented and adjusted well and rapidly;
- DEC's M2P2 approach is timely and effective;
- M2P2 is professionally challenging to staff members;
- A trust and partnership with industry will continue to develop;
- DEC is able to work within business time frames;
- Environmental decisions at facilities are being integrated into basic designs;
- Implementation of source reduction measures are difficult but challenging;
- Real reductions, especially source reductions, are possible;
- The M2P2 process requires a lot of effort from both the DEC and industry;
- DEC is able to accelerate its decision-making processes, thereby saving time and money for both the DEC and the regulated entity;
- The formation of the M2P2 Advisory Group has allowed to both the public and environmentalists to discuss this initiative with state agencies and the business community.

The multi-media pollution prevention initiative of the New York State Department of Environmental Conservation is an evolving process that will be modified and adjusted as needed. The process has worked thus far, but to continue to succeed, it will require time, patience and understanding from industry, environmentalists, and the public.

10.3
City of Cincinnati programs

Cincinnati City proper has a population of 1,744,125 in 1995. It is a urban and suburban type of city with an overall annual budget of approximately $718 million in FY95. Cincinnati is aspiring to create a model urban area pollution prevention program through the creation of a multifaceted strategy which focuses on city government operations, business and industries and outreach to the general public. This is being done through pollution prevention training, technical assistance, and promotional efforts.

Cincinnati's pollution prevention (P2) program began in August 1992 with the assistance from U.S. EPA, University of Cincinnati, Advanced Manufacturing Sciences, and other local resources. The initial goal of the program was to create a model urban area pollution prevention strategy for three sectors: local governments; business/industry; and the general public. The long range goal was to sustain local government P2 promotion and implementation efforts in all three sectors for community-wide pollution prevention.

The city adopted a P2 policy statement and implementation plan for "in-house" pollution prevention activities. All city departments and divisions have had P2 training and are responsible for implementing P2 practices on a daily basis. Baselines and goals are established along with measurements for improvements and cost savings. The city has also invited other local and state governmental agencies to participate in P2 training and promotional activities. The city has led the community in adopting an "Environmental Preference" purchasing ordinance for all city purchases. One particularly noteworthy example of in-house initiatives has been the conversion from lead, solvent-based highway line striping paints to lead-free, waterborne paints. While retraining employees and converting equipment has been challenging, management and employee commitment is beginning to produce P2 results.

The city is still in the initial stages of implementation of its urban area strategy. However, there have been some measurable successes in pollution prevention, either realized or projected. Ten small to medium sized industries have been assisted with P2 training, opportunity assessment, and implementation activities. Potential eliminations or reductions of pollution and wastes were measured in millions of gallons and tons per year. Potential cost savings were well over $2 million annually, in materials alone. Most of the industries are now in the process of implementing the identified P2 practices.

The city has performed its first departmental, or divisional, P2 waste reduction opportunity assessments. Two pilot assessments have just been completed: the City Printing Services and the Municipal Garage. Other city departments and divisions will soon follow using these two divisions as models and utilizing the cadre of trained P2 "cause champions" from their own departments and from the coordinating committee pool in forming interdepartmental P2 assistance teams. The approach will facilitate transfer of information and identification of possible P2 opportunities in similar city operations. The third sector of the urban area strategy related to the general public P2 activities will probably the most difficult to measure in terms of effectiveness. However, future success in waste reduction, recycle and reuse across the whole community, from all sources, are expected (NRC, 1997).

Lessons learned from Cincinnati P2 programs are that the city pays close attention to P2 activities because employee, division, and departmental competition and recognition activities and careful planning. The city faced some problems, but never gave up, and the employees believe that P2 is a way of life, an ongoing, never-ending process of improvement. They also believe that the process of P2 implementing is a challenging activity in waste management culture and P2 can be a building block toward sustainable city or community. As a result, there are now many businesses, industries, governments, and, most importantly, individuals who can attest that there are better, safer, cleaner, and more responsible ways of doing things in the city.

10.4
Local pollution prevention programs

In the United States, all local environmental programs including pollution prevention are managed under city and county governments in response to state laws and regulations. Most P2 programs in city and county governments emphasize on waste reduction with the assistance of the state and non-government organizations (NGOs). Such NGOs include the National Association of Counties (NACo), the US Conference of Mayors (USCOM), the National Recycling Coalition (NRC) and academic institutions.

Various research report findings indicate that while source reduction initiatives have traditionally focused on waste minimization and material reduction efforts in manufacturing, local governments are playing an increasingly prominent role in advancing source reduction. In recent years, communities utilize a variety of approaches to reduce waste generation. Educational initiatives, at-home composting, unit-based pricing and materials exchange and reuse programs are the most common source reduction programs practiced by the communities. Many communities have been successful with educational programs, focusing on the value of source reduction and it can be put into practice. Local governments have begun to form partnerships with multiple parties to effectively reach their target audience and gain the trust of constituents, and community-wide participation in their programs in manufacturing goods, using products and managing discarded materials. Communities have found a variety of both quantitative and qualitative ways to measure the results of their programs. However this is an area that requires continued attention and development. Communities are financing P2 programs in a variety of ways, including from the general fund, the recycling budget, disposal tipping fees, unit-based fees, and/or federal and state grants (NRC, 1997).

Hundreds of communities in the United States have found that piecemeal approaches to community issues have not been adequate for solving their environmental pollution problems, and many are lacking a new approach to developing long-term healthy communities based on the concepts of sustainability through pollution prevention. "Sustainable community" refers to community efforts to address problems by taking a systems approach to deal holistically with economic, social, and environmental concerns; adopting a long-term focus; and building consensus and fostering partnership among key stake-holders about community problems and solutions. With communities trying to evolve toward more sustainable practices, we, pollution prevention professionals, have an excellent opportunity to merge our efforts with sustainability activities (Lachman, 1997).

References

Lachman BE (1997) Linking Sustainable Community Activities to Pollution Prevention. Prepared for the Office of S&T Policy by Critical Technologies Institute, RAND, Santa Monica, Ca.

NRC (1997) Making Source Reduction and Reuse Work in Your Community – Executive Summary. Report by the National Recycling Coalition, Baltimore, Md.

NYSDEC (1993a) Directory of Pollution Prevention Unit, NYS Department of Environmental Conservation, Albany, N.Y.

NYSDEC (1993b) Pollution Prevention Guidance for Local Governments. NYS Department of Environmental Conservation, Albany, N.Y.

USEPA (1997) Pollution Prevention 1997: A National Progress Report. Office of Pollution Prevention and Toxics, Washington, DC., EPA 742-R-97-00, pp. 131–152, June

USEPA (1998) A Catalogue of The Agency's Partnership Programs. Office of the Administrator (1803), EPA-100-B-97-003, p. 44, Spring

11
Measuring pollution prevention progress

Large quantities of environmental quality data have been collected over the past 50 years by government agencies, industries, and non-governmental organizations. These databases have only limited usages to measure pollution prevention progress. There is a need to collect data on source reduction activities and their effects on waste and release quantities to measure pollution prevention progress. Pollution prevention is an increasingly popular subject, but different users have different measurement needs. Measuring pollution prevention progress sounds deceptively simple. Using a single measure to summarize pollution prevention will be applicable only in the simplest cases.

As pollution prevention becomes part of a facility's day-to-day operations, both industries and regulatory agencies must find a method to measure its progress toward reduction of hazardous wastes and releases to air and water. If facility environmental management practices are to shift towards a pollution prevention emphasis, regulatory agencies need to track pollution prevention progress and to know the sum of all or a great many individual pollution prevention actions, whether they represent successes or failures, and how this information relates to larger industrial, economic, and environmental issues and policies. In addition, many regulatory agencies in the United States are passing pollution prevention laws which require facilities to identify source reduction opportunities to be implemented over different time periods (Ryan and Schrader 1993).

In order to track the facilities' achievements and the overall effectiveness of this regulatory approach, pollution prevention related data must be available. Industry also needs independently verifiable data which effectively measure pollution prevention progress both to demonstrate compliance with newly evolving environmental laws and to promote an environmentally sensitive image (Pojasek and Cali, 1991).

A successful pollution prevention measurement needs to identify the specific objective, resource availability, and proper approach. If the industrial plant has the appropriate resources, the pollution prevention measurement can help develop a rapid understanding of the relationship between wastes and the manufacturing process.

This chapter covers pollution prevention data acquisition, data analysis, and methods of measuring pollution prevention progress with the emphasis on source reduction in various industrial processes. It also discusses toxic release inventories, material accounting data, and expansion of Community Right-to-Know program used by regulatory agencies to measure the progress of pollution prevention programs.

11.1
Purpose

The purpose of measuring pollution prevention progress is to evaluate the progress against the established goals which can be different in various settings. Measuring pollution prevention progress requires substantial commitment from both industries and regulatory agencies. we must minimize the data need-ed to measure progress and maximize the progress itself. Continuous improve-ment at facilities is more important than rigorous measurement and reporting. Fortunately, techniques for measurement of pollution prevention are evolving rapidly, and new methods are being implemented.

11.2
Data acquisition

Different data collection approaches might be required in different types of in-dustrial processes. They may be, for example, primary materials processing, sec-ondary manufacturing and product formulation, or packaging/container pro-ducers. We need to decide what kinds of data or specific wastes or pollutants to collect for measuring pollution prevention progress. Should we target specific "high risk" substances throughout their processes of extraction and use (e.g., chlorofluorocarbons, lead, and chlorine); or target particular stages of the waste generation process (extraction, manufacturing, commercial use, consumer use, and waste management); or target particular sectors, industries, or firms that are especially wasteful, especially hazardous, or especially attractive for oppor-tunistic waste reduction; or target product characteristics and specifications? We also need to determine what data shall be collected to help us measure pollution prevention progress (Andrew, 1989).

For data acquisition, we must select a quantity in terms of waste volume or waste toxicity. Quantity measurement must be normalized as necessary to cor-rect for factors not related to the pollution prevention. We must also realize that (1) the quantity selected to track performance must accurately reflect the waste/wastes in interest; (2) the quantity must be measurable with the resources available (material and energy balances will be helpful in organizing data and can help fill in some gaps in data); and (3) the data tracked must be decided be-fore collecting and normalizing.

The pollution prevention option may eliminate part of the target material but shift some of it to another plant stream, to another environment medium, or

into the product. It can be difficult to track the shift of a pollutant from one medium to another or to determine what new pollutants may be created by the procedure. Furthermore, the toxicity of the waste should be looked at, not just the
quantity produced. Reducing the sheer volume of a given waste product while
increasing its per-unit toxicity is a treatment option, but it is not pollution prevention. For example, adding lime to a waste-stream to precipitate metals reduces the volume of waste, but does not prevent pollution since the total quantity of
metal in sludge is not changed. Therefore, watch for shifts of wastes to other media and develop toxicity measures.

It may be necessary to normalize quantity comparisons to adjust for external
factors such as total hours the process operated; total employee hours; area,
weight, or volume of product produced; number of batches processed; area,
weight, or volume of raw material purchased; and profit from product. For continuous processes, the product output or raw material input can be a good nor-
malization factor. Flow processes may be measured by volume or weight, whereas plating or film-making may be better normalized by area. In batch processes,
production volume usually is related to waste production, but it may not be a lin-
ear relationship in all cases. For example, the quantity of solvent used at a printing plant is primarily a function of the total volume of stock printed and ink
used, but it is also significantly influenced by the number of color changes made
(USEPA, 1992).

A "loss tracking system" which provides a standardized procedure for tracking and recording material usage and various process losses is considered a useful tool in data collection for pollution prevention programs. A typical loss
tracking system includes:

- Using process flow diagram as the grid system;
- Accounting for all inputs (raw materials), losses (waste), and outputs (prod-
  ucts and by-products);
- Interfacing with existing manufacturing, production, accounting, and envi-
  ronmental compliance systems;
- Operating on a personal computer (PC) and using a database management
  system;
- Allowing multiple and end users to develop their own reports and queries.

The system should collect all the data needed to measure pollution prevention
progress. The data collected include:

- Material purchases and uses by unit operation;
- Material throughput for each unit operation;
- Generation of losses from each unit operation;
- Loss classification by medium (air, water, or land);
- Scrap and defective product generation; and
- Production outputs (Pojasek, 1991).

11.3
Data analysis

We must select the most useful methods for specific situations. Semi-quantitative methods are easier to prepare but have less utility. Lack of quantitative data also makes difficult to compare similar processes when looking at potential technology transfer. Data for analysis based on transfers should be easy to obtain. Shipping manifests and compliance reports provide data on quantities transferred. Changes in the quantities of materials brought on site, as determined from receiving records can be useful. Most facilities keep detailed, accurate records of material received from suppliers. These records provide a source of data to track changes in the types and volumes of materials brought into the facility. However, quantity purchased is an imprecise measure because it does not account for loss during processing.

Quantifying waste generated or used involves tracking the quantities of hazardous, toxic, and other materials flowing into and out of the facility. It uses data on the quantities of material purchased, produced and destroyed in the production process, and incorporated in products and by-products, as well as discharges to waste treatment and disposal. This approach gives an overall picture of material use but requires extensive data collection. Data on fugitive emissions are particularly difficult to track but can sometimes be estimated by calculating material balances.

The best way to measure pollution prevention is to determine the annual amount of pollutants that would have been released annually from a facility if prevention action had not been taken. In reality, very limited pollutants data are available that can be used to measure pollution prevention progress. Many factors contribute to reductions in pollutant releases, including increased waste treatment as well as decreased production. Therefore, progress should be measured by factoring in current production rates. However, the ability to normalize production quantities is often the most difficult aspect of measurement. Pollution prevention progress can be also measured and analyzed on a process-by-process basis by examining the production process in detail for changes due to pollution prevention activities. However, analyses strictly on processes will overlook facility-level waste, such as construction debris. Project analysis may be used to focus on measuring the results of each pollution prevention activity. It is more useful for production changes than for behavioral changes. Pollution prevention programs can also be measured by the change in total amounts of toxic materials released.

In the U.S., the change can be analyzed directly from SARA Title 313 Form R reporting (see Appendix E). The source of the toxicity can be identified by detailed testing. Process streams contributing to the plant waste effluents are sampled and partitioned into separate phases. The detailed toxicity testing allows identification and tracking of the actual toxicity of wastes from the plant. Toxicity testing requires sophisticated testing and data handling, however, and may not be feasible for all facilities.

Pollution prevention may have multi-faceted effects at an industrial facility. In order for a database to specifically distinguish the impacts that pollution prevention may have at an individual facility, it is critical to recognize that pollution prevention can have three distinct results:

(1) Reduction of toxics used per unit of product as primary or secondary input materials into a production process;

(2) Reduction of hazardous substances generated as non-product output per unit of product prior to recycling, treatment or disposal; and/or

(3) Reduction of hazardous substances in manufactured goods per unit of product.

Ideally, not only should analysis of a pollution prevention database be able to register variations in all three of these discreet areas, but information needs to be available to factor out non-source reduction influences. Furthermore, data analyses should include the economics of the pollution prevention activities by any of several techniques such as payback period, net present value, or return on investment.

11.4
Methods of measuring pollution prevention progress

The availability of Toxic Release Inventory (TRI) data collected under the Pollution Prevention Act in the computer database has greatly facilitated measuring progress of pollution prevention. The reported releases and transfers to TRI chemicals in the United States were 7.0 billion lbs. in 1987, 6.5 billion lbs. in 1988, and 5.7 billion lbs. in 1989 (USEPA, 1991). However, TRI does not contain information on why releases and transfers change from year to year and the pollution prevention information is optional. Thus, evidence is insufficient to conclude that there is a downward trend, or even that physical quantities are decreasing. In many cases, these apparent reductions are due to changes in reporting practices – accounting methods, estimation procedures and interpretation of the forms and instructions – rather than actual physical changes in quantities.

U.S. experiences indicate that changes in reporting requirements, and in respondents' understanding them, can introduce errors in analyses. Some changes in quantity reported are due to changes in the way the wastes were measured or the accounting practices used by the facility, rather than actual changes in the quantities generated. Substantial differences in reported quantities can result from changes in definitions of terms used in the reporting form. This can include changes in reporting criteria, changes in regulatory definitions, or clarifications to instructions.

It is essential that industries must reflect their reported pollution prevention data to their operations and their industries. The availability of information in the computer database will greatly facilitate comparisons between facilities with similar operations, as well as within industries, and geographic areas and facilities. Such facilities may not adequately consider the information. The information they are providing may find themselves at odds with other regulatory programs, or at a loss to explain discrepancies in future year's projections (Bolstridge, 1992).

Using 1988 TRI data as a baseline, EPA set two national goals: a 33% reduction using 1992 TRI data, and a 50% reduction using 1995 data. The 1995 goal translates to a targeted reduction of nearly 750 million lbs. of pollution from the nearly 1.5 billion lbs. reported to TRI for these chemicals. The efforts of 1,300 corporate participants (which own more than 6,000 facilities), surpassed its ultimate 50% reduction goal in 1994, a year ahead of schedule. By 1994 total releases and transfers of TRI chemicals were reduced by 757 million lbs. (USEPA, 1998a).

The new chemicals, collected in the annual Toxics Release Inventory data base, account for 237.7 million lbs. or 10% of all reported releases of toxic chemicals into air, land or water. 94% of the 286 newly added chemicals have demonstrated chronic health hazards and/or environmental effects, including cancer or reproductive disorders. For the core chemicals reported for 1995, the releases of pollution decreased by 4.9%, from 1.75 billion lbs. in 1994 to 1.66 billion lbs. in 1995. Reported air emissions were down by 88.8 million lbs., or 7%; reported discharges to surface water were down 4.1 million, or 10%. Releases to land were down by 17 million lbs., or 6%. However, underground injection releases increased, by 24.5million lbs., a 19.5% increase (USEPA, 1998b).

11.4.1
Toxic Release Inventory (TRI)

The Toxic Release Inventory (TRI), a publicly accessible database. TRI consists of relevant information on a set of toxic chemicals collected from a selected universe of facilities, put into a searchable computer database open to the public. It marked the entry of environmental data into the "information age", and created a national awareness that strategies based on information can create a powerful driver for environmental improvement. TRI motivates industries to reduce pollution without the prescriptive qualities of command and control regulations. TRI is widely viewed as a success story. Environmental and public interest stakeholders were also interested in the concept of facility-level chemical use data as a fundamental Right-to-Know issue, and recommended that EPA should initiate to include materials accounting data in the public domain.

It is essential to remember that TRI is only as valuable as the number of people who use the information. It's as important as ever to raise awareness of the value and availability of TRI. Individuals and organizations using TRI knit together concerned citizens with top corporate and government decision makers. The US-TRI Service provides general information about TRI and assistance in acquiring and using any of the TRI products, including assistance in searching the CD-ROM or TOXEN database. To contact TRI-US, call (202) 260-1531 or write: TRI-US, USEPA, 401 M St. S.W., Washington, DC 20460. Request for information and materials may also be faxed to (202) 401-2347. You may also contact TRI-US by email: tri.us@epamail.epa.gov

While many pollution problems can be addressed by preventive efforts, the emphasis to date has been on reducing industrial toxics. Currently, the best information on this subject is the Toxic Release Inventory (TRI). In 1990, EPA published the TRI data for the second time, covering 1988 releases of all manufacturing facilities in the U.S. The data include any individual toxic chemicals and/or 20 categories listed for reporting. The quantities produced, imported, or processed are 50,000 lbs. or more and/or used in any other manner 10,000 lbs. or more. Such chemicals vary widely in toxicity, frequency, amounts, and industrial processes in which they are used.

The TRI continues to evolve as a prime indicator of waste generation and as a focus for pollution prevention efforts. TRI data include releases to all environmental media – air, water, and land – in contrast to most previous single-media data collection efforts. However, the TRI has limitations. TRI describes only toxic emissions, which are a fraction of the broader waste problem and it is limited to those toxic chemicals listed by USEPA. Perhaps most significantly, TRI shows emissions or transfers after treatment, rather than the pollution generation data needed for prevention analysis. On the 1988 inventory form, only 10% of the respondents completed a voluntary section requesting data on waste minimization. This statistic led Congress to make reporting on waste minimization mandatory under the Pollution Prevention Act of 1990. Such additional data has helped refine the USA's evolving effort to prevent industrial toxics pollution.

11.4.2
Materials accounting data

Materials accounting involves determining the quantity of a chemical at key junctures in its progression through a facility, using readily derived information such as invoice and product composition data. It is distinct from engineering mass balance, which is a more technically rigorous exercise that includes measurement of process streams.

The idea of including materials accounting data in TRI has been an area of marked disagreement between stakeholders since the creation of TRI in 1986. EPA believes that chemical use information, while controversial, touches on many fundamental issues worthy of public dialogue. It relates to many important themes such as: the role of government as a provider of information to promote and empower efforts to improve environmental performance; of building problem-solving capability at the community level; of creating accountability that allows flexibility; and of providing the information basis for "cleaner, cheaper, smarter" approaches.

Toxic Release Inventory has been discussed that it may not be adequate to sufficiently measure pollution prevention. TRI describes only the emissions of manufacturers, which make up only a fraction of the total emissions of a range of small, dispersed sources such as motor vehicles, solid waste facilities, and small businesses that use hazardous chemicals. TRI can quantify changes in annual releases, which is not comparable to pollution prevention. If a materials accounting survey were added to TRI, which includes chemical throughout data to measure pollution prevention, TRI could become a powerful vehicle for promoting and measuring pollution prevention achievements (Freeman et al.,1992; Hearne and Ancott, 1992).

The DEQ-114 Forms of New Jersey State (see Appendix F) require TRI reporters to provide environmental release information as part of their throughput data and encourage them to conduct a "self-verification" exercise. Such exercise is intended to allow facilities to check the sum of starting inventory: quantity produced and quantity brought on site should approximately equal the sum of quantity consumed (chemical altered), quantity shipping off-site as (or in) product and as (or in) waste, quantity destroyed through on site treatment, air emissions, wastewater discharges, and ending inventory. It could also help determine if their information is approximately accurate and discover any apparent reporting errors.

Since pollution prevention can have varied effects on waste generation and total facility chemical use levels, both impacts should be assessed when measuring pollution prevention. Activities which increase process efficiency, decrease product composition of toxic substances or decrease overall use may not be recorded relying only on TRI data. Materials accounting survey data can potentially be used to confirm TRI data analysis findings when conflicting information arises. It is particularly important that significant data quality problems may exist with the TRI production index or activity index. Materials accounting survey can provide process efficiency values that reflect all historical improvements at the plant (Hearne, 1994).

The Toxic Release Inventory would be greatly enhanced for pollution prevention measuring purposes by the addition of materials accounting survey data to Form R of Section 313 data report as required by the Pollution Prevention Act. With this information, the industry, government, and public will have a better perspective of the impact source reduction activities have on total generation of non-product output, use, process efficiencies and product formulation of industrial facilities. Thus, more accurate pollution prevention progresses could be measured for program evaluation.

The existing state laws on materials accounting are (1) New Jersey Worker and Community Right-to-Know (PL 1984, c. 315) and (2) Massachusetts Toxics Use Reduction Act (YURA. M.G.L. c. 211).

(A) Lesson from the New Jersey experience
The New Jersey views materials accounting as an essential part of pollution prevention planning. It serves to highlight areas of potential cost savings and envi-

ronmental benefit. The data can be used to provide an estimate of production efficiency, and has also played an important role in facility-wide permitting project. The reported materials accounting (MA) benefits are: MA provides essential information such as: creating accountability to support flexible approaches, tracking source reduction progress, and improving community-level capability for looking at transportation-related issues.

(B) Lesson from the Massachusetts experience
The Massachusetts DEP summarized the lessons learned after four years of implementation in five major points:

1. The experience demonstrates that toxic chemical use data can be effectively collected from industrial facilities and managed by state agencies.
2. Toxic chemical use data provides significant benefits to state agencies, industrial firms, and the general public. These include: improved public and governmental understanding of the volumes and patterns of toxic chemical use; improved capacity to target state resources; better basis for measuring progress in pollution prevention; improved toxic chemical management in firms; and a better basis for promoting technology transfer among firms.
3. There is a public constituency for toxic chemical use data and both government and non-government organizations can use the data to promote public health and environmental protection.
4. While there are industrial costs to collecting and reporting this data, there are also savings that are associated with improved materials account, these benefits can outweigh the costs.
5. Confidential business information can be effectively protected by conventional trade secret protections (USEPA, 1996).

11.4.3
Evaluation

Evaluation is the administrative process of measuring and analyzing results in relation to specific goals, objectives, aims, or targets. It measures benefits against costs, or outputs against inputs. It can also determine the relevance of programs to various risk problems. Evaluation involves the following steps.

1. Statement of program objectives. An objective to be meaningful must:
   - define the problem
   - be measurable
   - dentify the target population
   - give the existing status of the program
   - specify the desired change
   - establish a timetable to accomplish the change.
2. Selection of indicators to measure achievement of the program objectives to which a program has achieved the stated objectives.

3. Comparison of program achievement with the stated objectives.
4. Measurement of program cost.
5. Measurement of productivity, with consideration of the expenditure and effectiveness or degree to which a program has achieved the stated objectives.
6. Analysis of the above findings, their overall program impact, and adjustment
   of the program and objectives as needed.

In order to perform an evaluation it is necessary that the data or information
collected the reliable and valid and that the achievement of goals be measurable.
To be valid the information must be accurate, sound, and correct, and measure
what it is supposed to according to established procedure. For example, the
number of inspections reported may be correct, but if the inspection were cursory, the information is not reliable. For an objective to be measurable it is necessary to have a means or yardstick to measure its achievement by a stated time.
  Objective evaluation of the pollution prevention program conducted by regulatory agencies serves many purposes. It can provide the basis for integrating,
adjusting, and balancing the program; it can be used to demonstrate the need for
obtaining and retaining competent personnel; it aids the administrator of the
program, and the supervisor of one or a group of activities, to determine whether available personnel are being utilized to perform the work considered most
important and if the established objectives are being accomplished. In addition,
it can provide cost data and facts for supporting program recommendations and
policy determinations. It is effective in showing which program activities need
more inspection time, which are receiving too much inspection time, and which
are not producing results.
  To be of value evaluation studies should consider the work load, work done,
quality of work, and its effectiveness. The data assembled must be interpreted in
the light of the thoroughness, competence, reliability and validity of the inspections or observations. Evaluation should start in the planning stage, before a program gets started. Goals are set, plans are made to reach feasible objectives toward the goals in a stated time, and analyses or studies are made to see how closely the goals and objectives have been met and what changes may be indicated.

References

Andrew RNL (1989) "Research Needs For Waste Reduction," background paper prepared
    for the National Research Council Workshop on Waste Reduction Research Needs, An-
    napolis, Md., May 8–9
Benforado D (1990) "Measuring Waste Reduction Progress," in: Environmental Challenge
    of the 1990's, Proceedings of the International Conference on Pollution Prevention:
    Clean Technologies and Clean Products, USEPA/600/9-90/039, September
Bolstridge JC (1992) "Pollution Prevention Act: Effects and Implications of Industry's Re-
    porting." Paper presented at AWMA Annual Meeting in Kansas City, Mo.
Freeman H et al. (1992) "Industrial Pollution Prevention: A Critical Review." J. A&WMA,
    Vol. 42, No. 5, pp. 636–654

Hearne SA and M Ancott (1992) "Source Reduction vs. Release Reduction: Why the TRI Cannot Measure Pollution Prevention," in: Pollution Prevention Review, Winter 1991/1992

Hearne SA (1994) Potential Modification to the USEPA Toxics Release Inventory to Better Track Pollution Prevention Achievements. Ph.D. dissertation, Columbia University, School of Public Health, New York, N.Y.

Pojasek R and LJ Coli (1991) "Measuring Pollution Prevention Progress," in: Pollution Prevention Review, Spring 1991

Ryan W and R Schrader (1993) "An Ounce of Toxic Pollution Prevention: State Toxics Use Reduction Laws." National Environmental Law Center, Center for Public Interest Research, and Center for Policy Alternatives. Washington, D.C.

USEPA (1991) "Pollution Prevention 1991: Progress on Reducing Industrial Pollutants." EPA 21P-3003, October

USEPA (1992) Facility Pollution Prevention Guide. Office of R&D, Washington, D.C., EPA/600/R-92/088, May

USEPA (1996) TRI-Phase 3: Expansion of the EPA Community Right-to-Know Program, Issue Paper #3. EPA Office of Pollution Prevention and Toxics

USEPA (1997) Press Release: EPA's 1995 Toxics Release Data, May 20

USEPA (1998) Toxic Release Inventory: Community Right-to-Know, 1995 TRI Public Data Release. EPA Office of Pollution Prevention and Toxics (http://www.epa.gov/opptintr/tri/national.htm)

12
The role of corporate management

One does not argue with the laws of nature. One either conforms or pays the penalty. The mathematics of compound interests is natural law. We are in the self-destruct stage. Our environmental quality is continuously getting worse. If we continue to seek socio-economic development and permit funds to accumulate, we are certain to have our economy destroyed and our people in revolt. Money, like everything else in the environment, must be recycled to prevent destructive pollution of the economic environment. Modern manager uses industry's capabilities and resources in a responsible and sustainable way. The fundamental problem lies in the fact that market prices for energy and raw materials tend to reflect only their direct costs. Prices reflect neither ecological costs, in terms of damage to the environment, nor the depletion of natural capital when a non-renewable resource is used. This problem must be resolved.

Corporate management has a great deal to offer to achieve sustainable development through industrial pollution prevention. Industry is the world's foremost creator of wealth, employment, trade, and technology, as well as the controller of tremendous human and financial resources. Industrial and business processes add value to resources and meet basic human needs. The multinational companies are powerful enough to influence international environmental and development problems. Some of them do take a more farsighted and international view than governments. Unfortunately, the value of environmental quality had not always been appreciated by business.

The history of industry's response to environmental concerns has four main stages: denial, data, delivery, and dialogue, or the four D's. In the 1970s, most companies adopted the attitude that what went on within their fences was no one else's business. They shrouded themselves in a veil of secrecy, claiming the need to protect confidentiality and proprietary interests. They denied that they had pollution problems. The second stage, roughly in the 1980s, was data collection and exchange. It was an improvement over the first stage in that people were at least communicating with each other on environmental data. But the public was still unimpressed. One important piece of learning from this period was that people simply do not care about what industries do. Caring means taking action. In the 1990s, message to business has been that fine words mean nothing if business cannot show continuous improvement of environmental performance and development of environmentally sound products and processes. Business leaders acknowledged this with the widely quoted maxim "don't trust us, track us." But experience shows that even delivery is not sufficient. There is a fourth stage of dialogue. It ensures that what is being delivered is what stakeholders want. It is also the key to getting recognition for success and tolerance for any temporary failures (Schmidheiny, 1994).

Industry and business committed to pollution prevention have to reexamine and reshape their entire enterprise, from the products they make, to the technologies they use to make them, to the raw materials they base production on, and even on how they market products and think about new acquisitions and investments. Modern corporate managers need to completely integrate pollution prevention thinking into absolutely every part of a company's management system rather than merely focus on regulatory compliance, the traditional environmental approach (Stevenson and Ling, 1991). Corporate managers must recognize that environmental groups are creating new demands. Environmental issues are not merely environmental strategies and the issues require a new business strategy, which is total environmental management (TEM). TEM translates a technological pollution prevention strategy into a new corporate culture which then drives a new business strategy. Paramount to this change and future industrial competitiveness will be measurable environmental quality and performance (Hirschhorn, 1994).

Industry and business have a crucial strategic choice to make: to stay defensive and reactive or to find innovative and proactive. The latter identifies and capitalizes on the unlimited business opportunities from growing global demand for environmental quality. Holding on to existing market shares for existing products made with existing technologies means waiting for failure. Industry and business in the 1990s face a variety of competing demands – maintaining high quality at low cost, staying competitive in a global marketplace, and meeting consumer preferences and regulatory demands for reduced environmental impacts. Pollution prevention has been recognized as an effective strategy for managing these challenging demands.

In general, industry and business appear to be motivated to adopt pollution prevention innovations by a combination of factors. Some are attracted by perceived economic benefits, either in the form of cost savings or increased market share. Some respond to the threat of government regulation, still others to the willingness of regulators to be flexible. Some companies are motivated by customer demand for "green" products; others by public attention to their polluting practices. However, implementation of pollution prevention in various sectors of industries is rather slow for various reasons such as financial, technical, institutional barriers, especially those medium and small business and industry (Shen, 1997).

This chapter discusses the role of industry's environmental responsibilities, management strategies, and environmental programs. It describes the need for total environmental management to achieve sustainable development with considerations of manufacturing green products and adequate plant design. It encourages industries to work closely with governments and international programs to achieve sustainable development.

12.1
Environmental responsibility

Business and industry should take the lead in providing its own solutions to environmental problems and world economic competition if our society is to enhance economic growth and environmental quality. Many corporate executive officers (CEOs), however, feel that there is a basic contradiction between the corporate objectives and the social responsibility of a corporation. And yet as one examines the responsibilities of the modern manager in relation to the human environment, one sees that it is perhaps only through proper concern about the environment and the avoidance of pollution that the modern manager can meet the classical and basic criteria for successful management.

Modern managers are responsible to solve environmental pollution problems with a preventive approach by controlling pollutants at the source. Pollutants can be valuable and expensive raw materials which are not fully used in production. In most cases, process modification, equipment redesign, material substitution, or other innovative approaches could greatly reduce or even eliminate pollution entirely. At the same time, this reduces the cost of valuable raw materials and can provide major savings in the cost of providing and operating pollution control facilities. As described in Chapter 2, when polluting materials cannot be removed at the source, they should be recycled, recovered, and reused in the manufacturing process or in other industrial processes. Companies that have taken this approach have seen handsome economic returns. Meanwhile, this approach will also reduce or eliminate potential future liability.

Most modern managers recognize that the public and government agencies increasingly demand comprehensive reduction and elimination of all kinds of nonproduct waste outputs, pollutants, and toxic chemicals. For industry, it means a fundamental rethinking of what are the environmental problems, laws, and regulations imposing on industry. As a result of continuing pressures from customers, investors, employees, legislators, and also from banks and insurance companies, changes of corporate management are necessary in order to response such pressures. Modern managers must know the importance of proactively pursuing resource productivity. This not only delivers more to society from less but also catalyzes more general innovation. Achieving it will become even more critical as the world becomes increasingly crowded and runs out of acceptable sinks for wastes and pollution, and as valuable resources become scarcer and increasingly costly.

In the past it was strictly costs and liabilities which were seen as threats to industry. Now there are new industrial opportunities in the global marketplace. The messages of the global pollution prevention movement in recent years are that preventive actions will improve protection of human health and the environment, as well as improve efficiency, profitability, and competitiveness in industry. The modern corporate executive officer (CEO) must realize that, in this complex world, he must consider many different ideas to guide his decisions. He is expected to get the best results from a system containing diverse and often seemingly contradictory factors. He is faced with a number of alternatives, including the trading-off of short-term against long-term interests, the striking of a balance between the interest of the corporation and that of the community, and taking a lead in national development as opposed to merely responding to legislative constraints and restrictions. The objectives of his corporation must include not only profit, growth and survival, but also human and social responsibilities (Royston, 1979; Ling, 1997).

Many business executives and some governmental officials fear that business excellence and environmental concern cannot be combined. Researchers found the opposite: they cannot be separated (Schmidheiny, 1994). We all agree, after studying worldwide business trends, that tomorrow's winners will be those who make the most and the fastest progress in improving their eco-efficiency and preventing their pollution. The reasons are that:

- Customers are demanding cleaner products.
- Banks are more willing to lend to companies that prevent pollution rather than pay for cleanup.
- Insurance companies are more amenable to covering clean companies.
- Employees prefer to work for environmentally responsible corporations.
- Environmental regulations are getting tougher.
- New economic instruments (taxes, charges, and tradable permits) are rewarding clean companies.

All these trends will accelerate as science offers more evidence of environmental damage. This means that investments in pollution prevention and eco-efficiency will help profitability, rather than hurt. Those industry and business, practicing pollution prevention and eco-efficient, will emerge more competitive as these trends take hold.

The CEO has a responsibility not only to the company which he manages (e.g., products, materials, manufacturing, pollution), but also to the society in which his company functions. Elements of corporate management environmental responsibility is presented in Fig. 12-1. Too often, in the corporate context, the decision makers and the policy makers are removed from public scrutiny and criticism. It is only when the corporation enters into conflict with the community that the decision makers become accountable to society for their actions. There lies the challenge for business and industry to be a part of the pollution solution, rather than part of pollution problem. Globally, industrialized countries should have the moral responsibility of helping the rapidly growing environmental management needs of the developing countries (Schmidheiny, 1994).

Industries must integrate business and environment, and decision-making systems in order to achieve pollution prevention goals. This requires a new model for business and industry to merge economic and competitive reality with environmental performance. Business and environment management systems have often been managed separately. By fully incorporating environmental goals into business goals, product by product, industries can make more informed de-

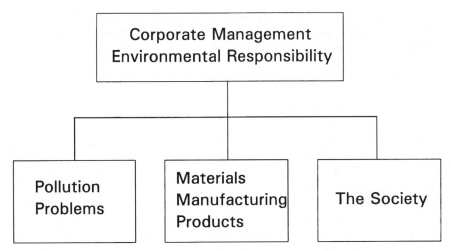

Fig. 12-1 Elements of corporate management environmental responsibility

cisions on where investments and resources are needed. Sustainability then becomes determinable on a cost basis. Environmental managers must work effectively with marketers. Corporate managers need critical third-party evaluations of corporate pollution prevention programs, environmental business strategies, and total environmental opportunities programs. Company programs need to be evaluated through the critical eyes of skeptical environmentalists. Finally, Business and industry need a new corporate culture and a major commitment to retooling work forces. In-house education and training programs are essential to teach every employee about environmental issues, pollution prevention, environmental marketing, and responsibility.

The world of corporate environmental management is changing. Whether under the auspices of corporate reengineering or ISO 14000, company after company is transforming the ways in which internal environment, health, and safety activities are managed, organized, and conducted. Element by element, environmental management systems are being developed. Policies and standards are being rewritten; responsibilities are being assigned; companies is being tracked; audit programs are being designed; best practices are being shared; performance is being measured (Brown and Larson, 1998).

The modern environmental manager needs to learn the total environmental management system or ISO 14000. It helps managers separate from profit motive toward management of the whole, the environmental issues have emerged as an integral part of the whole. The manager also needs to discover restructuring initiatives; leverage scare resources to maximize value and minimize costs; sell the benefits of responsible and proactive environmental behavior; and devise strategies to implement such actions. In addition to technical expertise in environmental areas, the manager will benefit from knowledge and skills in accounting, business strategy, marketing, finance, community organizing, and management ability.

12.2
Corporate environmental programs

Some industries may see no difference between an end-of-pipe pollution control approach and a front-end-pollution prevention solution. The danger is that those industries may not go beyond the first stage of waste reduction, in which opportunities spring from direct experience with naked-eye observations of waste generation. As the environmental concern deepens, industries have to move further up the production chain: first, end-of-pipe solution to wastes and pollutants; later, internal process modifications to reduce emissions and waste and eventually redesign the products to allow maximal recycling of raw materials and minimization of waste production after the products are used. At some point, industries may have to redesign the whole system that is providing the service in question if the environmental consequences of the present solution become unacceptable. The transport system has been cited as an area in search of new options. In this domain, industries need to enter a complex chain of new thinking in which the importance of the technical dimension decreases and the social and cultural dimensions grow in importance.

For industry, the most important part of the Pollution Prevention Act is to collect source reduction and recycling data. Facilities that file actual toxic chemical release forms must now file information on past and future source reduction and recycling. To make amounts of chemicals significant, the Act requires use of a production ratio, which can be very difficult for industries to develop, however. Industries must use new ways of defining their products and outputs to accurately and fairly describe their waste production performance, or face public criticism. The real issue is that reporting and planning requirements have become costly burdens imposed on all industries, as if all are equally guilty of too little commitment to pollution prevention. The challenge is that industries must work with government to modify reporting and planning requirements so that they can establish incentives and rewards for demonstrable industry's commitments and performance. Otherwise, limited resources diverted to planning and reporting deprive spending on actually implementing pollution prevention projects.

Because businesses and industry often need further information and technical resources in order to overcome various barriers to adopting pollution prevention practices, the Pollution Prevention Act provides state matching fund program, promoting the use of source reduction techniques. The Act also directs EPA to establish a Pollution Prevention Information Clearing House (PPIC) for the dissemination of pollution prevention information to federal, state, and local governments, industry and trade associations, public interest groups, academic and other public and private institutions.

Through practice of pollution prevention, many industries have enhanced their relationships with local communities, improved their products for their customers, and made real environmental improvements. Significantly, many industries have done all of this while improving their performance with profitable results. The positive results achieved so far are a good beginning, but much more remains to be done. The interesting and significant change that has occurred in industry since 1990 is not a new system or new technology, but rather a change in attitude. Industries are realizing that they can no longer afford to view the environment and business as two different topics, let alone two competing topics. Many large and small industries are establishing demanding waste and emissions reduction goals, enlisting in voluntary programs, and publicly reporting their progress.

The industrial pollution prevention (P2) program must integrate technology into a system which includes organization, operating procedures, and investment dollars. Corporate management plays the key role in developing a comprehensive program that reaches all levels of a company. The most successful organizations at preventing pollution and minimizing waste overall are those that employ multifaceted programs with the following elements (U.S.EPA, 1988):

- A formal commitment by management to the pollution prevention ethic, translated by management and employees into a company wide commitment in all divisions of the organization;
- Explicit program scope and goals;
- Accurate waste and cost accounting to assess costs, benefits, and programs;
- Periodic pollution prevention opportunity assessments to identify opportunities for improvement and evaluate progress;
- Periodic self-evaluation that keeps programs on track; and
- Educational workshops and training materials that share technical information and experience with affiliates, customers, and suppliers.

In industry, the technical people are concerned with innovative technology related to environmental quality or process development. Management on the other hand is concerned with profit, growth, and survival. The most successful ventures occur when the results of innovation, new product development, and process modification, with management support, provide substantial cost saving, gross reduction in or elimination of waste discharge, and preferably, a profit from the new system. All company employee will then have an incentive to participate in a program in which pollution prevention pays. If an industry can minimize discharge pollutants to the air, water or land, it will have a positive impact on preservation and enhancement of environmental quality or process development. For the company's general employee, pollution prevention program is a morale builder. It is good for the managers' conscience and for the environment.

Pollution prevention performance of a plant must be measured if it is to be improved, and it must be improved if it is to be praised. Performance is not merely regulatory compliance, endorsements of sustainable economic development, how much money has been spent on environmental compliance, the office paper recycling program, or participation in pollution prevention programs. Reliable facts and figures are needed for the reduction in the use of toxic chemicals and reduction in the generation of every form of nonproduct waste output. Performance must be measured in terms of the redesign, reformulation, and repackaging of existing products and the development of innovative new products. Products must be proven through accurate and credible life- cycle analysis to offer solid, measurable improvements for specific environmental problems.

Many industries must move swiftly to integrate the consideration of environmental impacts into all aspects of management decision including product cost, product price, product and process design, capital investments, and performance evaluation. However, industries are often unsure about how evaluations of changing regulations, changing environmental technologies, and changing costs of those technologies changes in decision making. In many cases, industries are being managed with a focus on regulatory compliance rather than environmental planning (WBCSD, 1998).

Three areas of industrial improvement that have significant positive impacts for both the environment and industrial profits are *capital investment decision making, cost management, and performance evaluation.*

(1) Capital investment decision making – Techniques are used to analyze uncertainty related to projections of sales, competition, production costs, and many other factors often exists and long time horizons are not uncommon. Nevertheless, risk and uncertainty are appropriately factored into the decision analysis and the decisions are made. These techniques are generally not used in environmental equipment decisions and the evaluation of quality improvements that benefit for both the environment and profits. Therefore, industries need to conduct product life cycle assessments for improving product design and process design that not only reduce waste and increase production yield, but also increase product marketability and sales.

(2) Cost management – Industries need to identify their total environmental costs, track those costs, and then determine the causes of those costs through a system like activity based costing. They need also to recognize that many of these costs can be controlled and reduced through strategic environmental management. The environmental cost accounting can dramatically improve environmental management and reduce environmental costs.

(3) Performance evaluation – Industrial managers are encouraged to consider the long term environmental and financial impacts of product and process decisions. They need to integrate environmental performance into the performance evaluation system. Many industries still do not recognize the benefits from proactive strategic environmental management.

Industry and business may encounter potential barriers to the adoption of pollution prevention and cleaner production as described in Section 2.4. Nine key elements can help overcome them:

- Leadership – top management support;
- Foresight – looking into the future;
- Culture – there must be a culture that creates commitment and action by everyone;
- Management tools – such as total management system, P2 performance measures, environmental value system and environmental accounting techniques;
- Life cycle management – understanding and improving the life cycle impact of products;
- Production and operations -beginning with waste minimization, material and energy saving;
- Market – improving products to meet the environmental labeling requirements and needs of consumers;
- Procurement – encouraging suppliers to improve their waste and product through P2 technologies; and
- After-sales service and disposal – considering the environmental impacts created by product use and disposal.

Progressive industries with superior management will understand the historical shift in emphasis from reaction and control to prevention. The global prevention movement includes the "green consumer" movement. It would be useful for industrial managers to understand that there are three levels to the global prevention movement, in order of descending importance:

(1) Products: changing their design, composition, and packaging to offer environmental benefits, and creating totally new products to replace old ones.

(2) Materials: changing materials used in industry to reduce use of toxic substances that may only be used in processing and that may reside in products.

(3) Manufacturing: making improvements in all processes, technologies, operations, and procedures to reduce and eliminate the generation of all wastes at their sources. Other sectors must also change, including energy production, agriculture, and transportation.

Presently pollution prevention is a top priority in the industrialized countries. Thus, government agencies encourage industries to develop source reduction and recycling strategies and technologies to prevent pollution, rather than simply relying on traditional strategies to manage and control it. While the traditional "end of pipe" and "command and control" regulatory requirements have significantly decreased overall releases of hazardous substances into the environment, limits on technology may limit the degree to which further reductions are possible. As technological controls become more difficult, they become more expensive. In addition, the diminishing availability of landfills and the increasing cost of incinerators make waste disposal an increasingly financial burden on society. The pressures of international industrial competition also point to the need to spend our limited environmental protection resources as cost-effectively as possible (Shen, 1997).

As discussed in the previous chapters, a successful industrial pollution prevention (P2) program must have a strong and clear policy, established at the highest level; otherwise, success with activities at the operational level will be only very occasional. The management policy level is under the influence of social change. Policies may thus be formulated or reformulated in a direct reaction to a specific external influence. In various countries, discussions have arisen about improvement and in particular the integration of laws covering all environmental media. The modern industrial managers are becoming more and more aware of the necessity to adopt a more integrated, preventive and anticipatory approach in their environmental policies.

12.3
Sustainable development

Sustainable development refers to socio-economical development that meets the needs of the present without degrading the needs of future generations. Development must be sustained not only environmentally, but also economically, socially, and politically. A successful sustainable development program requires: (1) a political system in which citizens participate in decision making; (2) an economic system that can generate surplus and technical knowledge on a self-reliant and sustained basis; (3) a social system that will resolve the tensions arising from disharmonious development; (4) a technological system that will continuously discover to new solutions; (5) an international system that fosters sustainable patterns of trade and finance; and (6) a flexible administrative system with the will and capacity to correct itself (Hinrichsen, 1987; Shen, 1996).

Sustainable development calls for industrial pollution prevention and clean technologies: the use of materials, processes, or practices that reduce or eliminate the creation of pollutants or wastes at the sources. Such practices would reduce the use of toxic and hazardous materials, energy, water or other resources through conservation or more efficient use. Two general methods of source reduction are product changes and process changes. They reduce the volume and toxicity of wastes and of end products during their life-cycle and at disposal (USEPA, 1992).

Fundamental realities relate to sustainable management of natural resources are that: (1) Basic needs must be met (i.e., food, clothing, shelter, and jobs); and (2) Limitation on development are not absolute, but rather are imposed by the present status of technology and social organization, by their impacts upon environmental resources, and by the biosphere's ability to absorb the effect of human activities. But, technology and social organization can be both managed and improved to allow for economic growth (Shen, 1993).

To pursue sustainable development, industry needs to apply five guiding principles:

(1) Anticipation and prevention: Industry must deal with the underlying causes, not just the symptoms of the problem. Only in this way will the industry lessen the environmental pressures generated by these economic activities;

(2) Full cost accounting: Natural assets must be fully valued to ensure proper use and allocation, and to make certain that the beneficiary of the activity pays the full price, including the cost of any environmental damage and use of the resource;

(3) Informed decision-making: Environmental and economic considerations must be integrated in decision making. The short- and long-term consequences of these decisions are important factors;

(4) Living off the interest: Our limited natural resources are part of the nation's capital wealth. Industry must do better with less; and

(5) Quality over nature and the rights of future generations: Industry must adopt its design and operation patterns that reduce energy and material requirements, reduce the amount of waste in its production and consumption activities, and choose materials and services with the minimum impact on the environment during their life-cycle (Hinrichsen, 1987).

In April 1991 at the World Industry Conference on Environmental Management in Rotterdam, 800 international business leaders endorsed the Business Charter for Sustainable Development. Under this document are 16 principles developed by the International Chamber of Commerce (ICC):

(1) Corporate priority. To recognize environmental management as among the highest corporate priorities and as a key determinant to sustainable development; to establish policies, programs, and practices for conducting operations in an environmentally sound manner.

(2) Integrated management. To integrate these policies, programs, and practices fully into each business as an essential element of management in all its functions.

(3) Process of improvement. To continue to improve corporate policies, programs and environmental performance, taking into account technical development, scientific understanding, consumer needs, and community ex-

pectations, with legal regulations as a starting point; and to apply the same environmental criteria intentionally.

(4) Employee education. To educate, train, and motivate employees to conduct their activities in an environmentally responsible manner.

(5) Prior assessment. To assess environmental impacts before starting a new activity or project and before decommissioning a facility or leaving a site.

(6) Products and services. To develop and provide products of services that have no undue environmental impact and are safe in their intended use, that are efficient in their consumption of energy and natural resources, and that can be recycled, reused, or disposed of safely.

(7) Customer advice. To advise and, where relevant, educate customers, distributors, and the public in the safe use, transportation, storage, and disposal of products provided; and to apply similar considerations to the provision of services.

(8) Facilities and operations. To develop, design, and operate facilities and conduct activities taking into consideration the efficient use of energy and materials, the sustainable use of renewable resources, the minimization of adverse environmental impact and waste generation, and the safe and responsible disposal of residual wastes.

(9) Research. To conduct or support research on the environmental impacts of raw materials, products, processes, emissions, and wastes associated with the enterprise and on the means of minimizing such adverse impacts.

(10) Precautionary approach. To modify the manufacture, marketing, or use of products or services or the conduct of activities, consistent with scientific and technical understanding, to prevent serious or irreversible environmental degradation.

(11) Contractors and suppliers. To promote the adoption of these principles by contractors acting on behalf of the enterprise, encouraging and, where appropriate, requiring improvements in their practices to make them consistent with those of the enterprise; and to encourage the wider adoption of those principles by suppliers.

(12) Emergency preparedness. To develop and maintain, where significant hazards exist, emergency preparedness plans in conjunction with the emergency services, relevant authorities, and the local community, recognizing potential transboundary impacts.

(13) Transfer of technology. To contribute to the transfer of environmentally sound technology and management methods throughout the industrial and public sectors.

(14) Contribution to the common effort. To contribute to the development of public policy and to business, governmental, and intergovernmental programs and educational initiatives that will enhance environmental awareness and protection.

(15) Openness to concerns. To foster openness and dialogue with employees and the public, anticipating and responding to their concerns about the potential hazards and impacts of operations, products, wastes, or services, including those of transboundary or global significance.

(16) Compliance and reporting. To measure environmental performance; to conduct regular environmental audits and assessments of compliance with company requirements, legal requirements, and these principles; and periodically to provide appropriate information to the board of directors, shareholders, employees, the authorities, and the public.

The 16 principles are designed to place environmental management high on corporate agenda and encourage policies and practices for carrying out operations in an environmentally sound manner. As part of the United Nations Conference on Environment and Development (UNCED) agenda, the delegates from 60 countries signed the 1992 Rio Declaration on Environment and Development. The delegates also committed to the "Agenda 21" setting the global environmental agenda for the next few decades. For industries and business, the major effect of UNCED is likely the hastening of a number of changes already in motion: the growing involvement of senior management, recent strides toward environmental excellence, development of management tools, attention to the need of corporate stakeholders, and a new focus on environmental issues across key business processes (Greeno, 1994).

Industry and business can provide leadership in the move toward sustainable production and consumption through:

- understanding fundamental customer needs and how these can be delivered through much more eco-efficient services and products;
- creating the preventive technologies that are critical to sustainable development;
- transferring technologies and ideas around the world; and
- using its marketing skills to inform consumers of the urgency and requirements of sustainable production and consumption (DeSimone and Popoff, 1997).

12.4
Plant design consideration

The design of new plants, or additions to existing plants offer excellent opportunities for pollution prevention through the concept and practice of design for the environment. When designing chemical manufacturing processes, it is important to select sequences of chemical reactions that avoid the use of hazardous chemical intermediates. It is necessary to find reaction conditions tolerant of transient excursions in temperature, pressure, or concentration of chemicals and to use safe solvents when extracting reaction products during purification steps. Finally it is important to minimize storage and in-process inventories of hazardous substances. One further consideration in the design of a new process or plant is whether it is going to generate polluting effluents or hazardous wastes. A good design should result in waste minimization and cleaner production in a manufacturing process.

Traditional analyses of process economics might prove that inherently safer and less polluting plants are less efficient in terms of energy of raw materials usage. Indeed, chemical plants have been designed in the past principally to maximize reliability, product quality, and profitability. Such issues as chronic emissions, waste disposal, and process safety have been treated as secondary factors. It has become clear, however, that these considerations are as important as the others and must be addressed during the earliest design stages of the plant. This is in part due to a more realistic calculation of the economics of building and operating a plant. When potential savings from reduced accident frequency, avoidance of generating hazardous waste that must be disposed of, and decreased potential liability are taken into consideration, inherently safer and less polluting plants may prove to cost less overall in expenses to build and operate.

From the time a new process is conceived until a plant begins production, there are many opportunities to consider pollution prevention. These include the periods when: the original laboratory and pilot-scale studies are conducted and the process is conceptually defined; the process is developed; the final technology is defined; the process unit is designed and engineering specifications are prepared; and the unit is constructed and started up. The emphasis at each stage should vary. In the early periods, pollution prevention shall focus on the actual product and the process. Once the process is defined and the facility is being scoped and designed, the focus should shift to the equipment needed. With the completion of the engineering design and the purchase of the equipment, the focus shifts to keeping pollution prevention in mind during equipment installation. Finally, the focus must emphasize on equipment startup and subsequent operations.

Exploratory researchers define a desired product. During the early stages of process development, source reduction opportunities can be most effectively considered. At this time, a key aspect of the waste minimization program is to assess the process waste for each stream. The following checklist should be addressed:

- How is the waste generated?
- Why is the waste generated?
- Why is it hazardous?
- How can its volume or quantity be reduced?
- How much will it cost to reduce its volume and toxicity?
- Is it economically practicable to reduce its volume and toxicity?

Source reduction should be emphasized at this stage of the operation. The process modifications recommended during this stage of technology development include:

- Improved controls;
- Minimizing water or solvent use;
- Waste stream segregation or concentration;
- Optimized reaction conditions;
- Improved catalyst;
- Change of reactants;
- Internal recycle; and
- Avoidance of contact between water and organics.

A key part of creating a vision of future products is paying attention to leading-edge thinkers. The following is a list of eight main ways of achieving eco-design, which broadly correspond to key stages in the product life cycle:

- Development of new product concepts
- Selection of low environmental impact materials
- Reduction in material use
- Optimization of production techniques
- Efficient distribution
- Reduced environmental impact in use
- Optimization of initial lifetime
- Optimization of end-of-life system (DeSimone and Popoff, 1997).

Sustainable production and consumption, where changing life styles and trends of individuals seem to become an increasingly important issue. To improve the lives of people in a nation and all parts of the world is the goal of economic and social activity. It should be done with proper regard for the potential environmental pollution. This requires that we learn how to prevent pollution and consume products differently, not less. To do this we need innovations in technology and ways of interacting. These will lead to products and services that improve the quality of lives in ways that are very different to those we have today. Achieving this vision requires continuous improvement in products, services, ways of working, ways of using and ways of any disposing of any wastes and resides. There will also be improvements in the tools used to assess the sustainability of products and services. Yet it seems likely that the principles against which such assessments are made will remain. Industrial products and services, across their life-cycle, must:

- Be safe for workers, users and environment
- Meet the legal requirements. Compliance is key, but going beyond is often right.
- Use resources wisely and minimize waste.
- Improve the quality of people's lives.

A sustainable community or a nation needs to have actors who understand and perform within the frames of sustainable development and pollution prevention, both as an employee and a consumer of goods and services. Future Industries will therefore need to have employees of all levels who understand these ideas in order to survive and grow. Business needs also examine how it shapes demand, through marketing and advertising, what responsibilities it should exercise in the global market place and what positive role it can play in fostering sustainable consumption in both developed and developing countries (WBCSD, 1998).

12.5
Partnership and international programs

In contrast to earlier years, the relations between government and industry are now less confrontational and more cooperative. Industry should work with government agencies and other concerned people to identify legislative and regulatory barriers to industrial competitiveness and make positive constructive recommendations for change which can improve environmental quality and enhance industrial competitiveness.

Policy decision making is very political. In contrast, problem solving is very technical. Development of costs follows from the technical solution. But timing is critical to have the technical and cost data ready. Industry must provide its input to regulatory decision-makers before decisions are made and public hearings are held. It is important technically and economically for us to encourage the right people to help decision makers to make the right decisions.

The concept and practice of partnership is essential for corporate management. The benefits of partnership can be summarized in this way:

- It can mobilize greater amounts, and a wider variety, of skills and resources than can be achieved by acting alone.
- It can address problems in a more integrated, multidisciplinary, and comprehensive manner.
- It can eliminate unnecessary duplication of cost and effort, which is especially important where there are shortages of financial resources or relevant skills.
- It can help traditional adversaries, or organizations which have had little cause to interact in the past, to broaden their perspectives and to respect each other's needs and capabilities.

This in turn can facilitate the dialogue, creativity and mutual trust toward common goals. The multiple face-to-face interactions which occur between partners can also facilitate the flow of information and promote technology transfer. There are many ways in which companies can achieve pollution prevention through individual actions or partnerships. Often this will be a powerful source of competitive advantage.

President Bush established the Partnership Award category within the President's Environment and Conservation Challenge Awards in 1991 and 1992. The Partnership Award was presented to organizations whose efforts of fostering cooperative approaches to environmental needs at the local, regional, or national level. As one of the technical reviewers, the author would like to present two recipients among others of the 1991 and 1992 Partnership Awards below:

(1) The 1991 Partnership Award was given to the McMonald's Corporation, Oak Brook, Illinois, and Environmental Defense Fund, New York City, New York. The award citation says that in a unique alliance, they developed one of the most comprehensive solid waste reduction plans ever – outlining 42 source reduction, reuse, recycling, and composting initiatives that have the potential for reducing McDonald's waste stream by 80 percent. The join task force also identified means for incorporating a commitment to reduce waste in the day-to-day operations of McDonald's restaurants, distribution centers and suppliers (WH, 1991).

(2) The 1992 Partnership Award was given to the New England Electric System, Westhoro, Massachusetts and the Conservation Law Foundation of New England, Boston, Massachusetts for their energy conservation collaborative effort. The award citation says that the unlikely union of an electric utility and an environmental group has resulted in two major achievement: the development of one of the nation's most successful energy conservation programs and regulatory approval for a utility earnings incentive. The power plant that conservation built significantly reduced the utility's air pollutants and the need for new capacity, while saving consumers and stockholders money (WH, 1992).

In recent years, environmental challenges have been a priority on the agenda for international cooperation. International environmental concerns include such diverse but interrelated issues as climate change, stratospheric ozone depletion, food security, health of the oceans, water supply, population change, conservation of biological diversity and natural resources, and pollution prevention and control. Recognizing that many of these problems cross political boundaries, they contribute to socio-political unrest, and are best addressed in a context of sustainable economic growth and political openness.

Held June 3–14, 1992, in Rio de Janeiro, Brazil, the United Nations Conference on Environment and Development (UNCED), also known as the Earth Summit, produced three principal documents and two framework conventions. One of the documents is the Agenda 21 Action Plan which contains 580-page charting a course toward global sustainable development. Its 40 chapters propose specific approaches to issues such as:

- Finance
- Forests
- Hazardous and solid wastes
- International institutions
- Oceans
- Technology cooperation
- Toxic chemicals
- Statement of forest principles
- Forests for the future
- Climate change convention
- Biodiversity convention
- UNCED follow-up in the United States

A high-level "Commission on Sustainable Development" was established, and the scope and nature of this commission's work are being discussed among the relevant UN agencies as well as among governments. One factor that will be given special consideration in a more sustainable development-oriented United Nations is the involvement of non-governmental groups, such as business and industry, in the decision making, and the implementation of Agenda 21. A special effort will be made to clarify the concept of sustainable development in relation to business (Hanson and Gleckman, 1994). In January 1992 President Bush announced the United States-Asia Environmental Partnership (US-AEP), involving U.S. and Asian businesses, governments, and community groups in a program that includes technical cooperation, environmental fellowships and training, and conservation of regional biodiversity.

The International Chamber of Commerce in Paris, and the World Bank in Washington, the International Office of USEPA in Washington have active environmental programs. The Organization for Economic Cooperation and Development in Brussels, the United Nations Environmental Program in Nairobe, and the United Nations Development Program in New York are good information sources on the activities in various countries and could provide industry with a forum to express itself (Stevensen and Ling, 1991). A wealth of literature on worldwide environmental conditions is available from sources such as the Organization for Economic Cooperation and Development (OECD), the United Nations Environmental Program (UNEP), the World Bank (WB), and private sources including the World Resources Institute (WRI), the World Environmental Center (WEC), the World Wildlife Fund (WWF), and the Worldwide Fund for Nature (WFN).

The UNEP Industry and Environment Center (UNEP/IEC) promotes partnerships to help strengthen institutional capacity and build the consensus needed to achieve socially responsible entrepreneurship. It networks with other United

Nation organizations, such as ILO, UNCTAD, UNDP, UNDIO, the World Tourism Organization, the World Bank and other development banks, providing the information and tools they need to integrate the environmental dimension in their worldwide activities. The goals of UNEP/IEC are to:

- build consensus for preventive environmental protection through cleaner and safer industrial production and consumption;
- help formulate policies and strategies to achieve cleaner and safer production and consumption patterns, and facilitate their implementation;
- define and encourage the incorporation of environmental criteria in industrial production; and
- stimulate the exchange of information on environmentally sound technologies and forms of industrial development (UNEP, 1998).

The UNEP, the International Chamber of Commerce (ICC), and the International Federation of Consulting Engineering (FIDIC) have produced a practical guide to environmental management system (EMS). This resource kit gives trainers and managers the tools necessary to conduct training courses in EMS for companies in their own region. Elements are included that are common to the International Organization of standardization (ISO) environmental management system standards, ISO 14001. The kits includes:

- Why introduce an Environmental Management System?
- What is an Environmental Management System?
- The A to Z of Environmental Management Systems.
- Getting stated.
- Integration with health, safety and quality management systems.
- Example of an initial environmental review protocol?
- Example of an EMS audit protocol.
- Case studies.

Industry should become more involved in environmental issues at the international level. Relatively few industries participate at present, even through other countries' decisions on environmental matters could have substantial impact on their business. It may be advantageous for all countries if the wealthier countries help the developing countries to avoid errors they themselves made with respect to a clean environment.

A failure to address developing countries' environmental problems can directly undercut development in two ways. First, environmental quality is itself an essential part of the improvement in welfare that development aims to provide. Development will therefore necessarily involve an enhancement of environmental quality: If the benefits from rising incomes are offset by the welfare costs of poor health and other non-pecuniary indicators of "quality of life," then it can make little sense to speak of development in such circumstances. Beyond this inherent connection, however, environmental degradation can undermine productivity and thus retard development. If soils are damaged, aquifers depleted, and ecosystems destroyed, the long-run prospects for development will be undermined. For development to be sustained, environmental quality must be maintained. This will be particularly the case for poorer countries with huge populations.

Sound environmental strategies should consider affordable costs, given their economic and environmental benefits, although some may carry a stiff political price. However, despite their evident cost-effectiveness over the long term, many sound strategies have substantial initial investment requirements. Exact costs will depend on standards chosen, the time path for reaching them, and the policy instruments used. In general, most developing countries cannot afford the costs for environmental protection including pollution prevention. Fortunately, some financial assistances are available from the United Nations, major financial institutions such as the World Bank, Industrial Development Bank, Asian Development Fund.

12.6
Future outlook

Many environmentalists and industrialists believe that the 1990s will be not only the decade of pollution prevention but also the historic time when consumer and industrial products drive the most substantial environmental improvements and industrial changes. Those who are willing to believe that the global pollution prevention movement will principally stimulates a global restructuring and modernization of industry, will profit from it.

An environmentally driven industrial revolution is beginning with inevitable winners and losers. Companies that delude themselves about their own environmental success, or retain existing market shares for existing products made with existing technologies, or underestimate environmental demands today will fail tomorrow. Environmental quality means reconfiguring the industrial economy to make it a protector of rather than a threat of planetary survival. Being proactive means using technological innovation for environmental progress that can be measured, communicated, and used effectively as a marketing tool.

At present, there is a major implementation gap between what many companies say and do about their environmental programs and their environmentally responsible products. Industrial managers and CEOs must recognize the new business ethics and opportunities. Clean technologies and clean products will bring competitive advantage. Companies that choose to wait for more certainty will face stiff competition from new startups based on true technological innovations and also from foreign firms.

The environmental crime of the future will be a failure to act preventively and proactively and a failure to measure performance, not failure to comply with regulations. The court of public opinion will issue the harshest judgments which will be translated into actions in the marketplace. Future competitive advantage will increasingly require company and product differentiation on the basis of measured environmental performance.

As corporate management moves into the 21st Century, it will continue to face economic and environmental challenges from public opinion to global competition. To remain a vital, thriving competitor, corporate management must gear up to meet the material and product needs of the next generation of consumers. Corporate executive officers should take the following steps:

(1) Restore public confidence and renew its faith in industry. It won't happen unless industry accepts responsibility for its past actions and voluntarily facilitates changes in its attitude and behavior toward a healthy environment.

(2) Develop a detailed in-house training program to improve the environmental literacy of the entire work force and build a new corporate culture that includes pollution prevention and a strong environmental ethic.

(3) Think seriously about changes in products, raw materials, and production systems that can concretely improve the life-cycle environmental performance of products and the whole company. Develop creative measurement systems to be accountable to the public with credible data. Better data on continuous pollution prevention performance also enhance management. Develop and implement manufacturing processes, new products and services that are congruent with the principle of pollution prevention.

(4) Assess totally new products which can satisfy both market needs and public demands for more effective solutions to environmental problems. Use new criteria and priorities for research and development or acquisitions to commercialize true technological innovations with environmental competitive advantages.

(5) Form partnership with government and become active in public policy discussions to ensure that government policies and programs reflect business needs and opportunities. Replace confrontation and litigation with proactive participation and cooperation while dealing with government officials and environmentalists.

(6) Aim for the global marketplace. There are constant new opportunities to become involved in public and private sector programs, such as ones focusing on pollution prevention and clean technologies for developing countries in the U.S. Agency for International Development. Be sensitive to different kinds of environmental problems and priorities in different countries. Take leadership role in our future environmental protection strategies and sustainable development (Hirschhorn, 1994; Ling, 1997).

In summary, the United Nations Environmental Program (UNEP) has suggested a list of management responsibilities. The UNEP advises the corporate management to:

- Commit to make the preventive strategy as preferred option to environment policy.
- Develop pollution prevention action plans and programs with clear, quantifiable and achievable targets and time frame to minimize waste, maximize resource use and avoid risks to human health, safety and environmental quality.
- Develop and implement manufacturing processes, new products and services that are congruent with the principle of pollution prevention.
- Demonstrate leadership in improving efficiency in resource consumption, reducing the use of toxic materials, reducing wastes and increasing the energy and material intensity of goods and services.
- Integrate preventive strategies into all relevant units of business and industry organizations and their management systems and all relevant operations.
- Conduct pollution prevention training activities and R&D innovative methods to overcome potential pollution prevention barriers in implementation.
- Encourage further adoption of pollution prevention practices and create a demand for environmentally compatible goods and services.
- Share pollution prevention experiences and disseminate information to others, worldwide (UNEP, 1998).

We challenge the industries to serve as the primary drivers in implementing pollution prevention and applying preventive technologies. It is industry that owns, develops, implements, markets, and uses the preventive technologies that protect the environment and help fuel economic growth. We also call upon industry to take the lead, along with non-governmental organizations, in building a more sustainable society through preventive technologies and cleaner production.

References

Brown H and T Larson (1998) "Making Business Integration Work: A Survival Strategy for EHS Managers," in: Environmental Quality Management, Vol. 7, No. 3, p. 1–8, Spring
DeSimone LD and F Popoff (1997) Eco-Efficiency: The Business Link to Sustainable Development. MIT Press, Cambridge, Ma., pp. 120, 132, 134, 167, and 246
Greeno JL (1994) "Corporate Environmental Excellence and Stewardship," in: Environmental Strategies Handbook, ed. by RV Kulluru, McGraw-Hill, Inc., pp. 43–64
Hansen MW and HR Gleckman (1994) "Environmental Management of Multinational and Transnational Corporations: Policies, Practices, and Recommendations," in: Environmental Strategies Handbook, ed. by RV Kulluru, McGraw-Hill, Inc. pp. 749–796
Hinrichsen D (1987) A Reader's Guide of Our Common Future. The International Institute for Environment and Development (IIED), with support of the Norwegian Government
Hirschhron JS (1994) "Business and the Environment," in: Environmental Strategies Handbook, ed. by RV Kulluru, McGraw-Hill, Inc., pp. 31–42

Ling JT (1997) "Next Stop: Designing for Sustainability," in: Pollution Prevention 1997 EPA National Progress Report, EPA 742-R-97-00, pp. 239–241

Royston MG (1979) Pollution Prevention Pays, Chapter 4. Pergamon Press, Oxford, pp. 41–54

Schmidheiny S (1994) "Looking Forward: Our Common Enterprise," in: Environmental Strategies Handbook, ed. by RV Kulluru, McGraw-Hill, Inc., pp. 1–10

Shen TT (1993) "Outlook For A Sustainable Development In China," Proceedings of the Second Mainland-Taiwan Environmental Technology Seminar, Taipei, Taiwan, December 21–26

Shen TT (1996) "Sustainable Development: The Role of Bankers and Investors." The Industry of Free China, Commission of Economical Development, Taipei, Taiwan, February, pp. 55–61

Shen TT (1997) "Industrial Pollution Prevention," Proceedings of the 1997 International Chinese Sustainable Development Conference in Los Angeles, California, July 4–5. ITRI of Taiwan

Stevensen AA, Ling JT (1991) "Environmental Quality and Industrial Competitiveness: Strategies For The Future." The Diplomate, Vol. 27, No. 3, p. 9

UNEP (1998) International Declaration on Cleaner Production, Recent Programme Highlights, April 1998. http:www.unepie.org/cp/cp rph.html

USEPA (1988) Waste Minimization Opportunity Assessment Manual, EPA Engineering Research Laboratory, Cincinnati, Oh.

USEPA (1990) Pollution Prevention News. USEPA Office of Pollution Prevention, Washington, D.C., August, p. 3

USEPA (1992) Facility Pollution Prevention Guide. Office of R&D, Washington, D.C., EPA/600/R-92/088, May

WBCSD (1998) The Sustainable Business Challenge Newsletter, January, http://challenge.bi./sbc/Newsletter.htm

White House (1991) The White House Office of the Press Secretary News Release, October 30

White House (1992) The White House Office of the Press Secretary News Release, December 12

13
The role of government

When the U.S. Congress began to pass environmental legislation in the early 1970s, it worked issue by issue, and often crisis by crisis. This symptom-by-symptom and crisis-by-crisis approach continued through the 1980s and early 1990s. The U.S. now has 16 major national environmental laws on the books, which are overseen by some 74 Congressional committees and subcommittees. There are thousands and thousands of pages of detailed, sometimes confusing environmental regulations. This fragmented approach has often resulted in pollution shifting. For example, pollutants have moved from air to water and water to land. It also has resulted frustration with the complex process of environmental regulation, and too little environmental protection at too high a cost. In recent years, EPA has been working in partnership with industry, government at the state and local level, and stakeholders representing environmental, community, and work force issues to prevent pollution at the source, prior to recycling, treatment, and disposal (USEPA, 1995).

In recent years, environmental policy and strategies have begun to focus upon preventive measures. Many government officials have started to realize the benefits of avoiding the production of industrial pollution. However, government implementation of pollution prevention policy has not progressed as expected. In the United States, for example, the source reduction approach is not becoming institutionalized as it should be with the state pollution prevention programs. In accordance with the Pollution Prevention Act, many state programs now provide technical assistance to industry in the areas of pollution prevention and technology transfer. However, the extent of this aid varies significantly. Regulatory programs emphasize mandatory facility planning of pollution prevention, while non-regulatory programs emphasize pollution prevention education and assistance. As a consequence of their divergent emphases, a majority of the state programs emphasize waste recycling, treatment, and disposal. The source reduction emphasis of the Pollution Prevention Act is inconsistently supported on a nationwide bases (GAO, 1994).

This chapter discusses the role of governments in developing pollution prevention policies and strategies, rules and regulations to achieve the goal of sustainable development. It also covers the issue of reallocating government agency resources and approaches of institutional arrangements, pollution prevention implementation plan as well as community actions.

13.1
Policies and strategies

Many countries have adopted sustainable development as their primary goal of environmental and economic policy. The most important and immediate target of environmental policy is to encourage the development and adoption of technologies compatible with sustainable development. All governments will face formidable environmental challenges in the 21st Century. This is because they will require moving well beyond the "identify and repair" approach of the past, and the "anticipate and prevent" of the 1990s, into an era of environmental management based on long-term strategic planning and closer international cooperation for sustainable development.

In the 1990s, governmental policies moving toward a new conceptual framework for managing the world environment based on "sustainable development". Global change processes, such as the enhanced greenhouse effect, the depletion to the ozone layer, and the threat of climate change have expanded the scope of environmental concerns to global dimensions. These change processes cannot be considered independently. And, there has been a worldwide movement toward market-based, democratic societies, with the political and economic reform process underway today in Asia, Eastern and Central Europe. This array of issues creates the new environmental management changes, but the changes also offer opportunities for environmental professionals seeking for improvement. Today, decisions on national environmental issues often have direct or indirect effects that impinge on environmental or economic interests beyond national boundaries and on international investment and trade. Interestingly, environmental issues and problems must be addressed on a regional basis and global scale that require global solutions (OECD, 1991; UNEP, 1998).

The Organization for Economic Cooperation and Development (OECD) proposed the following principles to guide action by OECD governments:

- There is a fundamental link between economic growth and the environment. Economic and environmental policies cannot be made and implemented in isolation;
- Environmental considerations must be brought to bear systematically on economic policy making;
- Conversely, sound economic analysis of costs and benefits and the distribution, coupled with scientific assessment of relative risk, is the optimal basis for setting priorities among environmental goals and choices;
- Compatibility between environmental and sectoral economic policies should be a central objective of policymakers; and subjected to continuous monitoring and evaluation;
- Economic instruments, used in conjunction with regulation, are important tools for achieving policy integration;
- International consultation and coordination is essential to ensure that national environmental policies, whether regulatory or market-based, do not

give rise to unwarranted or inappropriate constraints to national competitiveness and international trade.

The Industry and Environment Center of United Nations Environment Program (UNEP/IEC) was established to promote cleaner and safer industrial production and consumption patterns through consensus building, assistance in formulating policies and strategies, defining and encouraging the incorporation of environmental criteria in industrial production, and information exchange. It assesses trends, identifies emerging issues and encourages initiatives in the field of environmental management. UNEP/IEC also promotes partnerships to help strengthen institutional capacity and build the consensus needed to achieve socially responsible entrepreneurship. It networks with other UN organizations, such as ILO, UNCTAD, UNDP, UNDIO, the World Tourism Organization, the World Bank and other development banks, providing the information and tools they need to integrate the environmental dimension in their worldwide activities.

For sustainable development, government should play a role in helping "push" pollution prevention as a more appropriate production philosophy. Governmental pollution prevention programs can best counteract the pressure to invest in end-of-pipe pollution solutions by demonstrating the economic and environmental benefits of a source reduction approach, making technical information available, and providing technical assistance to those attempting to reduce their generation of pollutions. Michael E. Porter said that "government's proper role is as a catalyst and challenger, it is to encourage or even push companies to raise their aspirations and move to higher levels of competitive performance. Government policies that succeed are those that create an environment in which companies can gain competitive advantage..." (Porter, 1990), Pollution prevention as a production philosophy also conforms to the ideals of sustainable development.

Sustainable production and consumption, where changing life styles and trends of individuals seem to become an increasingly important issue. A sustainable community or a nation needs to have actors who understand and perform within the frames of sustainable development and pollution prevention. Problems such as climate change highlight the real impact of pollution and unsustainable patterns of consumption. Governments need to find ways to change the daily behavior of many millions of individuals. To do that governments need to work with business, industry and non-governmental organizations and to respect people's expectations for a high quality of life. Governments need to promote sustainable production and consumption, because governments can:

– ensure that all the policy signals point in the same direction (regulations, economic and fiscal measures, investment and planning policies, and social instruments);
– set the example by improving the environmental performance of their own activities such as procurement and consumption of environmental friendly products;

- promote debate to raise public awareness and encourage all stakeholders to set goals for managing the environmental impacts of consumption and production; and
- promote a framework for innovation (WBCSD).

Government action alone will not ensure a sustainable future. All stakeholders in society must play their part. Business and industry have a particular responsibility for making production process cleaner, designing more sustainable products, and adding value through greater emphasis on services (WBCSD).

Budgetary constraints on governments will require skillful use of human, financial and technical resources to ensure that, from the outset, they are targeted at the highest priority risks to human health and ecological stability. Costs and benefits of proposed environmental policies will have to be defined with more precision, over both the short and long terms, including the consequences of inaction. Full integration of environmental and economic policies will have to be pursued vigorously in all major economic sectors.

To pursue environmental goals, governments require public support and resources and require policies that are sensitive and attuned to the broader economic and social aspirations of people of each nation or region. It will also require demonstrating convincingly that a healthy environment and a healthy economy are fully compatible objectives, and that they are essential elements in a strategy for sustainable development (OECD, 1991).

In the United States, the Technology Innovation and Economics (TIE) Committee, a standing committee of EPA's National Advisory Council for Environmental Policy and Technology made an extensive review and analysis of the EPA's programs and new initiatives to improve the environmental quality. The TIE Committee found seven essential issues that changes are needed in manufacturing processes, feedstocks and operating procedures to reduce volumes and toxicity of pollutants prior to treatment and control. Each finding is not necessarily related exclusively to only one issue area, nor does a single finding fully express the conclusions drawn form exploration and review of a particular issue. The Committee recommended major areas for improvement by transforming environmental permitting and compliance policies to promote pollution prevention (USEPA,1993).

(1) Finding: The current system of single-medium permitting has achieved significant environmental gains primarily by stimulating a pollution control response, rather than by encouraging pollution prevention.
   Recommendation: Redesign permit procedures to foster and reward efforts by regulated facilities to expand their use of both multi-media management and pollution prevention approaches for environmental improvement.

(2) Finding: Existing permitting and compliance authorities at all levels of government lack flexibility necessary to encourage technology innovation for environmental purposes.

Recommendation: Accelerate the development of innovative pollution prevention technologies and techniques and encourage their use by implementing the recommendations on fostering technology innovation through permitting and compliance policy.

(3) Finding: Greater flexibility is required under federal, state, and local enforcement policies to allow and encourage facilities to use pollution prevention approaches as part of compliance and other environmental improvement activities.
Recommendation: Work with the states to encourage and develop pollution prevention enforcement initiatives.

(4) Finding: Facility-wide pollution prevention planning can be a valuable tool for encouraging regulated parties to comprehensively evaluate the cost and environmental implications of a range of product design and technical production alternatives, rather than simply focusing on mechanisms for environmental compliance.
Recommendation: Proactively support state initiatives in multi-media pollution prevention facility planning.

(5) Finding: The positive relationship between industrial productivity and environmental protection is not yet well understood or accepted by many leaders in industry and government. Pollution prevention can help achieve both industrial productivity and environmental improvement.
Recommendation: Create a culture change by working with federal, state, and local agencies, non-profit organizations, universities, trade associations, and environmental groups to facilitate the implementation of pollution prevention technologies and techniques by expanding training, educational, and technology diffusion efforts.

(6) Finding: Reward systems for EPA personnel place high value on the development, implementation, and enforcement of single-media, pollution-control regulations without creating strong incentives for activities that encourage and support pollution prevention.
Recommendation: Alter personnel reward systems to encourage EPA staff to champion pollution prevention.

(7) Finding: EPA leadership in rewarding pollution prevention efforts outside the agency provides significant encouragement for state and local regulatory personnel and for staff of regulated parties to explore the use of pollution prevention innovations.
Recommendation: Expand and publicize the system of national recognition and awards honoring outstanding pollution prevention research, training, and technology implementation.

These recommendations address changes are within the scope of USEPA's permitting and compliance system. Many can be accomplished with the agency's current statutory authorities. While there is consensus about many of the changes needed to reduce barriers and increase incentives, the Committee believes that much remains to be learned about the impacts of specific combinations of approaches to these permitting and compliance issues. The integration of pollution prevention into the environmental management system is a complex undertaking that will require a sustained commitment lasting several years. The USEPA's pollution prevention strategy is grounded in two additional principles: (1) a multi-media focus, one that looks at and avoids the potential transfer of risks from one medium to another; and (2) a comprehensive evaluation of the total environmental impacts of products over their entire life-cycle, from the development of raw materials through manufacturing (including energy use) to use and ultimate disposal (USEPA, 1993).

Experience indicates that consensus-seeking groups could work cooperatively to develop a mutually agreeable strategy. Such a strategy would integrate the environment with the economy at all government levels and across the private sectors by "round table" approaches. Environmental conservation and economic development not only can co-exist, but also must coexist, for one is a condition of the other. The Brundtland Commission Report (Hinrichsen, 1987) concluded that "those responsible for managing natural resources and protecting the environment are institutionally separated form those responsible for managing the economy. The real world of interlocked economic and ecological systems will not change; the policies and institutions concerned must".

Government is the policy maker and regulator that can exert influence through policies, education, regulations, and enforcement. Government is the single largest consumer and purchaser of goods and services as well as the largest property-owner. In its varied roles as purchaser of products, facility manager, regulator, and policy maker, the government is uniquely situated to encourage pollution prevention through the example of its own actions. Therefore, government can help create markets for environmentally-preferable products and preventive technologies.

13.2
Rules and regulations

Pollution prevention involves waste minimization, source reduction, design for the environment and other practices that reduce or eliminate the creation of pollutants through; (a) increased efficiency in the use of raw materials, energy, water, or other resources, or (b) protection of natural resources by conservation. Pollution prevention means any practice which: (a) reduces the amount of any hazardous substance, pollutant, used- or discarded-product entering any waste stream or otherwise released into the environment prior to recycling, treatment, or disposal; and (b) reduces the hazards to public health and the environment associated with the release of such substances, pollutants, or unwanted products.

Rules and regulations have to be continually reviewed and modified to keep pace with evolving approaches to environmental protection. These include the growing emphasis on comprehensive strategies, embodying such concepts as integrated pollution prevention, life-cycle management and integrated natural resource management. According to the findings of the Industrial Pollution Prevention Project (IP3), the four most important general motivators for pollution prevention in industry are economics, technical and financial assistance, open communications, and flexibility (especially regulatory flexibility). The economic benefits of pollution prevention have proven to be the most compelling argument for business to undertake prevention projects. The IP3 found that the key for pollution prevention is a stringent regulations or enforcement action. The desire to avoid being subject to regulations provided the most critical impetus for pollution prevention, not only motivating pollution prevention, but also ensuring their success in the marketplace (INFORM, 1994; USEPA, 1997).

It is important to reduce risks from man-made toxic chemicals and hazardous wastes in the environment. People must be ensured that chemicals and wastes are being tested effectively and efficiently and that measurement and monitoring methodologies for chemical risk assessment and reduction are being developed. People are expected that rules and regulations are being established to prevent pollution from toxic chemicals and hazardous wastes. Such work should be accelerated, in cooperation with industries and non-governmental organizations. Governments must stimulate and assist the private sector to introduce cleaner technologies to meet future environmental challenges.

In the United States, the Toxic Release Inventory has influenced state environmental policy in a number of cases, most notably increasing interest in pollution prevention. In response to 1988 TRI data, states developed statewide pollution prevention strategies. Many states are using TRI data in their permit programs to: (1) cross-check facilities for compliance with several different environmental programs; (2) develop state regulatory programs; (3) identify locations and chemicals for more detailed risk assessment; and (4) target technical assistance to TRI facilities (USEPA, 1991). From the Toxic Release Inventory (TRI) reporting under Title III of the Superfund Amendments and Reauthorization Act (SARA), to the Pollution Prevention Act of 1990, to a multitude of state pollution prevention laws, environmentalists have found the way to force industry providing the improved data quality. They ask for more data, especially about chemical use, manufacturing operations, products, and environmental claims. They raise more demands such as take back your waste, stop using chlorine, no more pollution, change your products and your packaging, change your production technology. Their demands are not merely regulatory compliance data, but also data that relate to pollution prevention, clean technologies, toxic use reduction, and environmentally responsible products. With this new environmentalism, any enlightened business strategy which builds on environmental issues must include a heavy emphasis on data if it is to be a successful strategy.

Regulations on international trade need special attention. In an increasingly interdependent world, environmental policies are likely to impact on levels and patterns of trade; and there is growing use of trade approaches for achieving environmental objectives. Other aspects to be considered include: management of government-owned or operated facilities; procurement policies for goods and services; product labeling programs; environmental impact assessment; and government interaction with private sector institutions and the public. The might lead to a set of common regulatory guidelines for good environmental practices by governments.

13.3
Institutional arrangements

Today, too high a percentage of government resources are spent in the administration of single-medium permitting actions, a high proportion of which are routine and provide little opportunity to have a significant environmental impact. For example, it is estimated that nearly 50% of a state's permitting resources are used for the routine reissuance of permits. The cost and time associated with the current permit application, negotiation, and renewal process is excessive, not only for permittees but also for the agencies which must process, draft, negotiate, review, conduct hearings, and finally approve or deny each new permit, permit modification, or permit renewal. Modifications to current permitting and compliance systems, such as increasing the use of permits-by-rule or reducing the time taken administratively to process and review standard permit renewals, could help free up resources so that agencies could not only focus on innovative pollution prevention alternatives across the board, but also shift their scarce resources to deal with those permits and enforcement cases which are the most environmentally pressing and the could benefit most from pollution prevention approaches.

In exploring the issue of reallocating agency resources, USEPA has been working on the following three approaches:

(1) Self-reporting: Increased use of self-reporting mechanism. Toxic release inventory reporting was cited as a model of what can be achieved through increased informational requirements outside of a permitting system.

(2) Facility siting contracts: Facility siting contracts with local governments and environmental public interest organizations. Such contracts could be negotiated in conjunction with initial permitting of a facility. As long as the facility keeps faith with the terms of the contract, subsequent permit modifications or renewals might be automatic or streamlined.

(3) Third party auditing: Third party auditing, paid for by the facility, to supplement state or federal agency inspections. It would be utilized as part of the

enforcement scheme and would not replace traditional enforcement activities. It is considered both the potential benefits of reducing resource demands on government resources, and potential problems associated with having a private entity fulfill an essentially governmental enforcement role.

Despite the encouraging signs of public and even producer support for higher levels of environmental protection, there is powerful resistance to change built into the existing systems. Institutional barriers to implement pollution prevention are discussed previously in Section 2.4. Many of the pollution prevention programs being implemented are at the cutting edge of science and information management, and especially of public program management. Designing and implementing diverse, flexible programs to produce sustained long-term results in a variety of sectors, for example, demands highly skilled environmental managers with the ability to tap much greater quantities of information not only about pollutants but about production processes and products. Yet, most developing countries have limited environmental professional to carry the work loads.

13.4
Pollution prevention implementation

Continued economic progress requires a broader way of thinking that encompasses concerns about economic development, global competitiveness, technological innovation, and a wider variety of environmental risks. Nonetheless, most governments' environmental pollution management strategies worldwide have been and still are emphasizing on waste management, adopting waste treatment and disposal technologies. Pollution prevention is still a relatively new way of practicing especially for those accustomed to pollution control through waste treatment and disposal practices. Regardless of the environmental strategy, progress relies on the effective use of the nation's managerial and technological resources. Fortunately, the technology for pollution prevention is developing. Economic competition, regulatory costs, and public concern, in concert with the systems approach, are driving institutional change toward pollution prevention strategies.

One of the critical problems for implementation is the money crisis. Industry and business can't function to accomplish the solution to the industrial pollution problems with a dollar that is rapidly becoming worthless. The trouble is that the wealth of the most nations has become concentrated in investment funds, the income from which is put back into the funds to make more money. The doubling rate for such funds is now somewhere in the 4- to 6-year range and the time is constantly getting shorter as interest rates go up. Financial crisis may occur in the next few years. We cannot avoid the strong measures and the economic reform necessary to establish a sound economy and a social order which makes peace instead of war.

All governments need a new implementing program on "Technology and Environment" which examines policies governments might use to stimulate and assist the private sector to introduce cleaner technology. Other areas where innovative technology could yield important benefits include: development and application of safe biotechnology to environmental management; prevention of accidents involving hazardous substances; waste minimization; energy efficiency; and the acquisition of improved environmental information by space technologies. All governments need to encourage companies to apply full-cost accounting for the environment which is slowly becoming a reality. Full cost accounting means a managerial cost accounting method to identify and quantify the direct (capital, operating, and regulatory), indirect (training and fines), and intangible (contingent liability, good will) costs of a product, process, or activity. Governments need also to encourage companies engaging the practice of merging and comparing environmental information with asset, resource, income, cost, managerial and financial data. Currently, USEPA, universities, the Global Environmental Management Initiative (Washington), and the Tellus Institute (Boston) among others are developing mechanisms and computer software for environmental cost accounting (Kirschner, 1994).

Environmental values must be integrated into the lifestyles of individuals and families as well as into the conduct of businesses, labor, and governments. People must have the knowledge, practical competence, and moral understanding to cooperate in building a sustainable civilization. The pursuit of environmental literacy will require curricular innovations from kindergarten through college, changes in teacher education programs, expanded graduate programs, and continuing education, both formal and informal. Enhanced science research will improve our knowledge of ecosystems, habitats, and public health and will add to environmental literacy. Well-educated people, consumers and citizens are crucial to successful environmental management in democratic societies. Governments should give high priority to expanding and strengthening environmental education at all levels, in particular to ensure that young people, and the future generations they represent, are sensitive to environmental values and risks.

Since pollution prevention is the best possible solution for environmental protection on both environmental and economic grounds, being potentially the most effective method for reducing risks to human health and the environment and for containing costs, we should call on all government to:

- Orient their existing environmental programs to emphasize pollution prevention.
- Develop and use compatible analytical methods to assess the costs and environmental impacts of the entire life-cycle management of products.
- Support the development and dissemination of better designs for industrial processes to reduce the use of energy and scarce raw materials, and toxic pollutants, and the release of pollutants.

- Lead in the adoption of pollution prevention techniques through government procurement practices the design and operation of government facilities, and the development of a mix of economic and regulatory incentives.
- Allow the maximum opportunity for flexibility and innovation in the design of pollution prevention approaches by industry and all other sectors of the economy.
- Support cooperative international ventures.
- Involve the public, as citizens and as consumers, in pollution prevention through education.
- Promote the use of pollution prevention impact statements.
- Establish through an international forum, an appropriate demonstration of pollution prevention (IUAPPA, 1991).

Governments need to help each other, and urgently. Working together, sharing information and performing cooperative projects is a way of enhancing the capability of individual government. Informed people with an ethical commitment to care for the environment is essential to the future we envision. Success with the technological, economic, and governmental changes is predicated on the understanding and wholehearted support of all people. International cooperation is critical to coping successfully with global-scale environmental problems and risks in the 21st Century. None can be solved by one nation working alone. The issues should be addressed through the United Nations Environment Program and in a variety of other international fora, as well as through their domestic programs, but agreed that these efforts must be strengthened.

13.5
Community actions

It is important that community must take its role to work with government and industry solving the community pollution problems. A basic aim is to make individuals and communities understand the complex nature of the natural and man-made environments resulting from the interaction of their biological, physical, social, economic and cultural aspects, and acquire the knowledge, values, attitudes and practical skills to participate in a responsible and effective way, in anticipating and solving environmental problems, and management of the quality of the environment.

The regulatory agencies have the obligation to designated competent authorities to be responsible for the supervision and administration of community operations for the disposal of toxic and dangerous waste. These authorities, responsible in given areas, must plan, organize, and supervise such operations. They are responsible for issuing permits for the storage, treatment, and/or deposit of toxic and dangerous waste, and for controlling such undertakings and those responsible for the transport of the waste.

If a community is to be effective in protecting itself from damage by pollution, it needs to have a basic community spirit. The community having such spirit can formulate its goals, its policy and its organizational structure and develop a strategy and tactics for determining its own-development and protecting itself from pollution. Each community needs a body such as an environmental committee which should be constantly monitoring the local environment, looking out for threats to it and educating citizens about its importance. As regards tactics, the community can involve itself in many activities, depending on the nature of the issue under consideration. The media can be exploited, and contacts with newspapers, radio and television maintained to keep the community's point of view well covered and specific environmental issues dealt with.

To raise public awareness and to pinpoint specific environmental issues, the local community can form alliances with government departments to fight pollution, particularly departments such as health, education, welfare, agriculture, fisheries, forestry and environment whose particular responsibility is touched by a given pollution issue. Clearly, the community needs technical assistance and it needs to be educated to take a tough technical line in dealing with the technocrats of government and industry. In doing so it will be helping government and industry to formulate better environmental projects.

The community can demand the preparation of an impact assessment of a given project or an environmental survey of damage caused by a particular activity. But above all, the purpose of this type of activity should be for spokesperson for the community to get together with representatives of the government and the promoters of proposed projects from both the private and the public sectors, to plan development projects on a fully participatory basis. Throughout the process of participatory planning the community is working from a position of strength, based on its knowledge of its own environment (Royston, 1979).

The general public must support and obey the established governmental rules and regulations and private initiatives for better environment projects. The general public must also accept the idea of modifying their lifestyle and consuming habits to save materials and energy uses and reduce their generation of wastes at home and workplace which cause immediate community pollution problems. The consumer behavior must gear to purchase recycled products and eco-labeling products whenever possible.

In summary, the government policies, strategies, rules, regulations and implementation plans must be continuously modified to fit the socio-economic, cultural and political situations of a given region or country.

The USEPA and UNEP have provided suggestions for all governments to:

– Commit to make the preventive strategy as preferred option to environment policy.
– Measure and monitor environmental quality for scientific research, engineering design, and management decisions such as applying the data and information for risk analysis in determining priority of environmental management.

- Build the principle of pollution prevention into decision making process to amend or create policies and regulations with the concept and practices of pollution prevention.
- Develop pollution prevention action plans and programs with clear, quantifiable and achievable targets and time frame to minimize waste, maximize resource use and avoid risks to human health, safety and environmental quality.
- Integrate preventive strategies into all relevant units of governmental agencies and management systems and all relevant operations.
- Develop innovative preventive techniques and encourage their use.
- Work with various government agencies and encourage them to develop pollution prevention enforcement initiatives.
- Create a cultural change to facilitate the implementation of pollution prevention technologies.
- Conduct pollution prevention training activities and R&D innovative methods to overcome potential pollution prevention barriers in implementation.
- Provide adequate funding for pollution prevention R&D and education and incentive programs.
- Disseminate various related environmental data and information to all sectors of the society, national and international.
- Modify the existing environmental permit systems and compliance requirements to facilitate pollution prevention implementation.
- Form partnership with industry in formulating simple, reasonable, implementable, and affordable rules and regulations for environmental protection.
- Expand and publicize the system of national awards and recognition.

References

GAO (1994) Pollution Prevention: EPA Should Reexamine the Objectives and Sustainability of State Programs. US General Accounting Office Publications, GAO/PEMD-94-8, Washington, D.C.

INFORM (1994) Stirring Up Innovation: Environmental Improvements in Paints and Adhesives, New York, N.Y.

IUAPPA (1991) Declaration on Pollution Prevention of the International Union of Air Pollution Associations, approved on September 4, 1991, at Seoul, Korea

OECD (1991) Press Release: Communiqué – Environment Committee Meeting in Paris on January 30–31, 1991, SG/PRESS (91)9

Porter ME (1990) "The Competitive Advantage of Nations", Harvard Business Review, March–April

Royston MG (1979) Pollution Prevention Pays, Chapter 11. Pergamon Press, Oxford, pp. 154–166

UNEP (1998) International Declaration on Cleaner Production, Recent Programme Highlights, April 1998. http:www.unepie.org/cp/cprph.html

USEPA (1993) Transforming Environmental Permitting and Compliance Policies to Promote Pollution Prevention: Removing Barriers and Providing Incentives to Foster Technology Innovation, Economic Productivity, and Environmental Protection. Office of the Administer (A-101 F6), EPA 100-R-93-004

USEPA (1995) EPA Pollution Prevention Accomplishments. Office of the Administrator
    (1102)
USEPA (1997) Pollution Prevention 1997: A National Progress Report. Office of Pollution
    Prevention and Toxics, EPA 742-R-97-00, June
WBCSD (1998) The Sustainable Business Challenge Newsletter, January, 1998, http://chal-
    lenge.bi./sbc/Newsletter.htm

14
Pollution prevention education and research

Our formal education is only the beginning of our lifelong learning process. Today, we can gain information and knowledge through the media, our workplace, and community activities. Non-formal education offers opportunities to gain more current information and knowledge through hands-on experiences as well as more modern modes of learning. Systematic approaches are needed to help educational consumers sort through and tie together the information resulting from everyday learning and from experiences.

Information about the local, regional, national or global environment is increasingly available and freely through television, print media, telecommunications, networks, and commercial software products. Educational tools such as CD-ROMs allow us to learn about the environment through text, audio, and video images. Such tools are designed to hold our interest and encourage creativity while convey information. Computer-aided environmental education that takes advantage of new interactive multimedia approaches will grow dramatically in the coming decade. However, educators, government, the scientific community, and the media should ensure that information provided to the public is accurate, useful and clearly presented. The vehicles by which information is furnished are continually changing and require ongoing training, skill acquisition and upgrades in equipment.

Education is a key to inform and involve citizenry having the creative problem-solving skills and cooperative actions. These actions will help develop public awareness and the capacity for pollution prevention. Changes of our concept in values, attitude, and behavior are required to take our personal responsibility for implementing pollution prevention and ensuring a healthy environment. This will mean that efforts must be directed to educate senior policy makers, scientists, engineers, economists, workers, and the general public.

Pollution prevention research provides innovative science, technology and management skills to improve our quality of living. Education and research are closely interrelated. If environmental pollution is to be prevented, we should take a leadership role in research, breaking new ground to prepare society for an age of accelerating change in a world of increasingly diverse and growing populations, an expanding economy and changing global environment.

This chapter consists of four sections: (1) public information and education; (2) the role of colleges and universities; (3) researches at universities; and (4) research at various organizations. The needs of quality information, broader education, and more researches for pollution prevention are discussed with justifications of why and how.

14.1
Public information and education

The need for public information and education of pollution prevention activities seems to be well recognized. It is essential to inform and educate people what will be done and is being done and why. It is even more essential to re-orient the direction of education in order to adapt to pollution prevention through enhanced public awareness, training, and an education system that enables people to implement pollution prevention principles in all aspects of their lives.

We must recognize that there really is not such a thing as a public – a single group of people with common interests and objectives. Rather, the people who determine public opinion comprise a great diversity of individuals, groups, and community forces. Three methods are usually followed in developing a community understanding: mass media, work with community groups, and person-to-person contact.

(1) Mass media approach include newspapers, radio, television, pamphlets, reports, exhibits, newsletters, posters, and films. When large numbers of people must be reached, and speed is essential, mass media are most commonly employed.

(2) Working with non-governmental organizations (NGOs), such as scientific and professional groups, trade and industrial associations, and academic institutes. They provide opportunities for involving citizens in the work of pollution prevention and also create an environment for changing opinions and actions.

(3) The personal contact with organization members permits more than merely reporting information, as is largely the case when mass media are used. Questions can be answered and data explained. These contacts are one of the best ways to develop support.

The United Nations Conference on the Environment and Development (UN-CRD) held in Rio de Janeiro in June 1992, led to specific education recommendations. The topics addressed in many chapters of Agenda 21 outline the platform from which information and education in the future must be launched. Many countries have embraced the themes of Agenda 21 as part of their programs in environmental education, global education, and development education. Lessons are being learned and the pace of progress continues on a global scale. To further these advances, countries should play an active role to ensure that pollution prevention themes cross curricula at all educational levels.

In view of the apparent appreciation of the need for pollution prevention education, it is surprising that so few of the environmental protection agencies and industries have staff members who are by education and experience skilled in pollution prevention information and education. The educators and decision-makers are responsible to carry out educational programs of various subjects on pollution prevention not only in academic institutions and industries, but also to general public by:

- Making use public media through activities such as: science and education films; compilation of a series of books about pollution prevention; publication of educational material in newspapers; development of television programs on pollution prevention activities; exhibitions about pollution prevention activities in science centers; and international workshops on increasing public awareness of pollution prevention.
- Developing college and university appropriate educational materials for use and training the qualified personnel for research on pollution prevention through educational courses as well as on-the-job training and extension activities.
- Increasing continuing education on pollution prevention for various professionals.
- Improving international cooperation and exchanging knowledge and experiences.

We reach our final opinion or conclusion from what we have seen, heard, and read. Public attitude is often passive and unreceptive to information not consistent with exiting concepts; so the attitudes of a large group of people are not quickly or easily changed. The public may be vitally concerned about pollution prevention at the time of a severe pollution control problem, and then have very little interest once these episodes have ended. The role and efforts of those attempting to educate the public are limited to making information available in a variety of ways. The goal of public information and education should be to give the facts to the public so the various groups can reach their own conclusions. It is unreasonable to promote, or seek support for a poor environmental program or a policy not to provide sufficient reasons and facts.

The U. S. EPA's pollution prevention educational programs emphasize four specific themes: wise use of natural resources, prevention of environmental problems, the importance of environmentally sensitive personal behavior, and the need for additional action a the community level to address environmental problems. EPA is working in partnership with State and local governments, industry, educational institutions, textbook publishers, and other entities to a project which would ultimately product pollution prevention education materials for students and teachers. This project will contribute to the establishment of an environmental ethic and work toward improved environmental quality (USEPA, 1996).

In summary, the immediate objectives of pollution prevention information and education are:

- To increase public awareness of pollution prevention;
- To introduce and reinforce educational institutes on pollution prevention;
- To develop higher education for decision-makers, scientists and engineers;
- To reinforce continuing education and training for pollution prevention through formal and non-formal means;
- To formulate educational programs and activities for pollution prevention; and
- To reinforce international cooperation and exchange for pollution prevention education.

In order to expedite finding essential sources of pollution prevention information, a general guide will be highlighted in Chapter 16. The guide will help readers search various P2 information which are provided by Internet, the United Nations, U.S. governments, Non-government organizations, as well as business and industries.

14.2
The role of high education

As pollution prevention becomes the dominant industrial and regulatory strategy for preserving environmental quality, the educational background of decision makers in industry and government will require more pollution prevention education. Educators will be the forefront in pursuing pollution prevention through their educational institutions, professional societies, government infrastructures, and local or national advocacy groups. As individual, educators are responsible for pursuing opportunities for professional training to incorporate the principles of pollution prevention in their courses. In addition, they can enlist the help of non-governmental organizations to ensure that their efforts embody diverse cultural perspectives. They can initiate innovations to bring the business sector and the community at large into the educational experience. They can participate in workshops and seminars to help other professionals find uses for advanced information and communication technologies for promoting pollution prevention. They can make the classroom serve as a model of pollution prevention for the community.

Colleges and universities play an important role not only in educating the next generation environmental leaders but also being as centers for research and development for creating and exchanging new pollution prevention technologies and management skills. Several educational centers go beyond engineering research and development and provide forums for regulators, businesses, and local communities to come together to resolve environmental issues through pollution prevention. Several well-known universities are in the forefront of this effort, including: the University of Michigan, University of California-Los Angels, Carnegie Mellon University, University of Wisconsin, and University of Massachusetts-Lowell.

Colleges and universities are also the principal providers of technical assistance to various sectors of our society. The growth of pollution prevention research at universities and colleges has been helped by partnerships between business/industry and various government organizations. In many instances, the business/industry community has been a significant funding source for university education and research efforts. Business, industry and universities are also working jointly on pollution prevention research and development projects. The goal of the partnership is to incorporate environmental concepts in design criteria and eliminate pollution from the manufacturing process. Partnership approach needs to be expanded among industry/business, government and university.

Many colleges and universities have established environmental engineering or science departments in order to meet the demand for trained environmental professionals. However, course work in environmental engineering or science was often not integrated with other disciplines. Concepts of source reduction and recycling were initially integrated in science and engineering department but have since spread to business schools and even to liberal arts programs. Prior to 1989, there was virtually no graduate or undergraduate business school program or course-work focused on environmental problems. Some business school faculty discussed environmental problems as a topic in "business, government, and society" or "business ethics" courses, but such treatment was inevitably cursory. Change began to occur in the late 1980 s when the National Wildlife Federation and Corporate Conservation Council (NWF/CCC) focused on business education as one of its outreach program projects. The first environmental business management course was offered at Boston University in the fall 1989 semester; the Loyola and Minnesota courses followed in the spring 1990 term. By the end of 1992, the three schools had offered the environmental management courses to several hundred business school students and shared course information, teaching materials, and environmental management case studies with hundreds of faculty from 15 other countries. The NWF/CCC curriculum project has proved to be the right idea at the right time in bringing environmental management into business education (Post, 1994).

In 1990, Tufts University President Jean Mayer convened 22 university presidents from 13 countries in Tallorires, France, to discussed the role of universities in working toward an environmentally sustainable future. The conference resulted in a declaration of actions to make environmental education and research a principal goal of universities around the world (Cortese, 1992). The Tallorires Declaration (excerpt) of University Presidents' Environmental Action Agreement states that: university heads must provide the leadership and support to mobilize internal and external resources so that their institutions respond to this urgent challenge. The 22 university presidents agreed to take the following actions:

1. Use every opportunity to raise public, government, industry, foundation, and university awareness by publicly addressing the urgent need to move toward an environmental sustainable future.
2. Encourage all universities to engage in education, research, policy formation, and information exchange on population, environment, and development to move toward a sustainable future.
3. Establish programs to produce expertise in environmental management, sustainable economic development, population, and related fields to ensure that all university graduates are environmentally literate and responsible citizens.
4. Create programs to develop the capability of university faculty to teach environmental literacy to all undergraduate, graduate, and professional school students.
5. Set an example of environmental responsibility by establishing programs of resource conservation, recycling, and waste reduction at the universities.
6. Encourage the involvement of government (at all levels), foundations, and industry in supporting university research, education, policy formation, and information exchange in environmentally sustainable development. Expand work with non-governmental organizations to assist in finding solutions to environmental problems.
7. Convene school deans and environmental practitioners to develop research, policy, information exchange programs, and curricula for an environmentally sustainable future.
8. Establish partnerships with primary and secondary schools to help develop the capability of their faculty to teach about population, environment, and sustainable development issues.
9. Work with the U.N. Conference on Environment and Development, the U.N. Environment Program, and other national and international organizations to promote a worldwide university effort toward a sustainable future.
10. Establish a steering committee and a secretariat to continue this momentum and inform and support each other in carrying out this declaration (Post, 1994).

Some universities are promoting environmental business education courses with pollution prevention principle for undergraduates, graduate students, and business executives. The Business-Environmental Learning and Leadership Program (BELL), a consortium of 25 business schools working with the Management Institute for Environment and Business (MEB), organized environmental courses on 25 campuses. The courses emphasize on pollution prevention, environmental accounting, design for environment, life-cycle analysis, and quality management. The BELL program is expected to incorporate environmental leadership, technology and economics, and science and policy (EnvironLink, 1996).

Recognition is growing that multicultural approaches to teaching and more inclusive content are needed in all forms of education. Educators in both formal and non-formal programs need special training to teach in settings that are increasingly diverse racially, culturally, and linguistically. Educators of all ethnic groups can benefits from preparation that assists them in performing effectively in diverse settings. The phrases "environmental Justice" and "environmental equity" are relatively new, but the underlying concepts are not. The goal of environmental justice is to ensure that all people, regardless of age, ethnicity, gender, social class, or race are "equally" protected from environmental hazards. Environmental justice expands the notion of environment from natural ecosystems to the landscapes where people live, work, and play (PCSD, 1996).

14.3
Researches at universities

Pollution prevention research activities in developing countries have only recently arrived at universities, and preventive concepts such as waste minimization and source reduction have taken hold, claiming an equal place with conservation, recycling, and environmental studies. Pollution prevention curriculum development has been underway, but is nowhere near as most universities in developed countries. Networking and exchange of curricula among university faculty of developing and developed countries are critically needed. Among many universities in the United States, this Section briefly introduces five relatively strong pollution prevention education and research programs.

14.3.1
University of Michigan

The University of Michigan offers many tools and strategies to incorporate pollution prevention concepts into the curricula of universities and colleges for faculty, students, and professionals. The National Pollution Prevention Center (NPPC) at the University of Michigan publishes Pollution Prevention Educational Resource Compendia in a variety of disciplines, including business law, chemical engineering, chemistry, accounting, industrial engineering/operation management, agriculture, architecture, and strategic management. Each compendium offers a discipline-specific resource list. The NPPC also publishes a Directory of Pollution Prevention in Higher Education: Faculty and Programs in order to help build a national network of pollution prevention educators in the United States who can contact each other to share information, ideas, and curricula (NPPC).

Faculty from the science and engineering departments of colleges and universities have prepared problem sets and new courses devoted exclusively to preventing pollution and have woven prevention concepts into existing courses. Engineering faculty teach students how to incorporate pollution prevention in process design, and also how to spot opportunities for waste reduction in unit operations. Science and engineering faculty have recognized the need for an interdisciplinary approach to environmental studies and also realized that they must be prepared to teach environmental issues both from an interdisciplinary perspective and with specific reference their own fields (Keniry, 1995).

14.3.2
University of California at Los Angeles

The Pollution Prevention Education and Research Center (PPERC) was established in 1991 by faculty members from the schools of engineering, public health and public policy. The Center's mission is to conserve resource and reduce or eliminate the use of toxic substances through an interdisciplinary program of education, research, publication and outreach. The Center has established itself among the leading academic pollution prevention programs, and has developed an impressive track record of accomplishments. PPERC faculty and associates have: collaboratively taught innovative, multi-disciplinary courses which examine pollution prevention opportunities in a wide variety of industry sectors; developed curricula, case studies and problem sets for students and professionals in diverse fields; sponsored public seminars and conferences to share pollution prevention information and stimulate discussion; written books and articles on technology, health and policy issues associated with pollution prevention; and have given presentations and participated in various roundtables working to reduce or eliminate the problems associated with toxics use.

PPERC's current research projects are summarized below:

- Wet cleaning demonstration project. The project searches effective alternative to prevent the problems associated with perchloroethylene (perc) use in dry cleaning and also find new technologies such as "wet cleaning" as a potential technology to reduce or even eliminate the use of dry cleaning chemicals.
- Reducing mercury use in semiconductor manufacturing. The project is to improve the efficiency of mercury use in the electronic materials industry and examine the feasibility of attaining zero discharge of mercury for selected semiconductor fabrication processes.
- Toxicology of bismuth and its inorganic compounds. It is a joint project with a faucet manufacturer to investigate the feasibility of replacing lead with the metal bismuth in brass faucet and other fixtures. The project addresses three issues: the toxicology of bismuth to ensure that an equally toxic material is not used to replace lead; how to manufacture brass containing bismuth, and whether the resulting product meets the needs of consumers; and how to meet the regulatory requirements of the state of California.
- Safe cleaning products for janitorial serve work. The project is to explore potential pollution prevention and risk evaluation decision-making modes to address the hazards of commercial cleaners impacting both the workplace and the environment.

- Community food security and pollution prevention. The project explores how direct marketing, specifically farmers' markets and community-supported agriculture programs, can be used to encourage reduced pesticide use and improve community food security.
- Case study development in industrial ecology. The project provides an analytical framework for assessing the environmental impacts of production and also an examination of materials flows in the industrial economy. It addresses such issues as material flow mapping, materials selection, public policy and industrial project prioritization.
- Size distribution and bioavailability of chromium aerosol in spray painting. The project studies the size distribution of chromium aerosol produced from spray painting and to provide research results for improving the quality of the cancer risk assessment for chromium in spray painting.
- Mortality study to determine excess risk to chemicals and radiation at Rockwell/Recketdyne facilities. The project is a cohort mortality study to determine if there is excess cancer mortality associated with employment and radiation/chemical exposure at the facilities in Southern California. It also tries to determine whether there is increased mortality and whether it can be attributed to specific chemical or radiation exposures.

14.3.3
Carnegie Mellon University

The Carnegie Mellon University developed a university-wide prevention research effort, the Green Design Initiative (GDI). The GDI consists of interdisciplinary teams of more than 30 faculty members are involved in research and education. The objective of the interdisciplinary teams is to prepare new environmental management and pollution prevention tools for product and process design, policy, and environmental management. In developing green technologies and policies, GDI pursues two main goals: (1) minimize and effectively manage the use of resources and (2) minimize toxic releases into the environment. Carnegie Mellon University also developed software tools to help engineers identify target areas for emission reductions, design environmentally conscious products, and clarify economic and environmental tradeoffs associated with design choices.

14.3.4
University of Wisconsin

The University of Wisconsin's Solid and Hazardous Waste education Center developed a list-serve on the Internet called P2TECH, that allows subscribers to exchange pollution prevention technical assistance. Subscribers can post questions to the e-mail address: *p2tech@great-lakes.net* and other subscribers can respond directly to the question sender. Similar list-serves are available for discussing pollution prevention regulations (P2REG); training (P2TRAINER); and pollution prevention for the printing industry (PRINTTECH). List-serves have proven to be an extremely useful method of sharing ideas on vendors, problem-solving approaches, and information sources.

The Internet has become a vital mechanism for exchanging information related to pollution prevention.

14.3.5
University of Massachusetts

The University of Massechusetts-Lowell has developed and maintains the Toxics Use Reduction Institute (TURI). It is a multidisciplinary research, education, and policy center that sponsors and conducts research, coordinates training programs, and provides technical support to promote education in the use of toxic chemicals. One of TURI's most ambitious projects is P2GEMS (http:www.uml.edu/turi). P2GEMS is an Internet search tool for facility planners, engineers, and managers who are looking for technical and process/materials management information. The web-site is full text searchable and includes documents, citations, names of experts, and other resource material designed to assist users in pollution prevention efforts.

14.4
Researches at other organizations

In the United States, the National Pollution Prevention Center (NPPC) research program has focused solely on life-cycle design, life-cycle assessment, and industrial ecology. The program's goal is to guide and enhance environmental decision making through effective metrics, identification and analysis of key stakeholder requirements, and selection of resource conservation and pollution prevention strategies. The NPPC also involved in demonstration projects for the testing and refinement of life cycle design techniques. These demonstration projects in cooperation with EPA and industrial partners have targeted a wide range of products, including automotive products, electronic products, and other systems ranging from milk and juice packaging to wet technologies for garment cleaning.

The Pacific Northwest Pollution Prevention Research Center (NWPPRC) has established a research project database on pollution prevention. The NWPPRC P2 research projects are listed below:

- Agronomic practices for reduced smoke and nitrate
- Analysis of P2 and waste minimization opportunities using total cost assessment
- Analysis of P2 investments using total cost assessment: A case study in the metal finishing industry
- Comparison of surge and conventional irrigation

- Economic analysis: Converting to UV-cured coating system
- Reducing the impacts of small petroleum discharges in marinas
- Transfer efficiency and VOC emissions from spray guns.

The Pollution Prevention Research Center (PPRC) provides detailed research projects database on Internet (http://www.pprc.org/pprc/p2tech/common96/reslist.htm). For example, the following aqueous and semi-aqueous cleaning research projects are listed on the PPRC research database as of April 1998:

- Automated aqueous rotary washer
- CFC alternatives testing
- Clean technology demonstrations for the 33/50 chemicals
- Development of an environmentally safe conversion coating system
- Evaluation of aqueous parts washer waste waters
- Fine-pitch cleaning
- Industrial health risk assessment in pollution prevention
- Ink and cleaner waste reduction evaluation
- Non-halogenated cleaning technology demonstrations
- Onsite solvent recovery with low emission vapor degreasing
- Power washer with wastewater recycling unit
- Replacement solvent cleaner/degreaser study
- Semi-aqueous cleaning
- Surface cleaning research
- Testing an aqueous solution through a filtering system for a part washer
- Ultrasonic tank cleaning
- Ultrasonic cleaning as a replacement for a CFC-based system.

The PPRC also publishes pollution prevention technology review series. The series is intended for manufacturers, researchers and others interested in the details of new cleaning technologies. These reviews are divided into several sections to make it easier for users to locate information of interest. Each review includes an overview of the technology, technical and economic performance, an identification of research that has been done and discussion of gaps in the existing research. The reviews offer hyperlink connections to projects listed in the Research Projects Database. The Research Projects Database provides the latest information on pollution prevention research activities in the United States. The database includes the most comprehensive and up-to-date information available from one source. The database currently includes nearly 400 projects, and will continue to grow as more research is conducted. The majority of the projects included in the database are those conducted by state and federal government agencies, and universities and nonprofit research institutions (NPPC, 1998).

References

Cortese A (1992) "Education for an Environmentally Sustainable Future." Environmental Science and Technology, Vol. 26, No. 617

EnvironLink (1996) A Newsletter for Educators in the Field of Business and the Environment, Spring 1996

Keniry J (1995) EcoDemia: Campus Environmental Stewardship at the Turn of the 21st Century. National Wildlife Federation, Washington, D.C., p. 194

NPPC (1998) The National Pollution Prevention Center for Higher Education. Program brochure via the web site: http://snre.umich.edu/nppc/

Post JE (1994) "Regulation, Markets, and Management Education," in: Environmental Strategies Handbook, ed. by RV Kulluru, McGraw-Hill, Inc., pp. 26–30

USEPA (1993) 1993 Reference Guide to Pollution Prevention Resources, prepared by Labat-Anderson Incorporated for the Office of Pollution Prevention and Toxics, Washington, D.C., EPA/742/B-93-001, February

USEPA (1994) EPA's Design for the Environment Program. Office of Pollution Prevention and Toxics, EPA 744-F-94-003, February

USEPA (1996) Partnerships in Preventing Pollution: A Catalogue of The Agency's partnership Programs, EPA 100-B-96-001, Spring

USEPA (1997) Pollution Prevention 1997: A National Progress Report, EPA 742-R-97-00

15
Pollution prevention in the U.S. Defense Department

Joseph Laznow, CCM, QEP, REA* and *Matthew P. Hanke,* CQE**
* Director, Environmental Programs, ADI Technology Corporation
** Environmental and Engineering Services Manager, ADI Technology Corporation

15.1
Introduction

The Department of Defense (DoD) has long relied on "end-of -pipe" solutions to control and mitigate the effects of using environmentally harmful materials and processes. Pollution prevention, which addresses the reduction of pollution at the source, provides a more efficient and effective means for DoD to protect the environment. Pollution prevention options include conservation of energy, water and other natural resources; elimination or reduced use of hazardous materials; and pollutant reduction and recycling techniques. Implementation of pollution prevention programs has been effective in supporting a primary mainstay of DoD's function, i.e., deterrence, by enhancing operational readiness, quality of life, and weapon system modernization goals.

Integrating pollution prevention at DoD military installations has resulted in reducing health and safety risks to personnel and surrounding environs while protecting facility natural and cultural resources. Pollution prevention activities have resulted in economic benefits by eliminating rather than controlling or cleaning up costly pollution problems and improving the effectiveness of other DoD operations, maintenance, and procurement budgets through more efficient use of materials and resources. Integrating pollution prevention into developing new or upgrading existing weapon systems has resulted in enhancing operational readiness. This has been accomplished by minimizing the environmental challenges associated with every stage of the life-cycle of a weapon system, in implementing process improvements to increase productivity and quality, in protection of human health, in improvement of operational performance, and in the significant reduction of the life-cycle costs.

In testimony before the Senate Armed Services Subcommittee on Readiness, Sherri W. Goodman, Deputy Under Secretary of Defense (Environmental Security) indicated that pollution prevention is the core of DoD's environmental protection efforts. "Only by reducing or eliminating hazardous materials and those processes that generate hazardous byproducts can DoD begin to lower overall compliance and cleanup costs."

15.2
DoD pollution prevention mission and implementation policies

The DoD's environmental strategic goal has been developed in accordance with the requirements of the Pollution Prevention Act of 1990 (PPA), the Federal Facilities Compliance Act (FFCA), the Emergency Planning and Community Right-to-Know Act (EPCRA), and Executive Orders (EO) 12088, 12856, 12873, and 13031. The DoD pollution prevention policy is published and implemented through DoD Instruction 4715.4 and DoD's pollution prevention goals.

In the past decade, DoD has established a new vision and program direction, making pollution prevention the preferred approach to environmental management across all DoD activities and in all phases of acquisition, design, operations, maintenance, support and the ultimate disposal of its weapon systems. This approach ensures that pollution prevention is implemented over the entire weapon system life-cycle. It is DoD's policy to reduce the use of hazardous materials, the generation and/or release of pollutants, and the adverse effects on human health and the environment caused by its operations and activities.

It is DoD's policy to prevent, mitigate, or remediate environmental damage caused by weapon system acquisition programs. DoD has concluded that prudent investments in pollution prevention can reduce life-cycle environmental costs and liability while improving environmental quality and program performance. Such pollution prevention activities include process efficiency improvements, material substitution, preventive maintenance, improved housekeeping and inventory control.

DoD program goal for pollution prevention
- Comply with all legal requirements by promoting pollution prevention as the preferred means of achieving environmental compliance.
- Protect human health and the environment by reducing the use of hazardous materials to as near zero as possible.
- Reduce costs by integrating cost-effective pollution prevention practices into all DoD operations and activities, while ensuring performance of DoD's mission.

DoD pollution prevention objectives
- Effectively promote and instill the pollution prevention ethic through comprehensive education, training and awareness in all mission areas.
- Incorporate pollution prevention into all phases of the acquisition/procurement process.

- Achieve and preserve environmental quality for all activities, operations, and installations through pollution prevention.
- Develop, demonstrate and implement innovative pollution prevention technologies.

Each branch of the Armed Forces has established specific pollution prevention goals and milestones. Pollution prevention applies to all phases of the life-cycle process, and is required to be fully integrated into an activity's or organization's thought processes and work habits. A carefully managed and effective pollution prevention program can produce tangible results including cost and labor-hour savings, and reduces the compliance challenges that activities and organizations face.

The DoD is required to comply with all the traditional end-of-pipe regulations required of the private sector. DoD is also guided by additional pollution prevention requirements in managing its environmental quality programs. The primary legal drivers for the DoD pollution prevention program are EPCRA, PPA, and FFCA. Taken together with the following international treaties and EOs, federal agencies are now required to conduct pollution prevention planning, practice source reduction, increase recycling, implement cost effective waste reduction practices, and make sound life-cycle decisions.

- International treaties, such as the 1978 International Convention for the Prevention of Pollution from Ships or Maritime Pollution Protocol (MARPOL) restrict disposal of shipboard wastes into the sea, and the Montreal Protocol bans most production of ozone-depleting halons, solvents, and refrigerants.
- Executive Order 12088, Federal Compliance with Pollution Control Standards, requires federal agencies to cooperate with the U.S. Environmental Protection Agency (EPA), States, and local agencies in the prevention, control and abatement of environmental pollution.
- Executive Order 12856, Federal Compliance with Right-to-know Laws and Pollution Prevention Requirements, requires federal agencies to comply with the provisions of EPCRA and PPA. DoD facilities and acquisition activities must put policies and practices in place which emphasize pollution prevention as the alternative of first choice in how they achieve compliance with existing regulations and requirements.
- Executive Order 12873, Federal Acquisition, Recycling, and Waste Prevention established requirements for the procurement of products containing recovered materials and environmentally preferable and energy-efficient products and services.
- Executive Order 13031, Federal Alternative Fueled Vehicle Leadership, which along with the Energy Policy Act of 1992, requires the purchase and use of alternatively fueled vehicles (AFV) for federal motor fleets.

DoD is also guided by State, regional, and local requirements, such as local recycling mandates that require reductions in solid waste generation, and state requirements to reduce or report on the use of toxic materials. Clean Air Act (CAA) mandated State Implementation Plans, which establish guidance for the reduction of air emissions in nonattainment areas, are also a consideration.

15.2.1
Pollution prevention objectives and metrics

In 1996 DoD issued a Pollution Prevention Instruction (4715.4) collating all existing DoD pollution prevention policies. This document provided the first comprehensive presentation of DoD's pollution prevention program's objectives, metrics, regulatory and legal requirements, roles and responsibilities.

Key objectives of the Instruction included a requirement for all DoD installations worldwide to have written, publicly available plans detailing how the installation would reduce pollution to air, land and water and requirements for all installations to operate recycling and composting programs. Most significantly, the document required pollution prevention to be incorporated into all phases of a weapon system's life-cycle.

The Instruction codified specific indicators to measure progress in meeting DoD's pollution prevention goals. When developing these metrics, DoD considered regulatory, legal and other requirements; significant environmental impacts; technological options; financial, operational and business requirements; and opinions and comments from stakeholders.

Pollution prevention metrics and objectives
- Toxic Release Inventory: Reduce toxic releases and off-site transfers 50% by Calendar Year 1999 (CY99), from a CY94 baseline.
- Hazardous Waste Disposal: Reduce the quantity of hazardous waste disposed of by 50% by CY99, from a CY92 baseline.
- Solid Waste Disposal: Reduce the quantity of nonhazardous solid waste disposed of by 50% by CY99, from a CY92 baseline.
- Solid Waste Recycling: Increase the recycling of nonhazardous solid waste by 50% by CY99, from a CY92 baseline.
- Alternatively Fueled Vehicles (AFV): Increase acquisition of nontactical (general purpose) AFVs to 75% by Fiscal Year 1999 (FY99).

15.3
DoD's unique pollution prevention attributes and considerations

DoD has developed a strategy to implement projects and actions to meet its pollution prevention goals and work toward its stated pollution prevention vision. In general, the DoD strategy is to thoroughly review available pollution prevention opportunities, identify alternatives, evaluate the life-cycle costs of all alternatives, focus on the opportunities that demonstrate the best short-term and long-term return on investment (ROI), and select opportunities that best balance cost and environmental performance.

15.3.1
Installation/facility operations

Executive Order 12856 and DoD's Pollution Prevention Instruction require all DoD installations worldwide to develop and routinely update their pollution prevention management plans. These plans call for an opportunity assessment, whereby the installation identifies all of its pollutant sources and defines opportunities (including the introduction of new materials, technologies, or management techniques) to reduce or minimize hazardous and nonhazardous wastestreams. The installation is also responsible for developing a prioritization roadmap and implementing pollution prevention projects to reach its goals. This planning effort provides the basis for DoD's pollution prevention budget requests, and also guides its research and development programs.

To assist installations in developing their pollution prevention plans, the DoD components have employed a variety of tools. For example, the Air Force has developed "Model Shop Reports," or prototype opportunity assessments for specific operations, such as Flightline Maintenance or Transportation shops. Each operations-specific model shop report can assist the applicable shop personnel in identifying typical raw materials used, potential wastestreams, processes to be examined, potential alternatives to be evaluated, and cost data. The report also provides guidance on how to implement the pollution prevention options. The Navy has also developed model pollution prevention plans for its industrial facilities, providing criteria, guidance and examples for installations to use in developing their pollution prevention plans.

The Navy's "P2 Afloat Program" has established prototype ships on each U.S. coast to demonstrate model pollution prevention programs. Each prototype establishes a hazardous waste baseline and demonstrates pollution prevention technologies and best management practices to reduce shipboard waste. The aircraft carrier USS CARL VINSON is one ship prototype for the program. The ship's crew plays a major role in helping reduce hazardous and other wastes by implementing new pollution prevention initiatives. The ship now recycles over 75% of its hazardous materials and has the lowest waste disposal cost of any aircraft carrier in the Pacific Fleet.

Toxic Release Inventory (TRI) data have become valuable in pollution prevention planning efforts. At Tinker Air Force Base (AFB), Oklahoma, for example, the installation's top TRI chemicals and the processes responsible for use of these chemicals were identified, pollution prevention opportunity assessments were conducted, and a roadmap to implement the findings of that assessment was created. This resulted in the most effective use of available funds, a positive impact on the mission, and increased worker health and safety.

The overall Tinker AFB TRI reduction roadmap projected an 88% reduction in TRI releases by CY99. Tinker's top TRI chemical, methylene chloride, is expected to be reduced by over 96% with the implementation of 6 new pollution prevention projects between 1994 and 1999. The first project, installation of high-pressure water blast robotic technology for aircraft component stripping (ACS), has eliminated about 15% of the use of the toxic solvent. Quantifiable benefits to the ACS operation thus far include: a financial savings of about $1.3 million per year, a 30% reduction in worker turnover rate, reduced turnaround time per aircraft component, reduced occupational illnesses, reduced personal protective equipment requirements, and reduced safety hazards.

The Navy has established the Fleet Assistance Support and Technology Transfer (FASTT) mission to assist in finding opportunities to reduce the cost associated with environmental compliance and improve maintenance work processes using the best technology and management practices available. The FASTT Team uses on-site surveys to bring together engineers, scientists, maintenance personnel, and process improvement specialists from a wide variety of commands to focus on unique problems at individual installations. This approach involves an assessment of maintenance processes, hazardous materials usage, hazardous waste generation and local environmental requirements. Based on on-site observations and interviews with process operators, pollution prevention opportunities are developed which can protect the environment and reduce workload and operational costs. The collective experience gained during site surveys, combined with continuous recruitment of new members, has lead to an ever-broadening pool of the Team's technical knowledge.

The FASTT Team has performed surveys at numerous naval activities and has expanded its role to include Air Force and Army installations. These visits have resulted in the formal preparation, documentation, and presentation of over 450 recommendations intended to increase the efficiency of a wide variety of maintenance processes that can assist the Navy and DoD in attaining its pollution prevention goals. FASTT efforts have resulted in a cumulative potential ten-year cost avoidance in excess of $90 million since the Team's inception in 1993. The overall ROI is in excess of 50 to 1. The net savings identified by FASTT and used to calculate the ROI includes the cost to procure, install, and provide training for any equipment required and also accounts for labor, maintenance, and other operational costs associated with implementing the recommended initiatives.

15.3.2
Weapon systems acquisition (life-cycle costing)

The DoD's ability to maintain readiness is linked to modernization and minimization of infrastructure costs. The strategic focus of the defense acquisition and technology program is on force modernization, fielding superior operational capability and reducing weapon system life-cycle costs. An important element of DoD's strategy to reach these goals is ensuring that consideration of environmental, health, and safety (EH&S) issues is fully integrated into a weapon system's entire life-cycle. DoD implements this strategy by ensuring that each program's EH&S life-cycle effects and costs are understood and considered throughout the process.

DoD has been successful in reducing costs of environmental compliance and cleanup through sound pollution prevention initiatives. By conducting systematic assessments and cost data analyses early in the planning stages and the engineering and manufacturing phases of the weapon system life-cycle, DoD has been able to decrease the life-cycle costs and develop a more effective environmental management program. DoD studies indicate that the Operation and Support (O&S) phase of the life-cycle process accounts for at least 80% of the hazardous materials generated by weapon systems. The opportunities for applying pollution prevention efforts at this phase are significant. Additional studies indicate that environmental costs can encompass up to 15–30% of a weapon system's life-cycle costs.

The DoD has made significant progress in implementing acquisition reform initiatives that enhance pollution prevention in the weapon system acquisition process. Actions that reduce hazardous materials or that eliminate the need to follow command and control rules and regulations will produce savings in acquisition and O&S costs. These initiatives include the following:

1. Streamline acquisition management. DoD has issued new, streamlined acquisition policies. In addition, DoD has developed the Defense Acquisition Deskbook. The Deskbook is an automated reference tool that provides acquisition information for all DoD components across all functional disciplines. It contains standard practices and practical advice for environmental professionals. The Deskbook provides access to current mandatory directives, guidance, advice, and software tools. For example, one section describes the purpose and typical content of the "programmatic EH&S evaluation" that each weapon system's program must develop. Another section offers recommendations on how to include a hazardous materials management program in a weapon system's contract. A third provides advice on how environmental studies conducted to meet the requirements of the National Environmental Policy Act (NEPA) can enhance pollution prevention efforts.

2. Integrated Product Teams. DoD environmental staff actively participates on Integrated Product Teams (IPT) for many of the major acquisition programs. An IPT is a multidisciplinary team of experts from all DoD communities and includes customers and suppliers. This participation ensures that weapon system program managers provide early and continuous consideration of environmental issues, rather than after-the-fact oversight that had been common in the past.

3. Use cost as an independent variable. The Army has developed an "activity-based costing" methodology as a tool for identifying a weapon system's life-cycle environmental costs. The methodology helps program managers better predict what environmental issues drive significant O&S costs and make better decisions in designing and developing weapon systems.

4. Reform military specifications. DoD initiated the Toxic Reduction Investment and Management (TRIM) pilot program to identify and revise the military specifications that require the greatest use and ultimate release of toxic chemicals. DoD plans to expand the TRIM program to include all DoD installations that submit TRI reports.

5. Single Process Initiative (SPI). SPI is an important tool that will afford the opportunity for contractors to eliminate separate production lines for their military and commercial products. Separate production lines have increased DoD's costs to produce weapon systems and often require the use of greater quantities of hazardous materials per military specifications and standards. The SPI is an expedited process for testing and validating new alternative processes so that they can become acceptable to multiple program managers. The DoD Joint Group on Acquisition Pollution Prevention (JG-APP) is partnering with major defense contractors to reduce the use of toxic chemicals used in weapon system manufacturing processes through the SPI program.

The Environmental Technology Team at the U.S. Army Missile Command Research, Development and Engineering Center, for example, has incorporated pollution prevention into the systems engineering process. This was accomplished by developing new source selection evaluation criteria, participating on Integrated Product Teams, reviewing trade-off studies and recommending alternative materials and processes that will result in reduced pollution. A similar environmental team from the Air Force's Kelly AFB, Texas won the 1996 Secretary of Defense Weapon System Pollution Prevention Team Award for revising aircraft maintenance technical orders which had required the extensive use of hazardous materials and for incorporating pollution prevention into its program management business practices.

15.4
Installation/facility pollution prevention accomplishments

In FY97, DoD invested about $244 million in pollution prevention. About $64 million or 26% of the pollution prevention investment was spent on recurring costs, or those routine activities required to support the mission and maintain compliance at an installation, such as running a recycling program or TRI reporting. The remaining $180 million or 74% was spent on one-time, nonrecurring projects such as the purchase of new pollution prevention technologies.

Hazardous material, hazardous waste, solid waste, clean air and clean water requirements are DoD's biggest cost drivers. These are also the areas of greatest emphasis in the pollution prevention program. FY97 pollution prevention expenditures by media (see Fig. 15-1) indicate that about 40% of the nonrecurring funds were used to reduce the generation of hazardous material, including releases reported under the TRI program. About 13% of the funds went to reduce

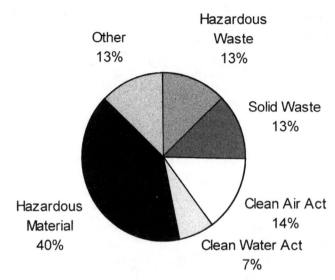

Fig. 15-1 FY 97 nonrecurring pollution prevention investment by media

hazardous waste and about 14% were used to reduce CAA pollutants, including the use of ozone-depleting substances (ODS). Additionally, about 12% of the funding were used to reduce generation of solid waste and establish recycling programs and about 7% went to reduce Clean Water Act pollutants. The remaining 13% went for other miscellaneous efforts.

15.4.1
Hazardous materials and toxic chemicals

The DoD has made significant progress in reducing the use and release of toxic chemicals and hazardous materials to all media. One of DoD's key pollution prevention metric is the reduction of hazardous waste because of the high costs and potential liability associated with its handling and disposal. DoD has made significant progress toward its goal to reduce the amount of hazardous waste it disposes. From CY87 through CY92, DoD met its first goal to reduce hazardous waste disposal by 50%. After CY92, DoD expanded its definition of hazardous waste counted in the metric and established a new 50% reduction goal, using CY92 data as the baseline. Since 1992, hazardous waste disposal within DoD declined from approximately 415 million lbs. to about 283 million lbs., or some 32% (see Fig. 15-2). Information on the annual hazardous waste disposal by individual DoD component is provided in Table 15-1.

Table 15-1 DoD components: hazardous waste disposal

(Millions of lbs.)	CY 92	CY 93	CY 94	CY 95	CY 96
Air Force	49	38	33	31	25
Army	60	83	67	41	44
Navy	207	223	220	181	136
Marine Corps	79	114	76	79	75
Defense Logistics Agency	20	12	15	4	3
Total	415	470	411	336	283

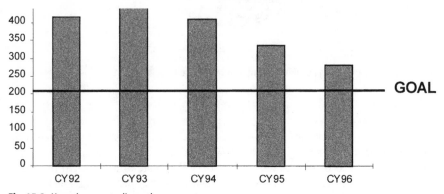

Fig. 15-2 Hazardous waste disposal

The DoD components have developed numerous cost-effective tools and techniques to reduce the use and release of toxic chemicals. They generally fall into two categories: improved business practices and innovative technologies.

15.4.1.1
Improved business practices

One of DoD's most successful pollution prevention tools has been the use of improved management techniques based on centralized control and cradle-to-grave management of hazardous materials. In short, these common-sense practices of managing hazardous materials can be summed up by the phrase, "Buy what you need, use what you buy, and manage what you use."

Each of the DoD components has established significant hazardous material "pharmacy" programs at most of their major installations. The pharmacy concept is a program that establishes a single point of control and accountability for the requisition, receipt, and distribution of hazardous materials. Additional management practices include "just-in-time" delivery, exact quantity product distribution, and distribution to authorized users only. These practices have greatly reduced purchase and handling costs of hazardous materials, reduced the amount of expired shelf materials that require disposal, reduced worker exposure to hazardous materials and potential liability from improper handling, and improved supply support to customers.

For example, under the Navy's Consolidated Hazardous Material Reutilization and Inventory Management Program (CHRIMP), every shore facility and ship inventories the locations, kinds, and amounts of its hazardous materials and establishes a central point for procuring, storing, issuing, reissuing, and eventually disposing of hazardous materials as waste. Cost avoidance throughout the Navy from the reduction of hazardous waste disposal and hazardous material procurement has been as high as $18 million in a single year.

The most significant reductions and savings are found at the large depots, such as the Tobyhanna Army Depot, Pennsylvania. As the largest communications-electronics facility in DoD, Tobyhanna produces large quantities of hazardous waste. A new hazardous materials management program established at Tobyhanna is projected to save the depot over $1.6 million from FY96 to FY02 as a result of a 30% reduction in hazardous material purchases alone. This is in addition to previous accomplishments in which the depot had reduced hazardous waste generation 82% and solid waste generation 74% from FY95 to FY96.

The pharmacy concept has even realized positive results from smaller activities. Four aviation maintenance battalions located at Fort Campbell, Kentucky, saved over $100,000 in the first six months of operating a new hazardous materials control center (HMCC). The initial focus of the HMCC was to reduce inventories that contained dozens of unique and hazardous chemicals, thereby reducing the amount of hazardous materials that need to be managed and disposed of.

The benefits of establishing a hazardous materials "pharmacy" at Hill AFB, Utah, included the following:

- Elimination of 1000 tons of hazardous waste and 170 tons of air emissions
- Reduction of annual hazardous materials acquisition from $14 million to $4 million
- Reduction of on-hand inventory from $2.3 million to $0.2 million
- Reduction of supply processing time from 1 day to 15 min
- Reduction of delivery time from 25 days to 10 days
- Reduced potential for spills and worker exposure to hazardous materials

The Defense Reutilization Marketing Service (DRMS), operated by the Defense Logistic Agency (DLA) provides a major contribution to DoD's pollution prevention program. DRMS recovered and recycled over 16% of all hazardous material dispositions in FY96. The acquisition value of this property was about $48 million and resulted in disposal cost avoidance for DoD of about $3.6 million. A new DRMS initiative, the "return to manufacture" program, which offers surplus hazardous property to the original manufacturer or other manufacturers in lieu of waste disposal, will further reduce the disposal of solid and hazardous wastes.

The DoD is also deploying a standard software system for installations to track hazardous materials from procurement to disposal. The Hazardous Substance Management System (HSMS) tracking system provides environmental managers with a detailed picture of how and where installation personnel use hazardous materials. It supports the rapid identification of processes and materials that are potential candidates for alternative pollution prevention opportunities. It also enhances DoD's ability to comply with regulatory and legal requirements and communicate important health and safety information to its personnel and the local community.

15.4.1.2
Innovative technologies

Another keystone of DoD's pollution prevention program is the utilization of innovative technologies. DoD has recognized significant savings from both the use of relatively simple and low-cost technologies as well as the application of state-of-the-art technologies.

The DoD has widely implemented water jet technologies to reduce the quantity and toxicity of wastestreams from many cleaning and depainting operations, two of its major waste-generating activities. For example, a high-pressure water jet system is being used by the Navy to remove paint, corrosion and marine growth from underwater mines. The previous system used an abrasive media that created a hazardous wastestream and posed potential health and safety problems to workers. This off-the-shelf high-pressure water jet technology will improve the work place environment and reduce the quantity of hazardous waste generated by approximately 48,000 lbs. per year. It has resulting in a pay back of the initial investment in less than a year and a projected saving of over $2.5 million during the 10-year lifetime of the equipment.

Tinker AFB also uses high-pressure water stripping in lieu of toxic solvents to depaint aircraft, eliminating more than 140,000 lbs. of methylene chloride, 100,000 lbs. of solid waste, and 8.3 million gallons of waste water each year. This application reduces potential worker health hazards, reduces processing time by about 21%, and has resulted in savings of over $1 million a year when compared to the previous process.

The DoD has been, and continues to be, on the cutting edge of developing and transferring the use of new pollution prevention technologies. One of DoD's largest wastestreams is used lead acid batteries from both tactical and nontactical vehicles. Using a new technology called "sweeping pulse," DoD can extend the life of its batteries by nearly 80%. The technology is solar powered, and can be permanently mounted on vehicles, providing a safe, energy-efficient and environmentally sound alternative to the replacement and disposal of spent batteries. Army implementation of this technology at Fort Hood, Texas, and for its entire III Armored Corps is projected to reduce its battery procurement and disposal costs by more than $2 million a year.

The DoD is also developing novel approaches to pest control and management without using toxic or ozone-depleting pesticides. An ancient Egyptian tactic, heat, has been particularly successful. Because insects have no way of per-spiring, heat quickly and effectively kills them. At Fort Knox, Kentucky, Fort Bragg, North Carolina and Fort Belvoir, Virginia, the Army heated 14 buildings to 120 degrees to kill roaches. The DLA has also successfully used heat to control termites in infested crates and boxes, in lieu of using fumigants containing ozone-depleting chemicals.

15.4.2
Nonhazardous waste reduction and recycling

Solid waste reduction and recycling programs have helped DoD cut waste, re-duce costs and generate income from the sale of recyclables. Recycling programs operate at every DoD installation. Many also have composting programs in place. The DoD set goals to reduce solid waste disposal 50% and increase recy-cling 50% by CY99. Since 1992, solid waste disposal at DoD declined from about 10.3 billion lbs. to 6.1 billion lbs., or about 41%. Recycling increased to nearly 50%, from about 868 million lbs. to about 1.3 billion lbs. (see Figs. 15-3 and 15-4). Information on annual nonhazardous solid waste disposal and nonhazard-ous solid waste recycling by individual DoD component is provided in Tables15-2 and 15-3 respectively.

The Marine Corps Recruit Depot Parris Island, South Carolina, is an example of a comprehensive recycling program at a DoD installation. The installation's recycling center recovers grease from the mess halls, used tires, used oil, paint,

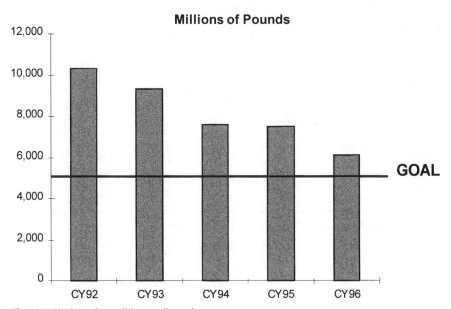

Fig. 15-3 Nonhazardous solid waste disposal

steel cans, aluminum cans, ethylene glycol, scrap metal, brass shell casings, ex-
pended grenade fuses, plastics, glass, office paper, newspaper, cardboard and
pallets. The program has reduced the Depot's wastestream by about 52% in less
than 2 years and reduced the demand on the installation's operating budget by
about $435,000 through reduced disposal and maintenance costs and the gener-
ation of revenue from the sale of recyclables.

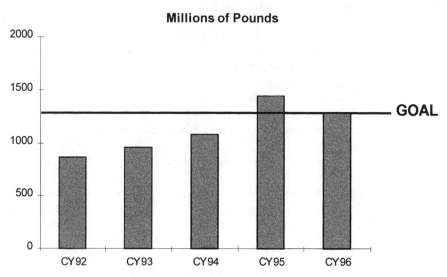

Fig. 15-4 Non-hazardous solid waste recycling

Table 15-2 DoD components: non-hazardous solid waste disposal

(Millions of lbs.)

	CY 92	CY 93	CY 94	CY 95	CY 96
Air Force	1,943	1,376	1,295	1,159	1,016
Army	6,415	6,079	4,415	4,453	3,478
Navy	1,531	1,459	1,522	1,532	1,224
Marine Corps	409	399	367	322	326
Defense Logistics Agency	1	1	29	26	76
Total	10,299	9,314	7,628	7,492	6,120

Table 15-3 DoD components: non-hazardous solid waste recycling

(Millions of lbs.)

	CY 92	CY 93	CY 94	CY 95	CY 96
Air Force	213	186	272	345	370
Army	97	65	134	215	172
Navy	509	637	556	745	573
Marine Corps	477	55	97	121	114
Defense Logistics Agency	2	15	20	20	54
Total	868	958	1,079	1,446	1,283

The Navy is also focusing on finding ways to reduce its use of plastics in order to reduce solid waste from ships. Faced with a significant threat to the readiness of its fleets, the Navy implemented an aggressive affirmative procurement program, Plastic Reduction In Marine Environments (PRIME), to find alternative plastics or plastic-free packaging in order to comply with the MARPOL international protocol. The protocol prohibits disposal of plastics into the sea. The PRIME program has reviewed over 670,000 shipboard consumables and identified nearly 375,000 for alternative products and processes for use in the Fleet.

For example, paper substitutes are being used instead of polystyrene peanuts or bubble wrap; 100% biodegradable rayon scrims are replacing nylon scrims used in cleaning machinery; new biodegradable paper cups are replacing polystyrene; water-soluble packaging are replacing plastic bottles; and wet-strength paper trash bags are replacing plastic bags. Since the PRIME program was initiated, the Navy has reduced shipboard generation of plastic waste by about 70%. Any remaining plastic waste generated on Navy ships is returned to shore for disposal.

The reuse and/or recycling of construction and demolition wastes has also been a growing concern for DoD. At McClellan AFB, California, new rock and rubble crusher equipment was installed to crush demolition rubble and recover the material as useful construction aggregate. About half of the demolition material has been used for on-base construction projects. It took approximately four months to recover the initial $554,000 investment for the crusher through reduced landfill disposal costs. Additional benefits included the elimination of approximately 56 tons of air emissions that would have resulted from hauling the wastes to a distant landfill.

15.4.3
Environmentally preferable products

The DoD has also made significant progress in purchasing recycled goods and other environmentally preferable products. The purchase of these products has assisted DoD in reducing its life-cycle environmental impacts. DoD policy requires that all contract documents eliminate preferences for use of virgin materials and encourage the use of recovered, recycled, reused and other environmentally preferable products. It also requires the procurement of "EPA guideline products," products EPA has designated as having recycled content, such as paper products, re-refined lubricating oil, retread tires, certain building insulation, and concrete and cement that contain fly ash. In addition to changing its policy, DoD has successfully encouraged procurement of environmentally preferable products and services through better awareness, marketing and the use of innovative contracting arrangements.

The DLA's Defense Supply Center in Richmond, Virginia, publishes an Environmental Products Catalog marketing over 800 environmentally friendly products stocked by DLA and purchased across the federal government. This simple tool has resulted in almost doubling the sales of these products in just the past few years.

15.4.4
Alternatively fueled vehicles

The DoD has been instrumental in increasing federal purchasing of AFVs as required by the Energy Policy Act and in ensuring that these clean vehicles support other important objectives such as reducing dependence on foreign oil and complying with the CAA's air pollution emission reduction requirements. The vehicles are generally assigned to installations located in air quality nonattainment areas because they can assist installations in achieving their emission reduction goals. For example, McClellan AFB received emission reduction credits for its fleet of AFVs.

The DoD operates nearly 8,000 AFVs nationwide. DoD is also purchasing 150 hybrid electric vehicles to replace gasoline-fueled vehicles at Navy and Air Force installations in CAA nonattainment areas, including Sacramento, California; Atlanta, Georgia; and Washington, DC. The Navy is managing data collection for performance and life-cycle costs at the demonstration sites and will operate about one-half of the hybrid AFVs. The Navy also operates over 2,700 AFVs and has installed 21 compressed natural gas stations to support its vehicles.

15.5
Weapon system programs pollution prevention accomplishments

The DoD acquisition pollution prevention program has made substantial progress in reducing pollution associated with developing, testing, operating, and disposing of its weapon systems. This effort includes new, developing and mature systems. The LDP 17 (amphibious-transport docking ship) program is a good example of what DoD is accomplishing in the acquisition of new systems. The Navy plans to replace 4 entire amphibious ship classes with 12 new ships. The new ships are designed to minimize many of the environmental problems associated with older ships. For example, the LDP 17 will have a dry bilge (i.e., no deck or other drains leading to the bilge that can cause water contamination).

"Blackwater," or sewage, will be handled by the latest vacuum transfer systems. The refrigeration and fire suppression systems will be ODS free. Plastics will be recycled and the most advanced solid waste processing equipment will be installed.

Further along in the development cycle, the Air Force's F-22 program successfully eliminated a requirement for cadmium plating, a highly toxic and strictly regulated heavy metal, on the aircraft's nose and main landing shock strut assemblies and replaced it with nonhazardous plating materials. This change will greatly reduce the Air Force's use of cadmium plating in the future.

The Navy's F/A-18 E/F and V-22 aircraft programs had each won EPA Stratospheric Ozone Protection Awards for their pioneering work in developing and using non-ODSs for fire suppression. The fire suppression alternatives developed will not compromise aircrew safety or the defense mission and are environmentally acceptable. The F/A-18 E/F will employ a revolutionary inert gas generation system and the V-22 will use a combination of inert gas generation and hydrofluorocarbon-125 (HFC-125). These systems will be the first new aircraft to enter DoD service that do not use halon-1301 for fire suppression in engine nacelles and dry bays.

During production of the B-2 Bombers ODSs were reduced from 180,000 lbs. of emissions to less than 50 lbs. and the cost of the hazardous waste management system was reduced from $3.7 million to less than $1000.

The DoD has also succeeded in preventing pollution from systems that have been in production for many years. The Army's Bradley Fighting Vehicle System continues to be improved both operationally and environmentally. In FY96, a Bradley Pollution Prevention Program was established, requiring every prime and major subsystem contractor to establish pollution prevention programs based on National Aerospace Standard (NAS) 411, Hazardous Materials Management. The program has already achieved significant reductions in the use of zinc chromate, methylene chloride, methyl ethyl ketone, acetone, ethylene glycol, and many other volatile organic compounds (VOC). The Air Force, working with its contractor, reduced the number of Class I ODSs used on the F-15 aircraft from five to one.

The Navy's New Attack Submarine is being designed and constructed so that its operation, deployment, maintenance, overhaul, and ultimate disposal meet or exceed all current environmental requirements. To accomplish that goal, teams representing all aspects of a submarine's life-cycle were formed to make environmentally conscious decisions. Key elements in the approach of the program included identifying applicable compliance requirements and performing environmental analysis on systems, components, materials, and wastestreams. It also included investigating alternatives if items currently used do not meet environmental requirements; assessing alternative materials and processes; and planning to control and manage those hazardous materials that must be used. The New Attack Submarine won the 1997 Secretary of Defense Weapon System Acquisition Team – Pollution Prevention Award. The benefits of the New Attack Submarine's life-cycle EH&S program thus far include the following:

- Reduction of hazardous waste by 90%
- Reduction of high level radioactive waste by 50%
- Utilization of low VOC paints
- Recycling of lead ballast and chromated water
- Elimination of ODSs in air conditioning and refrigeration
- Deployment of fewer oil systems
- Elimination of dredging requirements at operating bases due to the boat's smaller size

The DoD has made substantial progress in improving its ability to share information about innovative pollution prevention technologies and processes. For example, the Air Force and Army Joint Surveillance and Target Attack Radar System (STARS) program has taken advantage of the Navy's efforts to eliminate the use of chlorofluorocarbon 114 (CFC-114). By applying the results of the Navy's work, the program will be able to reduce the costs of eliminating CFC-114 from $22 million to $2 million.

Reducing the volume of hazardous materials used to support fielded weapon systems is an important element of DoD's weapon systems pollution prevention program. Demonstrated successes include Robins AFB, Georgia. The AFB used over 35,000 lbs. of methylene chloride per year to strip F-15s for repainting in 1994. By the end of 1996, methylene chloride use was eliminated and replaced with a closed loop plastic media blasting system in which the spent plastic media is completely recycled. In 1996, over 80 F-15s were stripped and almost no waste was generated. Over 80,000 lbs. of the used plastic media were recycled into bathroom fixtures and highway pavement.

A medium pressure/bicarbonate of soda stripping (BOSS) system being used by the C-141 Aircraft Directorate at Robins AFB reduced the use of methylene chloride by an estimated 1.5 million lbs. for paint stripping of these aircraft. The number of barrels of methylene chloride per aircraft has been reduced from 54 to near zero with a resulting annual savings of about $750,000. Robins AFB won the 1997 Secretary of Defense Pollution Prevention – Industrial Installation Award.

The Army's Tank-Automotive Research, Development, and Engineering Center (TARDEC) conducted a test program to identify halon-1301 alternatives for use in the fire suppression systems of the engine compartments of seven combat vehicles. The program is expected to eliminate the need for 1 million lbs. of halon used in 15,000 combat vehicles. TARDEC was awarded a 1996 International Stratospheric Ozone Protection Award for its work.

The development of tools and models for improving decision making is an important component of DoD's efforts to incorporate pollution prevention into weapon system programs. To facilitate better understanding of life-cycle costs associated with weapon systems, the Army's Sense and Destroy Armor (SA-DARM) program undertook an activity-based costing study that estimates system life-cycle environmental costs. The study used activity-based costing to collect environmental cost data in three major phases of the program: production, testing, and O&S. The study showed that 10% of production, 16% of testing, and 30% of O&S costs for the system are related to environmental requirements. By understanding the costs and knowing the factors from which they arise, program managers are better able to reduce these costs during the design of new systems or modification of existing systems.

15.6
Future DoD pollution prevention initiatives

The DoD has been increasing its emphasis on promoting investments in pollution prevention to meet regulatory and legal requirements. DoD's long-range strategy is to focus on pollution prevention opportunities that demonstrate the best short-term and long-term ROI. At the same time, DoD is faced with decreasing or flat environmental budgets and increasing regulatory and legal requirements. The DoD's challenge is to continue to protect human health and minimize the impact of its activities on the environment given these real world constraints. Over the past several years, DoD has revised its budget and programming policies, metrics, strategic plans, and oversight approach to meet this challenge.

The future direction for DoD's pollution prevention program is expected to focus on smart business decisions and the exploitation of the most promising pollution prevention opportunities, not simply meeting regulatory and legal requirements and waste reduction quotas. Additionally, to make pollution prevention a DoD-wide investment strategy, DoD will be engaging in an intensive effort to achieve total and committed buy-in of its pollution prevention vision from senior management to shop-level personnel.

Pollution prevention adopted metrics have yielded DoD enormous benefits in achieving its pollution prevention goals. DoD recognizes that metrics need to be evolutionary and visionary or otherwise programs lose momentum as their goals are achieved. Furthermore, these indicators must provide better incentives to focus on potentially higher return on pollution prevention investments; consider pollution prevention as the first choice in meeting regulatory and legal requirements; and address DoD's most significant environmental risks first. Future metrics could be integrated with performance criteria for management and functional organizations, so there is a connection between personnel actions and pollution prevention goals, costs and benefits. The DoD is currently considering such improvements for its indicators. For example, a waste diversion goal instead of a waste disposal goal to better encourage source reduction and recycling efforts; a clean air metric based on the DoD's highest toxic air releases; and an economic metric which would demonstrate ROI from source reduction and recycling initiatives.

All DoD installations have pollution prevention plans. These plans are intended to assist an installation commander in identifying funding and technology requirements. Many installations, however, need tools and/or technical assistance to make these plans meet operational requirements on an annual basis. One tool that DoD is developing and promoting is a consistent methodology to identify life-cycle costs and benefits for individual pollution prevention projects. Such a tool would ensure that pollution prevention projects are competitive with end-of-pipe solutions in the budget process and enable installations to implement the most cost-effective and best performing environmental management programs.

Technical assistance is also an important resource element for many installations. DoD intends to expand to other state regulatory agencies, a similar partnership it began in 1995 with the Texas Natural Resource Conservation Commission. This partnership involves state regulators in nonregulatory pollution prevention Site Assistance Visits (SAV) at DoD installations along with staffs from other participating DoD components.

The Acquisition Pollution Prevention Initiative (AP2I) will further streamline the ability of acquisition programs to adopt new materials and processes that can reduce the use of hazardous materials at manufacturing facilities. The AP2I will be based on the joint acquisition pollution prevention process that was developed and validated by the JG-APP. The AP2I will expand DoD's SPI by encouraging contractors and the government to partner in identifying and evaluating alternative materials and processes that can be used at both contractor facilities and DoD installations.

The greatest opportunity to identify and implement pollution prevention projects may lie with the shop-level personnel. However, they generally do not have the appropriate level of knowledge about environmental requirements, costs and benefits, or how to implement new pollution prevention options. The Air Education and Training Command at Randolph AFB, Texas, is piloting a new means for bringing pollution prevention training to the shop level. As discussed previously, the Air Force has developed "Model Shop Reports," or prototype opportunity assessments for specific operations, such as aerospace systems and components. The training initiative is intended to help shop-level personnel understand basic pollution prevention concepts, regulatory and legal requirements, how to identify existing or generate new pollution prevention options, and how to acquire project funding.

References

Army Environmental Center (1998) P2 Initiatives Cut Weapons Systems Costs. Environmental Update, Spring

Department of the Air Force (1993) Pollution Prevention. Air Force Policy Directive, November 30

Department of the Air Force (1996) Environmental, Safety, and Health (ESH) Management and Cost Handbook. United States Air Force Space and Missile Systems Center, Los Angeles, AFB, Ca., September 13

Department of the Air Force (1997) Pollution Prevention Award Industrial Installation, 1996–1997. United States Air Force, Robins Air Force Base, Ga.

Department of Defense (1991) Environmental Considerations During Weapons Systems Acquisition, Report to Congress. Deputy Assistant Secretary of Defense (Environment), Washington, D.C., September

Department of Defense (1993) Executive Order 12856, Federal Compliance with Right-To-Know Laws and Pollution Prevention Requirements. Memorandum, The Under Secretary of Defense, Washington, D.C., December 10

Department of Defense (1994) Comprehensive Pollution Prevention Strategy. Memo, The Secretary of Defense, Washington, D.C., August 11

Department of Defense (1995) Pollution Prevention. DoD Progress Report, Washington, D.C., December 11

Department of Defense (1996) Defense Environmental Quality Program, Annual Report to Congress for Fiscal Year 1995, Washington, D.C.

Department of Defense (1996) Pollution Prevention. DoD Instruction 4715.4, June 18

Department of Defense (1997) Defense Environmental Quality Program, Annual Report to Congress for Fiscal Year 1996, Washington, D.C.

Department of Defense (1997) Staff Reporter: Technology, Pollution Prevention Called "Pillars" of DOD Environmental Program. The Bureau of National Affairs, Inc., Vol. 27, No. 50, Washington, D.C., p. 2662

Department of the Navy (1997) Pollution Prevention – Weapon System Acquisition Team, NSSN Program. Secretary of Defense Environmental Security Award

Department of the Navy (1998) Pollution Prevention. OPNAVINST 5090.1B CH-1, Office of Chief of Naval Operations, Chapter 3

Goodman SW (1996) Pollution Prevention is Good Business. In: Environmental Management, p. 34–36

Morris P (1995) Pollution Prevention Funding: What It's For and How to Get It. Langley Air Force Base, Global Environmental Outreach, November 6

Papatyl T (1994) Institutionalizing Pollution Prevention – A Step in the Right Direction. American Defense Preparedness Association, 3rd Annual Air Force Worldwide Pollution Prevention Conference and Exhibition, San Antonio, Tx., August 29–September 1

Porth A (1998) Personal communication – EQ/Pollution Prevention, Office of the Deputy Undersecretary of Defense (Environmental Security), July

Public Works Digest (1998) Pollution Prevention Requirements, April

Stephans E (1994) Pollution Prevention in System Acquisition. American Defense Preparedness Association, 3rd Annual Air Force Worldwide Pollution Prevention Conference and Exhibition, San Antonio, Tx., August 29–September 1

Sullivan M (1997) Pollution Prevention in Weapon Systems Acquisition, Briefing Slides. Office of the Deputy Under Secretary of Defense (Environmental Security), January 29

16
Sources of pollution prevention information

The environmental information has been increasing rapidly and freely in recent years. This information is being stored in and manipulated by computers. Many of the data measuring status and change in the environmental management are being gathered by automated systems (such as satellites) and fed electronically into various networks. The need for information arises at all levels, from that of senior decision makers at the national and international levels to the grass-roots and individual levels. The ability to use these data to inform decisions about pollution prevention management and sustainable development depends on ability to locate and access, combine, compare and collate them, and on collaboration and communication between the data gathers and data users.

Often there is apathy about environmental problems which have been present for a long time, but in most instances the majority of the public wants to solve the problems of the community in which they live. Efforts to resolve them, however, will be supported only when the public is well informed, sufficiently knowledgeable, and sufficiently concerned to desire a solution. Not only must the public be made aware of pollution prevention, it must know something about what actions the community must take if environmental quality is to be protected. It is obvious that adequate sources of pollution prevention information are critical to successful pollution prevention programs and education must be tailored to the objectives.

This chapter provides a general guide for readers to find sources of pollution prevention information. It is divided into six major categories: (1) the Internet, (2) the United Nations, (3) U.S. federal agencies, (4) state and regional governments, (5) Non-governmental organizations, and (5) business and industries. The purpose of this chapter is to assist readers in locating relevant information sources on pollution prevention and sustainable development in a rapid manner. Some useful World Wide Web addresses, telephones and fax numbers related to those listed organizations are provided to aid the readers in accessing the vast amount of pollution prevention information across the world.

16.1
The Internet

The Internet is a global network of computer networks that provides the user of a connected computer with access to information sources all over the world. It is an excellent tool in the search for pollution prevention information and sustainable development. Accessing the Internet is becoming an easier, more efficient process every year. Today's Internet provides access to a number of basic pollution prevention services via the World Wide Web, electronic mail, information resources, network news, and the ability to transfer files. There are also a number of tools available to guide the user through the library-like system of the Internet. Electronic mail is a communication tool that enables a user to send and receive messages or mail. The World Wide Web is an information system based on "hypertext." It provides a means of obtaining information. The Web has become a major part of the Internet because of its hypertext or hypermedia capabilities. Through the use of software called a "browser," the Web user can jump from web page to web page by selecting coded links with a mouse or by typing in a web address.

The Internet is a powerful tool to search and evaluate information on pollution prevention. It also provides users information on various pollution topics, such as laws and regulations, technologies, environmental data, and case studies. When looking for pollution prevention ideas and information, an e-mail list-serve could be a great place to communicate with fellow professionals. An e-mail list-serve is a system that allows users to send e-mail to a large group of individuals who subscribe. Users simply send e-mail to the list-serve address and a copy of that message is forwarded to each subscriber's mail box. In addition, each subscriber's response to that message will wind up in the other subscribers' mailboxes, inclusive of the original sender's.

The Pollution Prevention Global Environmental Management (P2GEM) maintains a database of P2-related Internet resources. Each entry in the P2 GEM's database consists of a keyword, a URL (also known as an Internet address), and a description which provides us with two ways to search this database: by keyword and by category. Searching by keyword is like using the index of a book. We think of a word, find it in the index, and then look at each of the references listed under the word. Search by category is like using the table of contents of a book. We think of a topic, find chapter that covers that topic, look at the list of sub-topics in the chapter, and go to the sub-topic that contains the references we are looking for.

16.2
The United Nations

The United Nations Environmental Programme (UNEP) is the responsible agency to provide the environmental functions and information for the United Nations.

UNEP information can be located under its website home page (http://www.un-ep.org/). Its information in the website is frequently updated. Four important UNEP global environmental information exchange networks are listed below:

16.2.1
The Mercure Telecommunication System

The Mercure Telecommunication System, donated by European Space Agency, is a suite of 16 earth stations providing global telecommunications via the Intelset system. It is designed to improve the dissemination of environmental information, especially to developing countries and countries with economies in transition. By use of off-the-shelf technology and non-proprietary standards, UNEP is ensuring that public and private organizations, as well as individuals, can reach UNEP's information. Mercure contributes significantly speedier and more cost-effective communication among UNEP offices and other organizations around the world. Its website is (http://www.unep.org/unep/eia/eis/mercure/home.htm).

16.2.2
INFOTERRA

INFOTERRA operates through a system of government-designed national focal points of 176 countries and facilitates access to environmental information by a diverse spectrum of users. INFOTERRA also provides a wide range of environmental information products and services, including environmental bibliographies; directories of sources of information; query-response serves; environmental awareness leaflets; training manuals; source books; and access to Internet services. Its website is (http://www.unep.org/infoterra/overview.htm).

16.2.3
UNEPnet

UNEPnet is the international environmental Internet and uses cost-effective modern data communications which is designed to meet the needs of developing countries for timely and comprehensive environmental information. It employs Internet standards (TCP/IP, HTTP, FTP, SMTP, Gopher, Telnet) to provide its users with flexible access and transport, common services and common communications conventions. UNEPnet serves all UNEP Offices and other organizations worldwide, especially developing countries where the information gap remains great. Its website is (http://www.unep.org/unep/eia/eis/unepnet/home.htm).

16.2.4
Pollutant Release & Transfer Register (PRTR)

PRTR is maintained by the International Register of Potentially Toxic Chemicals. International bodies, environmental groups, industrial firms and associations, and other non-governmental organizations are involved in developing the register system. More PRTRs are underway, stimulated by the recommendations of the 1992 UN conference on Environment and Development in Rio de Janeiro. The Conference affirmed the right of communities and workers to know about toxic chemicals and the importance of chemical inventories to meet that right-to-know. PRTR's website (http://irptc.unep.ch/prtr/) offers information about PRTR from many sources. It also provides links to other websites containing PRTR data.

16.3
U.S. federal agencies

The U.S. federal government is uniquely situated to promote and implement pollution prevention through its various roles. Major environmental information and data can be obtained directly from the U.S. Environmental Protection Agency (EPA). However, there are other federal agencies which provide environmental information and data, such as the White House, Department of Agriculture, U.S. Agency for International Development, Department of Commerce, Department of Energy, Department of Defense, General Services Administration, Department of the Interior, National Aeronautics and Space Administration, U.S. Post Service, and Department of Transportation.

Spurred by the Pollution Prevention Act, the National Energy Policy Act, and a series of Executive Orders dealing with recycling, acquisition, procurement, energy efficiency, reporting of releases to the toxic release inventory, and other pollution prevention issues, federal agencies have begun the laborious process of rethinking all the various ways in which their actions impinge upon the environment. Using International Standard Organization's Environmental Management Standard (ISO 14000) or EPA's Code of Environmental Management Principles, federal facilities must adopt facility-wide environmental management systems that will make pollution prevention a day-to-day reality.

16.3.1
White House

Pollution prevention information and activities in the White House can be obtained through the President's Council on Sustainable Development (PCSD) in Washington, DC or from its website (http://whitehouse.gov/WHEOP/pcsd/).

16.3.2
U.S. Environmental Protection Agency (EPA)

EPA's website (http://www.epa.gov/) contains a vast array of information. The home page provides a list of categories that link users to more specific menus. Through the home page users can: get in-depth information about EPA's projects and programs; find out about laws and regulations; locate EPA offices, laboratories, and regions; browse through EPA publications; get the latest news and upcoming events; discover new databases and software tools; or see what grants and fellowships are available. In addition, the home page provides links to a range of other resources both inside and outside the agency. EPA also provides pollution prevention programs and services across the United States through its various organizations and programs. Some major program information services are listed below:

Office of Pollution Prevention & Toxics (OPPT)
The mission of OPPT is to promote pollution prevention; risk reduction; public understanding or risks of chemicals; and safe chemicals, processes, and technologies in design, development, and application. Activities include (1) developing voluntary partnerships; (2) working cooperatively with customers and interested parties to further environmental protection; (3) protecting children and other vulnerable populations from environmental risks; (4) targeting solutions to specific situations instead of trying to come up with a "one-size-fits-all" answer; and (5) reinventing government through regulatory flexibility and innovation (USEPA, 1998).

OPPT's home page (http://www.epw.gov/opptintr) has seven broad categories that users can click on to link to sites that provide more in-depth information on topical areas in each category. Its programs and projects, publications, databases and software can all be accessed through the home page. Users can also link to other information resources for specific concerns such as dockets, clearinghouses, libraries, and hotlines.

OPPT's Pollution Prevention home page (http://www.epa.gov/oppintr/p2home/index) links users to information about pollution prevention programs and activities both inside and outside of EPA. Specifically, users can choose from the following categories: EPA's P2 programs and projects; publications, such as OPPT's PPN newsletter; the latest announcements on conferences, training and Federal Register notices; grant programs for P2 activities at the state, local, and tribal level; and other informational resources and links.

The OPPT's Pollution Prevention Homepage website contains information on EPA pollution prevention grants and initiatives. It also provides links to numerous EPA pollution prevention programs (e.g., Environ$ense and Design for the Environment). Several pollution prevention publications are also available through this website (http://www.epa.gov/opptintr/opptp2.htm).

Pollution Prevention Information Clearinghouse (PPIC)
PPIC is primarily a distribution center for EPA documents and fact sheets dealing with source reduction and pollution prevention information. PPIC also provides a telephone hotline for document orders and to refer callers to other information resources. PPIC maintains a collection of documents relating to pollution prevention, waste minimization, and alternative technologies. Documents may be ordered by phone, fax, e-mail, or from the P2 Web site. Tel: 202-260-1023; Fax 202-

260-0178; E-mail: ppic@epamail.epa.gov; and website (http://www.epa.gov/oppin-tr/p2home). The collection of pollution prevention documents is available in the EPA Headquarters Library and through EPA's Online Library System. The OPPT Library at EPA Headquarters in Washington, DC maintains a collection of books, journals, newspapers, and government documents in support of the Toxic Substance Control Act and Emergency Planning and Community Right-to-Know Act.

The Toxics Release Inventory (TRI)

TRI is a database of toxic chemicals maintained by EPA under mandate of Section 313 of the Emergency Planning and Community Right-to-Know Act. Manufacturing facilities are required to report on releases of toxic chemicals to air, water, and land and off-site transfers. TRI provides information to the public on releases and other waste management information for more than 600 chemicals and chemical categories from certain industry sectors. The information includes the amounts of each listed chemical released to the environment at the facility; amounts of each chemical shipped off-site for recycling, energy recovery, treatment, or disposal, burned for energy recovery, or treated at the facility; and maximum amounts of the chemical present on site at the facility during the year. With this information, communities know what toxic chemicals are present in their neighborhoods, and facility managers can identify opportunities for source reduction and compare their progress to other facilities. The TRI program is available on CD-ROM, microfiche diskette, reports, directors, and on the Internet via PPIC's website. Tel: 202-260-1512; User support: 202-260-1531.

Integrated Risk Information System (IRIS)

The IRIS operates a Web site (http://www.epa.gov/ngispgm3/tris) which provides an electronic database containing information on human health effects that may result from exposure to various chemicals in the environment. It is intended for those without extensive training in toxicology, but with some knowledge of health sciences. IRISD is a tool that provides hazard identification and dose-response assessment information. Combined with specific exposure information, the data in IRIS can be used for characterization of the public health risks of a chemical in a particular situation that can lead to a risk management decision designed to protect public health. To aid users in accessing and understanding the data in the IRIS chemical files, the system provides extensive supporting documentation.

TSCA Assistance Information Service (TSCA/AIS)

TSCA/AIS provides information and technical assistance about programs implemented under TSCA, ASHAA, and AHERA. Its hotline typically handles questions involving the handling and disposal of PCBs, asbestos in schools and public buildings, registration of new chemicals, import certification, and reporting requirements under TSCA. Documents available through the hotline (202-554-1404) include Federal Register notices, asbestos guidebooks, Chemical Hazard Information Profiles, and the Chemicals in Progress Bulletin. Fax: 202-554-5603.

Environfacts
Environfacts is a national information system with a website (http://www.epa.gov/enviro) that allow users to retrieve environmental information from seven major EPA databases on Superfund sites, drinking water, toxic, and air releases, hazardous waste, water discharge permits, and grants information. In addition, there are three integrated databases: the Facility Index System, the Master Chemical Integrator, and Locational Reference Tables. The databases are accessible via Envirofacts. The Environfacts system also provides query forms that retrieve information from the various databases and then generates facility-based reports using the information. Results of queries can also be mapped via Environfacts' "Map On Demand" feature.

Partners for the Environment
EPA's partnership programs build cooperative partnerships with a variety of groups, including small and large businesses, citizen groups, state and local governments, universities and trade association. Partnership means that EPA is working cooperatively with the private sector to provide stakeholders with effective tools to address environmental issues. Thousands of organizations are working cooperatively with EPA to set and reach environmental goals such as conserving water and energy, and reducing greenhouse gases, toxic emissions, solid wastes, indoor air pollution, and pesticide risk. EPA views these partnership efforts as key to the future success of environmental protection (USEPA, 1996).

Currently, EPA operates 28 voluntary pollution prevention programs which are reported to have achieved measurable environmental results often more quickly and with lower costs than would be the case with regulatory approaches. These programs are briefly described in Appendix D: A list of EPA's Voluntary Initiatives under the P2 Partnership Program.

Enviroene
Enviroene program, supported by the USEPA, is a very comprehensive site that attempts to provide a single repository for pollution prevention, compliance assurance, and enforcement information and databases. Its website (http://es.imel.gov/index.html) also provides P2 case studies, technologies, and compliance and enforcement policies and guidelines.

Pollution Prevention Incentives for States (PPIS)
PPIS builds and supports state pollution prevention capabilities. It funds the institutionalization of multimedia pollution prevention as an environmental management priority. PPIS grants fund other pollution prevention activities such as providing direct technical assistance to businesses, collecting and analyzing data, conducting outreach and funding demonstration projects for testing and evaluating innovative pollution prevention approaches and methodologies. States and federally recognized tribes are eligible for awards. Tel: 202-260-2237; Fax: 202-260-0178.

National Pollution Prevention Center for Higher Education (NPPC)
The EPA created the NPPC in 1991 to collect, develop, and disseminate educational materials on pollution prevention; the University of Michigan was selected as the site. The NPPC represents a collaborative effort between business and industry, government, non-profit organizations, and academia. Materials developed by the NPPC include resource lists, annotated bibliographies, problem sets, case studies, teaching notes, syllabi, and videos in the following disciplines: accounting, architecture, business law, chemical engineering, chemistry, environmental studies, industrial engineering, materials and logistic management, operations research and industrial ecology, and sustainable agriculture. NPPC is emphasizing the Internet in the outreach activities and is planning to focus on sustainability in its research over the coming years. The mission is to promote sustainable development, create educational materials, provide tools and strategies for addressing relevant environmental problems, and establish a national network of pollution prevention educators. The mission is carried out through its program activities. E-mail: nppc@umich.edu; Fax: 313-936-2195; Web site: (http://www.umich.edu/(nppcpub/info.html).

Regional Pollution Prevention Coordinators
The EPA's Office of Pollution Prevention and Toxics (OPPT) has Regional Pollution Prevention Coordinators to provide information and other services to public and private organizations and individuals in each of the EPA's 10 Regional Offices. The locations, telephones and faxes of the OPPT's pollution prevention coordinators are as follows:

Region 1 (CT, MA, ME, NH, RI, VT) in Boston, MA
   Tel: 617-565-4523; Fax: 617-565-3346

Region 2 (NJ. NY, Puerto Rico, Virgin Islands) in New York, NY
   Tel: 212-637-3584; Fax: 212-637-0545

Region 3 (DC, DE, MD, PA, VA, WV) in Philadelphia, PA
   Tel: 215-597-0765; Fax: 215-597-0765

Region 4 (AL, GA, FL, KY, MS, NC, SC, TN) in Atlanta, GA
   Tel: 404-347-3555; Fax: 404-347-1943

Region 5 (IL, IN MI, MN, OH, WI) in Chicago, IL
   Tel: 312-886-0180; Fax: 312-886-0957

Region 6 (AR, LA, NM, OK, TX) in Dallas, TX
   Tel: 214-665-6444; Fax: 214-666-7466

Region 7 (IA, KS, MO, NE) in Kansas City, KS
   Tel: 913-551-7065; Fax: 913-532-5985

Region 8 (CO, MT, ND, SD, UT, WY) in Denver, CO
   Tel: 303-293-1471; Fax: 303-391-6216

Region 9 (AZ, CA, HI, NV) IN San Francisco, CA
   Tel: 451-744-2190; Fax: 415-744-1796

Region 10 (AK, ID, OR, WA) in Seattle, WA
   Tel: 206-223-1151; Fax: 206-223-1165 (USEPA, 1997b)

16.3.3
U.S. Department of Energy (DOE)

The DOE has institutionalized pollution prevention in several important ways:

- Establishing a top-level Pollution Prevention Executive Board, chaired by the Under Secretary, to set priorities and assist in achievement of goals;
- Creating an Office of Pollution Prevention within the Office of Environmental Management;
- Applying pollution prevention program managers in other DOE Secretarial organization that generate wastes, and installing pollution prevention coordinators at field sites;
- Decentralizing program implementation, thereby allowing each site to develop its own goals and to fund activities to achieve these goals, in a manner consistent with the best practices at that site; and
- Elevating pollution prevention to the status of a "national program" so that it will no longer have to compete with mission activities in site budgets (USEPA, 1997b).

The DOE's Office of Transportation Technologies and Office of Utility Technologies contain pollution prevention information describing current efforts to expand the use alternative fuels. The DOE's pollution prevention programs include:

- Alternative fuel vehicle fleet buyer's guide
- Alternative fuels data center
- Bioenergy information network
- Biofuels information center
- Biomass resource information clearinghouse
- BioPower
- Clean cities
- Energy information administration
- Energy science and technology database
- Green power network
- Hydrogen infoNet
- Photographic information exchange
- Regional biomass energy program
- Renewable electric plant information system
- DOE Chicago regional support office
- DOE Denver regional support office
- DOE golden field office

The DOE Pollution Prevention Information Clearinghouse was developed under a joint effort with the EPA to enhance the exchange of pollution prevention information among federal, state, and local government agencies, as well as with industries, academic institutions and the general public.

16.3.4
U.S. Department of Agriculture (USDA)

The USDA contributes to pollution prevention through a number of programs. Many of these programs have emphasized reducing pollution from excessive agricultural chemicals or soil nutrient management and soil conservation issues.

16.3.5
U.S. Agency for International Development (USAID)

The USAID's pollution prevention program is a global initiative focused on creating and supporting locally sustainable pollution prevention projects to address industrial and urban waste problems in developing countries. The program provides technical assistance to help participating countries understand how pollution prevention can be used to address environmental problems. This assistance consists of four general categories: diagnostic assessments and other technical assistance; training; information dissemination; and assistance in developing sustainable government and non-governmental pollution prevention programs. The HQ Clearinghouse contains approximately 1,000 items compiled into database. Through its linkage with INFOTERRA, USAID has access to several hundred data bases, EPA documents, and other information sources.

16.3.6
U.S. Department of Commerce (DOC)

The DOC's Manufacturing Extension Partnership (MEP) is a partnership of federal and state organizations working together to address the needs of small and medium-sized manufacturers. The MEP has a network of regional manufacturing extension centers in all 50 states and Puerto Rico. The centers provide direct service to smaller manufacturers, helping them address their most critical needs in areas such as pollution prevention, production techniques, technology applications, and business practices.

16.3.7
U.S. General Service Administration (GSA)

The GSA's pollution prevention strategy focuses on purchasing alternative products that do not contain or have reduced amounts of hazardous chemicals and educating other federal agencies on purchasing decisions. Although GSA does not manufacture or process toxic chemicals, it does use and transfer toxic chemicals and hazardous wastes off-site. Therefore, GSA has established goals to reduce toxic chemical purchase and use as well as to handle hazardous waste transport and disposal.

16.3.8
U.S. Department of the Interior (DOI)

The DOI pollution prevention strategy commits to pursuing a hierarchical approach to pollution prevention, beginning with source reduction. It identifies pollution prevention as the primary approach to managing waste activities on all Interior-managed lands and facilities. DOI has issued general guidance on pollution prevention and Right-to-Know. The National Park Service has developed its own pollution prevention strategy for achieving the pollution prevention goal.

16.3.9
U.S. Post Service (USPS)

The USPS's pollution prevention strategy includes using environmental considerations among the criteria by which projects, products, processes, and purchases are evaluated. All managers are required to participate in waste reduction initiatives, including source reduction, reuse, and recycling activities.

16.3.10
U.S Department of Transportation (DOT)

The DOT's pollution prevention strategy provides detailed tables summarizing the applicability, major requirements, key deadlines, and responsible DOT offices. In addition, the strategy directs each covered facility to develop a facility-specific pollution prevention plan to include: facility-specific goals for toxic chemical release reductions; an inventory of products used and waste streams containing extremely hazardous substances and listed toxic chemicals; evaluation and selection of pollution prevention alternatives; procedures and a schedule for implementation, communication and training needs; consideration for involving the community; and procedures for measuring success.

DOT is also utilizing innovative pollution prevention technologies including materials substitution, process reengineering, and alternative waste disposal options. Examples include participation in the alternative fuel vehicle program, development of a model consolidated hazardous materials management program, development of improved products for highway structure and materials, and use of effective chemical (USEPA, 1997b).

16.3.11
National Technical Information Services (NTIS)

The NTIS, located in Springfield, Va., is the central source for the public sale of U.S. Government-sponsored research, development, and engineering reports and federally generated machine-processible data files. NTIS's information services include various reports, such as air pollution, acid rain, water pollution, marine pollution, land use planning, solar energy, solid wastes, and radiation monitoring. Tel: 703-487-4650.

16.3.12
Standard Industrial Classification (SIC) Codes

The SIC Codes' website (http://www.osha.gov/oshatats/sicser.html) is one of several sources of Standard Industrial Classifications. The website is maintained by the U.S. Department of Labor Occupational Safety and Health Administration.

16.3.13
The National Round Table on the Environment & the Economy (NRTEE)

The NRTEE is an independent agency of the federal government that seeks to provide objective views and information regarding the state of the debate on the relationship between the environment and the economy.

16.3.14
U.S. Department of Defense (DoD)

The DoD's Environmental Quality Program protects DoD personnel and surrounding communities from exposure to hazardous materials and reduces pollution to the air, land, and water. It focuses on source reduction, reuse, and recycling. DoD's pollution prevention mission and implementation policies, attributes and considerations, accomplishments and future initiatives can be very valuable information and a model for international defense professionals and consultants. Such pollution prevention program information and practical experience with factual data are highlighted as a case study in Chapter 15.

16.4
State and regional governments

State-based environmental programs have a unique contribution to pollution prevention through their direct contact with industry and awareness of local needs. Since 1991, most of state programs were focused on teaching small and medium-size businesses about pollution prevention through information, outreach and technical assistance. State and regional government have jurisdiction over activities that create direct and indirect impacts of various pollution prevention activities. Readers may wish to contact the following state and regional governments for their specific pollution prevention programs and activities:

16.4.1
State governments

Alabama Dept of Environmental Management
   Tel: 205-250-2782; Fax: 204-250-2795

Alaska Depart of Environmental Conservation, P2 Office
   Tel: 907-463-6529; Fax: 907-562-4026

Arizona Dept of Environmental Quality
   Tel: 602-207-4210; Fax: 602-207-4236

Arkansas Dept of Pollution Prevention and Ecology, Hazardous Waste Division
   Tel: 501-570-2861 or 682-7325; Fax: 501-682-7341

California EPA, Dept of Toxic Substances Control
   Tel: 916-322-3670: Fax: 916-327-4494

Connecticut Department of Environmental Protection, Office of P2
   Tel: 203-566-8476 or 5217; Fax: 203-566-4924

Delaware Dept of Natural Resources and Environmental Control
   Tel: 302-739-6400; Fax: 302-739-5060

Florida Dept of Environmental Protection, P2 Office
   Tel: 904-488-0300; Fax: 904-922-4939

Georgia Hazardous Waste Authority, P2 Division
   Tel: 404-651-4778; Fax: 404-651-5778

Hawaii Environmental Management Division
   Tel: 808-586-8143; Fax: 808-586-4370

Idaho Division of Environmental Quality
   Tel: 208-334-5879; Fax: 208-334-0576

Illinois Environmental Protection Agency, Office of P2
   Tel: 217-524-1849; Fax: 217-524-4959

Indiana Dept of Environmental Management, Office of P2
   Tel: 317-232-8172; Fax: 317-232-8564

Kentuchy Dept for environmental Protection
   Tel: 502-563-2150; Fax: 502-584-4245

Louisiana Dept of Environmental Quality
   Tel: 504-765-0720; Fax: 504-765-0742

Maine Department of Environmental Protection, Office of P2
  Tel: 207-287-3811; Fax: 207-287-2814

Maryland Department of the Environment
  Tel: 410-631-4122; Fax:410-631-3936

Massachusetts Department of Environmental Protection, Bureau of waste Prevention
  Tel: 508-767-2775; Fax: 508-792-7621

Michigan Office of Waste Reduction Services
  Tel: 517-373-1871; Fax: 517-335-4729

Minnesota Pollution Control Agency
  Tel: 612-296-8643; Fax: 612-297-8676

Mississippi Dept of Environmental Quality, Waste Minimization Unit
  Tel: 601-961-5241; Fax: 601-354-6612

Nevada division of Environmental Protection
  Tel: 702-667-4870; Fax: 702-885-0888

New Hampshire Department of Environmental Services
  Tel: 603-271-6398 or 2867; Fax: 603-271-2867

New Hampshire WasteCop in Concord, NH
  Tel: 603-224-5388; Fax: 603-224-2872

New Jersey Department of Environmental Protection, Office of P2
  Tel: 609-292-3600 or 777-0518; Fax: 609-777-1330

New Mexico Environment Department, P2 Program
  Tel: 505-827-0197; Fax: 505-827-2836

New York State Dept of Environmental Conservation, P2 Unit
  Tel: 518-457-7267; Fax: 518-457-2570

North Carolina Office of Waste Reduction
  Tel: 919-571-4100; Fax: 919-571-4135

Ohio Environmental Protection Agency, Office of P2
  Tel: 614-644-3469; Fax: 614-644-2329

Oregon Dept of Environmental Quality, Hazardous & solid Waste Division
  Tel: 503-229-5913; Fax: 503-229-6124

Pennsylvania Dept of Environmental Resources, Office of Air & Waste Manatgement
  Tel: 717-783-0540 or 787-7382; Fax: 717-787-8926

Rhode Island Department of Environmental Management, Hazardous waste Reduction Section
  Tel: 401-277-4700; Fax: 401-277-2591

South Carolina Dept of Health and Environmental Control, Center for Waste Minimization
Tel: 803-634-4715; Fax: 803-734-5199

Tennessee dept of Environment and Conservation,P2 Division
Tel: 615-532-0736; Fax: 615-532-0231

Texas Water Commission, Office of P2
Tel: 512-475-4580; Fax: 512-375-4599

Vermont Department of Environmental Conservation, P2 Division
Tel: 802-241-3629 or 3888; Fax: 802-244-5141

Virginia Department of Environmental Quality
Tel: 804-371-8712; Fax: 804-371-0193

Washington Wate Department of Ecology, Toxic Substance Section
Tel: 206-493-9380; Fax: 206-438-7789

Wisconsin Dept of Natural Resources, Office of P2
Tel: 608-267-9700 or 3763; Fax: 608-267-5231

16.4.2
Regional governments

The National Association of Counties (NACo)
The NACo is playing a role in advancing local pollution prevention efforts. As a national organization, NACo provides a link across counties and between counties and national pollution prevention organizations. Counties are well positioned to lend P2 efforts by providing information and assistance. County workers such as fire inspectors, permitting and licensing officers, health officials, and zoning and planning board members have numerous opportunities to promote P2. County governments can incorporate P2 into internal operations and provide the benefit of their experience to the community (USEPA, 1997a).

NACo published "Preventing Pollution in our Cities and Counties: A Compendium of Case Studies." In conjunction with the National Recycling Coalition, NaCo co-wrote "Making Source Reduction and Reuse Work in Your Community: A Manual for Local Governments." NACo is developing a "starter kit" to help counties nationwide embark on similar efforts (NACo, 1995).

Northeast Waste Management Officials' Association (NEWMOA) in Boston, MA
Tel: 617-367-8558; Fax: 617-367-0449

Solid Waste Association of North America (SWANA)
The SWANA, located in Silver Spring, Md., operates the Solid Waste Assistance Program (SWAP) and the Peer Match Program. SWAP is a technical information hotline designed to collect and distribute materials and provide assistance to all

interested parties on solid waste management. The Peer Match Program aids state and local governments by connecting knowledgeable municipal solid waste professionals with communities in need of assistance. Tel: 800-677-9424 (SWAP) and 301-589-2898 (SWANA); Fax: 301-589-7068.

Waste Reduction Resource Center (WRRC)
WRRC, located in Raleigh, N.C., helps coordinate P2 with other agencies, promotes economic development, helps deploy new technology, and provides direct assistance to reduce pollution generation and business compliance. Tel: 800-476-8686; Fax: 919-571-4135.

16.5
Non-governmental organizations

The Air & Waste Management Association (AWMA)
AWMA has a website (http://www.awma.org/). The site is maintained by AWMA, which is a not-for-profit technical, scientific, and educational organization. It provides quality environmental information on pollution prevention, publications, meetings, key links, public outreach, news items, education and certification.

National Center for Manufacturing Sciences (NCMS)
The NCMS's website (http://www.ncms.org) is a consortium of U.S. and Canadian corporations promoting next-generation technologies. It supports a web page for the National Metal Finishing Resource Center.

World Resources Institute (WRI)
The WRI, created in 1982, is an independent center for policy research and technical assistance on global environmental and development issues. WRI is dedicated to helping governments and private organizations of all types cope with environmental, resource, and development challenges of global significance. Detailed information of WRI can be located from its website (http://www.wri.org/wri).

Battelle Seattle Research Center
The Battelle's website (http://www.seattle.battelle.org/) provides a list of P2 and environmental links, a page entitled "Finding and Using P2 Information on the Internet." It includes pollution prevention technology resources, design for environment information, recycling information, state pollution prevention programs, regulations and policy information, and trade and professional associations.

Center for Technology Transfer and Pollution Prevention (CT2P2)
The Center, located at Purdue University, provides the users with the tools necessary to transfer technical information about the environment and P2 worldwide. The website of CT2P2 is (http://ingis.can.purdue.edu:9999/cttpp/cttpp.html).

National Center for Environmental Publications and Information (NCEPI)
This Center focuses on scientific/technical and public-oriented environmental information. Approximately 2,500 new titles are added annually to the NCEPI system database. Services are provided to federal, state, and local agencies, businesses, civic and environmental groups, academia, and the public. Tel: 513-891-6561; Fax: 513-891-6685.

Los Alamos National Laboratories (LANB)
The LANB's website (http://perseus.lanl.gov/), maintained by the Materials Substitution Resource, provides access to pollution prevention publications and a comprehensive list of environmental links.

Minnesota Technical Assistance Program (MnTAP)
The MnTAP's website (http://es.inel.gov/techinfo/facts/macts/mpca/mpca.html) facilitates access to a large number of fact sheets that provide pollution prevention research materials.

New Jersey Technical Assistance Program (NJTAP)
The NJTAP's website (http://www.njit.edu/njtap) facilitates access to a large amount of P2 reference material. It also supports a search engine, EnviroDaemon, which is designed to concentrate the P2 data searching process.

– North Carolina Division of Pollution Prevention and Environmental Assistance (DPPEA)

The DPPEA's website (http://owr.ehnr.state.nc.us/) provides information on pollution prevention, technical assistance, and recycling.

The P2-Tech Archives
The Archives' website (http://gopher.great-lakes.net:2200/1 s/mailarc/p2tech), supported by GLIN, provides access to archives from the P2 Tech list-serve and a variety of other references.

The Tellus Institute
The Tellus Institute has a website (http://www.tellus.org/), supported by GLIN, which provides access to archives from the P2Tech list-serve and a variety of other references. The site contains a great deal of information on pollution prevention and environmental accounting, as well as other environmental topics.

Center for Hazardous Materials Research (CHMR) in Pittsburgh, PA
This Center offers pollution prevention workshops for industrial representatives, consultants, engineering students, and regulatory personnel. Other features include on-site technical assistance, a technical assistance hotline, a quarterly newsletter, and industry-specific manuals and fact sheets. Tel: 412-826-5320; Fax: 412-826-5552.

16.6
Business and industry

16.6.1
Canadian Chemical Producer's Association (CCPA)

The CCPA's website includes copies of the "Reducing Emission" report. It details the CCPA's efforts in the field of Responsible Care. The website hosts a comprehensive set of links to safety and health organizations, governments, associations and companies that deal with issues related to the Canadian Chemical Industry.

16.6.2
Center for Waste Reduction Technologies (CWRT)

The CWRT, located in New York, N.Y., is an industry-driven, non-profit organization dedicated to sponsoring and developing new and innovative waste reduction technologies and methodology transfer, and enhanced education in a collaborative effort among industry, government, and academia. The Center is sponsored by major manufacturing and chemical/petroleum companies. Technology transfer activities include publications, and the development of broad, interlinked databases for identification and application of waste reduction technologies. Tel: 212-705-6462; Fax: 212-752-3297.

16.6.3
Green Business Resource Center

The Center's website (http://www.enn.com/green.htm) is a one-stop information source for environmental managers and others in business, government, and institutions interested in learning how to integrate environmental thinking throughout their organizations. At the heart of the Center is the Best Practices Database, which will allow users to search through hundreds of company environmental initiatives, including energy efficiency and renewable, solid waste reduction, pollution prevention, toxic reduction, and recycling and recycled procurement.

16.6.4
Chicago Board of Trade (CBOT)

The CBOT's Recyclable Exchange in Chicago, Ill., offers: (1) on-line (via Internet) posting and trading of various grades of glass and paper as well as PET and HDPE plastics (other recyclable commodities will be added in the future); (2) a miscellaneous category available for trading other materials; and (3) testing, dispute resolution, and specifications for materials traded. Tel: 312-341-7955.

16.6.5
Northeast Waste Management Officials' Association (NEWMOA) in Boston, Mass.

The NEWMOA operates a clearinghouse of information on P2, conducts training sessions for state officials and industry representatives on source reduction, coordinates an interstate roundtable of state pollution prevention programs, and researches source reduction strategies. A quarterly newsletter is also published. Tel: 617-376-0449; Fax: 516-376-0449.

16.6.6
Pacific Northwest Pollution Prevention Research Center (PNPPRC) in Seattle, Wash.

The Center is a public-private partnership formed to identify opportunities for and overcome obstacles to pollution prevention. Activities include identifying research and project needs, facilitating transfer of pollution prevention information, and providing research support. The Center acts as a referral service and is establishing an Industrial Liaison Project to transfer the results of nonproprietary pollution prevention research from large to small companies. A bimonthly newsletter is published and the Center hosts seminars, and maintains several database to help technical assistance programs. Tel: 206-223-1151; Fax: 206-223-1165.

16.6.7
Gulf Coast Hazardous Substance Research Center (GCHSRC)

The GCHSRC is located in Lamar University, Beaumont, Tex. It is a consortium of eight universities which conducts research to aid more effective hazardous substance response and waste management. The Center focuses on waste minimization and alternative technology development with an emphasis on the petrochemical and microelectronics industries. Tel: 409-880-8707; Fax: 409-880-2397.

16.6.8
National Materials Exchange Network

The network, located in Spokane, Wash., is an electronic linking of over 40 industrial waste exchanges across North America that allows users to locate available and wanted materials. The materials, which are organized into 17 categories, include waste by-product, surplus, off-spec, over-stock, obsolete, and damaged materials. Tel: 509-325-0551; Fax: 509-325-2086.

References

NACo (1995) Preventing Pollution in Our Cities and Counties. Co-published by National Association of Counties, 440 First Street NW, Washington, DC 20001

NPPC (1998) Educational Materials Available From the National Pollution Prevention Center for Higher Education, NPPC Document List, June

PPIC (1998) Pollution Prevention Information Clearinghouse, USEPA, Office of Pollution Prevention and Toxics, EPA/742/F-98/004, Summer

USEPA (1994) Directory of Pollution Prevention in Higher Education: Faculty andPrograms. Regents of the University of Michigan, EPA742/B-94/006

USEPA (1996) Partnerships in Preventing Pollution: A Catalogue of Th Agency's Partnership Programs. EPA-100-B-96001, Spring

USEPA (1997a) Pollution Prevention News, Office of P2 and Toxics, September–October issue, EPA-742-N-97-004

USEPA (1997b) Pollution Prevention 1997: A National Progress Report. EPA742-R-97-00, pp. 105–106; 117–119

USEPA (1998) Office of Pollution Prevention and Toxics: 1997 Annual Report. EPA 745-R-98-003, January

Appendix A: Pollution Prevention Assessment Worksheets

The worksheets in this appendix were designed to be useful at various points in the development of a pollution prevention program. TableA-1 lists the worksheets and describes the purpose of each.

Since these worksheets are intentionally generic, you may decide to redesign some or all of them to be more specific to your facility once you have your program underway. The checklists in Appendix B contain information that you may find helpful in deciding how to customize these worksheets to fit your situation.

TableA-1. List of Pollution Prevention Assessment Worksheets

Phase	Number and Title	Purpose/Remarks
Assessment Phase	1. Assessment Overview	Summarizes the overall program.
	2. Site Description	Lists background information about the facility, including location, products, and operations.
	3. Process Information	This is a checklist of process information that can be collected before the assessment effort begins.
	4. Input Materials Summary	Records input material information for a specific production or process area. This includes name, supplier, hazardous component or properties, cost, delivery and shelf-life information, and possible substitutes.
	5. Products Summary	Identifies hazardous components, production rate, revenues, and other information about products.
	6. Waste Stream Summary	Summarizes the information collected for several waste streams. This sheet can be used to prioritize waste streams to assess.

TableA-1. (continued)

Phase	Number and Title	Purpose/Remarks
	7. Option Generation	Records options proposed during brainstorming or nominal group technique sessions. Includes the rationale for proposing each option.
	8. Option Description	Describes and summarizes information about a proposed option. Also notes approval of promising options.
Feasibility Analysis Phase		
	9. Profitability	This worksheet is used to identify capital and operating costs and to calculate the playback period.

Firm _____	**Pollution Prevention Assessment Worksheets**	Prepared By _____
Site _____		Checked By _____
Date _____	Proj. No. _____	Sheet ___ of ___ Page ___ of ___

WORKSHEET
1

ASSESSMENT OVERVIEW

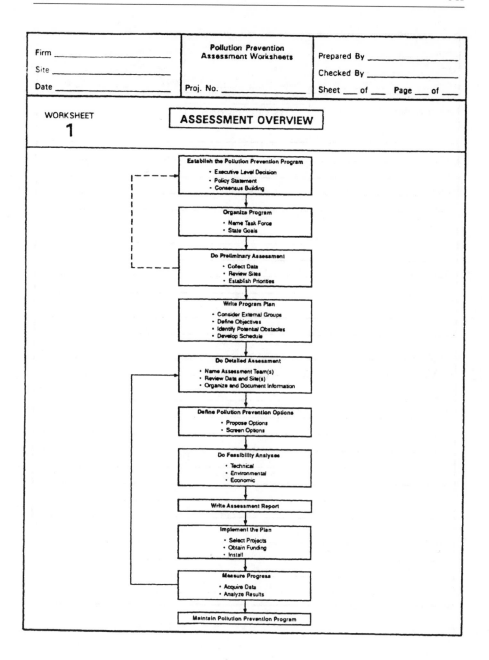

Establish the Pollution Prevention Program
- Executive Level Decision
- Policy Statement
- Consensus Building

Organize Program
- Name Task Force
- State Goals

Do Preliminary Assessment
- Collect Data
- Review Sites
- Establish Priorities

Write Program Plan
- Consider External Groups
- Define Objectives
- Identify Potential Obstacles
- Develop Schedule

Do Detailed Assessment
- Name Assessment Team(s)
- Review Data and Site(s)
- Organize and Document Information

Define Pollution Prevention Options
- Propose Options
- Screen Options

Do Feasibility Analyses
- Technical
- Environmental
- Economic

Write Assessment Report

Implement the Plan
- Select Projects
- Obtain Funding
- Install

Measure Progress
- Acquire Data
- Analyze Results

Maintain Pollution Prevention Program

Firm _____	**Pollution Prevention Assessment Worksheets**	Prepared By _____
Site _____		Checked By _____
Date _____	Proj. No. _____	Sheet ___ of ___ Page ___ of ___

| WORKSHEET 2 | SITE DESCRIPTION | |

Firm: _____

Plant: _____

Department: _____

Area: _____

Street Address: _____

City: _____

State/Zip Code: _____

Telephone: (   ) _____

Major Products: _____

_____

SIC Codes: _____

EPA Generator Number: _____

Major Unit: _____

Product or Service: _____

Operations: _____

_____

_____

_____

_____

_____

Facilities/Equipment Age: _____

_____

_____

_____

Firm _____	**Pollution Prevention Assessment Worksheets**	Prepared By _____
Site _____		Checked By _____
Date _____	Proj. No. _____	Sheet ___ of ___ Page ___ of ___

WORKSHEET **3**	PROCESS INFORMATION

Process Unit/Operation: _____

Operation Type: ☐ Continuous   ☐ Discrete

☐ Batch or Semi-Batch   ☐ Other _____

Document	Status					
	Complete? (Y/N)	Current? (Y/N)	Last Revision	Used in this Report (Y/N)	Document Number	Location
Process Flow Diagram						
Material/Energy Balance						
Design						
Operating						
Flow/Amount Measurements						
Stream						
Analyses/Assays						
Stream						
Process Description						
Operating Manuals						
Equipment List						
Equipment Specifications						
Piping and Instrument Diagrams						
Plot and Elevation Plan(s)						
Work Flow Diagrams						
Hazardous Waste Manifests						
Emission Inventories						
Annual/Biennial Reports						
Environmental Audit Reports						
Permit/Permit Applications						
Batch Sheet(s)						
Materials Application Diagrams						
Product Composition Sheets						
Material Safety Data Sheets						
Inventory Records						
Operator Logs						
Production Schedules						

Firm _____	**Pollution Prevention Assessment Worksheets**	Prepared By _____
Site _____		Checked By _____
Date _____	Proj. No. _____	Sheet ___ of ___ Page ___ of ___

WORKSHEET **4**	INPUT MATERIALS SUMMARY

Attribute	Description		
	Stream No. ____	Stream No. ____	Stream No. ____
Name/ID			
Source/Supplier			
Component/Attribute of Concern			
Annual Consumption Rate			
Overall			
Component(s) of Concern			
Purchase Price, $ per _____			
Overall Annual Cost			
Delivery Mode[1]			
Shipping Container Size & Type[2]			
Storage Mode[3]			
Transfer Mode[4]			
Empty Container Disposal Management[5]			
Shelf Life			
Supplier Would			
— accept expired material? (Y/N)			
— accept shipping containers? (Y/N)			
— revise expiration date? (Y/N)			
Acceptable Substitute(s), if any			
Alternate Supplier(s)			

Notes:
1. e.g., pipeline, tank car, 100 bbl tank truck, truck, etc.
2. e.g., 55 gal drum 100 lb paper bag, tank, etc.
3. e.g., outdoor, warehouse, underground, aboveground, etc.
4. e.g., pump, forklift, pneumatic transport, conveyor, etc.
5. e.g., crush and landfill, clean and recycle, return to supplier, etc.

Firm _____	**Pollution Prevention Assessment Worksheets**	Prepared By _____
Site _____		Checked By _____
Date _____	Proj. No. _____	Sheet ___ of ___ Page ___ of ___

WORKSHEET	
5	**PRODUCTS SUMMARY**

Attribute	Description		
	Stream No. _____	Stream No. _____	Stream No. _____
Name/ID			
Component/Attribute of Concern			
Annual Production Rate			
Overall			
Component(s) of Concern			
Annual Revenues, $ _____			
Shipping Mode			
Shipping Container Size & Type			
Onsite Storage Mode			
Containers Returnable (Y/N)			
Shelf Life			
Rework Possible (Y/N)			
Customer Would			
— relax specification (Y/N)			
— accept larger containers (Y/N)			

Firm _____	**Pollution Prevention Assessment Worksheets**	Prepared By _____
Site _____		Checked By _____
Date _____	Proj. No. _____	Sheet ___ of ___ Page ___ of ___

WORKSHEET **6**	**WASTE STREAM SUMMARY**

Attribute	Description		
	Stream No. _____	Stream No. _____	Stream No. _____
Waste ID/Name:			
Source/Origin			
Component or Property of Concern			
Annual Generation Rate (units _____)			
Overall			
Component(s) of Concern			
Cost of Disposal			
Unit Cost ($ per: _____)			
Overall (per year)			
Method of Management[1]			

Priority Rating Criteria[2]	Relative Wt. (W)	Rating (R)	R x W	Rating (R)	R x W	Rating (R)	R x W
Regulatory Compliance							
Treatment/Disposal Cost							
Potential Liability							
Waste Quantity Generated							
Waste Hazard							
Safety Hazard							
Minimization Potential							
Potential to Remove Bottleneck							
Potential By-product Recovery							
Sum of Priority Rating Scores		Σ(RxW)		Σ(RxW)		Σ(RxW)	
Priority Rank							

Notes: 1. For example, sanitary landfill, hazardous waste landfill, on-site recycle, incineration, combustion with heat recovery, distillation, dewatering, etc.

         2. Rate each stream in each category on a scale from 0 (none) to 10 (high).

Firm _____	**Pollution Prevention Assessment Worksheets**	Prepared By _____
Site _____		Checked By _____
Date _____	Proj. No. _____	Sheet ___ of ___ Page ___ of ___

WORKSHEET

7

OPTION GENERATION

Meeting format (e.g., brainstorming, nominal group technique) _____

Meeting Coordinator _____

Meeting Participants _____

_____

_____

_____

List Suggestion Options	Rationale/Remarks on Option

Firm _____	**Pollution Prevention Assessment Worksheets**	Prepared By _____
Site _____		Checked By _____
Date _____	Proj. No. _____	Sheet ___ of ___ Page ___ of ___

| WORKSHEET **8** | OPTION DESCRIPTION | |

Option Name: _____

Briefly describe the option: _____
_____
_____
_____
_____
_____
_____
_____

Waste Stream(s) Affected: _____
_____

Input Material(s) Affected: _____
_____

Product(s) Affected: _____
_____

Indicate Type: ☐   Source Reduction
           _____ Equipment-Related Change
           _____ Personnel/Procedure-Related Change
           _____ Materials-Related Change

          ☐   Recycling/Reuse
           _____ Onsite   _____ Material reused for original purpose
           _____ Offsite   _____ Material used for a lower-quality purpose
                           _____ Material sold

Originally proposed by: _____ Date: _____

Reviewed by: _____ Date: _____

Approved for study? _____ yes _____ no   By:_____

Reason for Acceptance or Rejection _____

Firm _____	**Pollution Prevention Assessment Worksheets**	Prepared By _____
Site _____		Checked By _____
Date _____	Proj. No. _____	Sheet ___ of ___ Page ___ of ___

| WORKSHEET **9** | **PROFITABILITY** |

Capital Costs

Purchased Equipment _____

Materials _____

Installation _____

Utility Connections _____

Engineering _____

Start-up and Training _____

Other Capital Costs _____

    Total Capital Costs _____

Incremental Annual Operating Costs

Change in Disposal Costs _____

Change in Raw Material Costs _____

Change in Other Costs _____

    Annual Net Operating Cost Savings _____

$$\text{Payback Period (in years)} = \frac{\text{Total Capital Costs}}{\text{Annual Net Operating Cost Savings}} = \underline{\hspace{4cm}}$$

Appendix B: Industry-Specific Checklists

This appendix tabulates information that may be helpful to you if you decide to customize the worksheets in Appendix A for your own company's needs. Some ideas for achieving pollution prevention through good operating practices are shown in Table1. Approaches to pollution prevention in material receiving, raw material and product storage, laboratories, and maintenance areas are shown in Table2. Information in these two tables can apply to a wide range of industries. Industry-specific checklists for five example industries are presented in Tables 3 through 7.

Table1. Pollution Prevention Through Good Operating Practices
Table2. Checklist for All Industries
Table3. Checklist for the Printing Industry
Table4. Checklist for the Fabricated Metal Industry
Table5. Checklist for the Metal Casting Industry
Table6. Checklist for the printed Circuit Board Industry
Table7. Checklist for the Coating Industry

Table 1 Pollution Prevention Through Good Operating Practices

Good Operating Practice	Program Ingredients
Waste Segregation	Prevent mixing of hazardous wastes with non-hazardous wastes
	Store materials in compatible groups
	Segregate different solvents
	Isolate liquid wastes from solid wastes
Preventive Maintenance Programs	Maintain equipment history cards on equipment location, characteristics, and maintenance
	Maintain a master preventive maintenance (PM) schedule
	Keep vendor maintenance manuals handy
	Maintain a manual or computerized repair history file
Training/Awareness-Building Programs	Provide training for
	–Operation of the equipment to minimize energy use and material waste
	–Proper materials handling to reduce waste and spills

Table 1 (continued)

Good Operating Practice	Program Ingredients
Training/Awareness-Building Programs	– Emphasize importance of pollution prevention by explaining the economic and environmental ramifications of hazardous waste generation and disposal
	– Detecting and minimizing material loss to air, land, or water
	– Emergency procedures to minimize lost materials during accidents
Effective Supervision	Closer supervision may improve production efficiency and reduce inadvertent waste generation
	Centralize waste management. Appoint a safety/waste management officer for each department. Educate staff on the benefits of pollution prevention. Establish pollution prevention goals. Perform pollution prevention assessments.
Employee Participation	"Quality circles" (free forums between employees and supervisors) can identify ways to reduce waste
	Solicit and reward employee suggestions for waste reduction ideas
Production Scheduling/Planning	Maximize batch size to reduce clean out waste
	Dedicate equipment to a single product
	Alter batch sequencing to minimize cleaning frequency (light-to-dark batch sequence, for example)
Cost accounting/Allocation	Charge direct and indirect costs of all air, land, and water discharges to specific processes or products
	Allocate waste treatment and disposal costs to the operations that generate the waste
	Allocate utility costs to specific processes or products

Table 2 Checklist for All Industries

Waste Origin/Type	Pollution Prevention and Recycling Methods
Material Receiving/Packaging materials, off-spec materials, damaged container, inadvertent spills, transfer hose emptying	Use "Just-in-Time" ordering system. Establish a centralized purchasing program. Select quantity and package type to minimize packing waste. Order reagent chemicals in exact amounts. Encourage chemical suppliers to become responsible partners (e.g., accept outdated supplies). Establish an inventory control program to trace chemical from cradle to grave. Rotate chemical stock. Develop a running inventory of unused chemicals for other departments' use. Inspect material before accepting a shipment. Review material procurement specifications. Validate shelf-life expiration dates. Test effectiveness of outdated material. Eliminate shelf-life requirements for stable compounds.

Table 2 (continued)

Waste Origin/Type	Pollution Prevention and Recycling Methods
Material Receiving/ Packaging materials, off-spec materials, damaged container, inadvertent spills, transfer hose emptying	Conduct frequent inventory checks. Use computer-assisted plant inventory system. Conduct periodic materials tracking. Properly label all containers. Set up staffed control points to dispense chemicals and collect wastes. Buy pure feeds. Find less critical uses for off-spec material (that would otherwise be disposed). Change to reusable shipping containers. Switch to less hazardous raw material. Use rinsable/recyclable drums.
Raw Material and Product Storage/Tank bottoms; off-spec and excess materials; spill residues; leaking pumps, valves, tanks, and pipes; damaged containers; empty containers	Establish Spill Prevention, Control, and Countermeasures (SPCC) plans. Use properly designed tanks and vessels only for their intended purposes. Install overflow alarms for all tanks and vessels. Maintain physical integrity of all tanks and vessels. Set up written procedures for all loading/unloading and transfer operations. Install secondary containment areas. Instruct operators to not bypass interlocks, alarms, or significantly alter setpoints without authorization. Isolate equipment or process lines that leak or are not in service. Use sealless pumps. Use bellows-seal valves. Document all spillage. Perform overall materials balances and estimate the quantity and dollar value of all losses. Use floating-roof tanks for VOC control. Use conservation vents on fixed roof tanks. Use vapor recovery systems. Store containers in such a way as to allow for visual inspection for corrosion and leaks. Stack containers in a way to minimize the chance of tipping, puncturing, or breaking. Prevent concrete "sweating" by raising the drum off storage pads. Maintain Material Safety Data Sheets to ensure correct handling of spills. Provide adequate lighting in the storage area. Maintain a clean, even surface in transportation areas. Keep aisles clear of obstruction. Maintain distance between incompatible chemicals. Maintain distance between different types of chemicals to prevent cross-contamination. Avoid stacking containers against process equipment.

Table 2 (continued)

Waste Origin/Type	Pollution Prevention and Recycling Methods
Raw Material and Product Storage/Tank bottoms; off-spec and excess materials; spill residues; leaking pumps, valves, tanks, and pipes; damaged containers; empty containers	Follow manufacturers' suggestions on the storage and handling of all raw materials. Use proper insulation of electric circuitry and inspect regularly for corrosion and potential sparking. Use large containers for bulk storage whenever possible. Use containers with height-to-diameter ratio equal to one to minimize wetted area. Empty drums and containers thoroughly before cleaning or disposal. Reuse scrap paper for note pads; recycle paper.
Laboratories/Reagents, off-spec chemicals, samples, empty sample and chemical containers	Use micro or semi-micro analytical techniques. Increase use of instrumentation. Reduce or eliminate the use of highly toxic chemicals in laboratory experiments. Reuse/recycle spent solvents. Recover metal from catalyst. Treat or destroy hazardous waste products as the last step in experiments. Keep individual hazardous waste streams segregated, segregate hazardous waste from non-hazardous waste, segregate recyclable waste from non-recyclable waste. Assure that the identity of all chemicals and wastes is clearly marked on all containers. Investigate mercury recovery and recycling.
Operation and Process Changes Solvents, cleaning agents, degreasing sludges, sandblasting waste, caustic, scrap metal, oils, greases from equipment cleaning	Maximize dedication of process equipment. Use squeegees to recover residual fluid on product prior to rinsing. Use closed storage and transfer systems. Provide sufficient drain time for liquids. Line equipment to reduce fluid holdup. Use cleaning system that avoid or minimize solvents and clean only when needed. Use countercurrent rinsing. Use clean-in-place systems. Clean equipment immediately after use. Reuse cleanup solvent. Reprocess cleanup solvent into useful products. Segregate wastes by solvent type. Standardize solvent usage. Reclaim solvent by distillation. Schedule production to lower cleaning frequency. Use mechanical wipers on mixing tanks.
Operation and Process Changes Sludge and spent acid from heat exchanger cleaning	Use bypass control or pumped recycle to maintain turbulence during turndown. Use smooth heat exchange surfaces. Use on-stream cleaning techniques. Use high pressure water cleaning to replace chemical cleaning where possible. Use lower pressure steam.

Table 3 Checklist for the Printing Industry

Waste Origin/Type	Pollution Prevention and Recycling Method
Image Processing/Empty containers, used film packages, outdated material	Recycle empty containers. Recycle spoiled photographic film.
Image Processing/Photographic chemicals, silver	Use silver-free films, such as vesicular, diazo, or electrostatic types. Use water-developed litho plates. Extend bath life. Use squeegees to reduce carryover. Employ countercurrent washing. Recover silver and recycle chemicals.
Plate Making/Damaged plates, developed film, outdated materials	Use electronic imaging, laser plate making.
Plate Making/Acids, alkali, solvents, plate coatings (may contain dyes, photopolymers, binders, resins, pigment, organic acids), developers (may contain isopropanol, gum arabic, lacquers, caustics), and rinse water	Electronic imaging/laser print making. Recover silver and recycle chemicals. Use floating lids on bleach and developer tanks. Use countercurrent washing sequence. Use squeegees to reduce carryover. Substitute iron-EDTA for ferrocyanide. Use washless processing systems. Use better operating practices. Remove heavy metals from wastewater.
Finishing/Damaged products, scrap	Reduce paper use and recycle waste paper.
Printing/ Lubricating oils, waste ink, cleanup solvent (halogenated and non-halogenated), rags	Prepare only the quantity of ink needed for a press run. Recycle waste ink and solvent. Schedule runs to reduce color change over. Use automatic cleaning equipment. Use automatic ink leveler. Use alternative solvents. Use water-based ink. Use UV-curable ink. Install web break detectors. Use automatic web splicers. Store ink properly. Standardize ink sequence. Recycle waste ink.
Printing/ Test production, bad printings, empty ink containers, used blankets	Install web break detectors. Monitor press performance. Use better operating practices.

Table 3 (continued)

Waste Origin/Type	Pollution Prevention and Recycling Method
Printing/ (Continued)	Use alternative fountain solutions. Use alternative cleaning solvents. Use automatic blanket cleaners. Improve cleaning efficiency. Collect and reuse solvent. Recycle lube oils.
Finishing/ Paper waste from damaged product	Reduce paper use. Recycle waste paper.

Table 4 Checklist for the Fabricated Metal Industry

Waste Origin/Type	Pollution Prevention and Recycling Methods
Machining Wastes/ Metalworking Fluid	Use of high-quality metalworking fluid. Use demineralized water makeup. Perform regularly scheduled sump and machine cleaning. Perform regularly scheduled gasket, wiper, and seal maintenance. Filter, pasteurize, and treat metalworking fluid for reuse. Assigning fluid control responsibility to one person. Standardize oil types used on machining equipment. Improve equipment scheduling/establish dedicated lines. Reuse or recycle cutting, cooling, and lubricating oils. Substitute insoluble borates for soluble borate lubricants.
Machining Wastes/ Metal wastes, dust, and sludge	Segregate and reuse scrap metal.
Parts Cleaning/ Solvents	Install lids/silhouettes on tanks. Increase freeboard space on tanks. Install freeboard chillers on tanks. Remove sludge from solvent tanks frequently. Extend solvent life by precleaning parts by wiping, using air blowers, or predipping in cold mineral spirits dip. Reclaim/recover solvent on- or off-site. Substitute less hazardous solvent degreasers (e.g., petroleum solvents instead of chlorinated solvents) or alkali washes where possible. Distribute parts on rack to allow good cleaning and minimize solvent holup. Slow speed of parts removal from vapor zone. Rotate parts to allow condensed solvent drop-off.
Parts Cleaning/ Aqueous Cleaners	Remove sludge frequently. Use dry cleaning and stripping methods. Use oil separation and filtration to recycle solution.

Table 4 (continued)

Waste Origin/Type	Pollution Prevention and Recycling Methods
Parts Cleaning/ Abrasives	Use of greaseless or water-based binders. Use an automatic liquid spray system for application of abrasive onto wheel. Ensure sufficient water use during cleaning by using water level control. Use synthetic abrasives.
Parts Cleaning/ Rinsewater	Improve rack and barrel system design. Use spray, fog, or chemical rinses. Use deionized water makeup to increase solution life.
Surface Treatment and Plating/Process Solutions	Use material or process substitution e.g., trivalent chromium. Use low solvent paint for coating. Use mechanical cladding and coating. Use cleaning baths as pH adjusters. Recover metals from process solutions.

Surface Treatment and Plating/Rinsewater	Reduction in drag-out of process chemicals:
	   Reduce speed of withdrawal
	   Lower plating bath concentrations
	   Reuse rinsewater
	   Use surfactants to improve drainage
	   Increase solution temperature to reduce viscosity
	   Position workpiece to minimize solution holdup
	 System design considerations:
	   Rinsetank design
	   Multiple rinsing tanks
	   Conductivity measurement to control rinse water flow
	   Fog nozzles and sprays
	   Automatic flow controls
	   Rinse bath agitation
	   Countercurrent rinse.

Table 5 Checklist for the Metal Casting Industry

Waste Origin/Type	Pollution Prevention and Recycling Methods
Baghouse Dust and Scrubber Waste/ Dust contaminated with lead, zinc, and cadmium	Identify the source of contaminants, e.g., coatings on scrap, and work with suppliers to find raw materials that reduce the cotaminant input. Install induction furnaces to reduce dust production. Recycle dust to original process or to another process. Recover contaminants with pyrometallurgical treatment, rotary kiln, hydrogen reduction, or other processes. Recycle to cement manufacturer.

Table 5 (continued)

Waste Origin/Type	Pollution Prevention and Recycling Methods
Production of Ductile Iron/ Hazardous slag	Reduce the amount of sulfur in the feedstock. Use calcium oxide or calcium fluoride to replace calcium carbide as the desulfurization agent. Improve process control. Recycle calcium carbide slag.
Casting/ Spent casting sand	Material substitution, e.g., olivine sand is more difficult to detoxify than silica sand. Separate sand and shot blast dust. Improve metal recovery from sand. Recover sand and mix old and new sand for mold making. Recover sand by washing, air scrubbing, or thermal treatment. Reuse sand for construction if possible.

Table 6 Checklist for the Printed Circuit Board Industry

Waste Origin/Type	Pollution Prevention and Recycling Methods
PC Board Manufacture/ General	Product substitution: Surface mount technology Injection molded substrate and additive plating
Cleaning and Surface Preparation/Solvents	Materials substitution: Use abrasives Use nonchelated cleaners Increase efficiency of process: Extend bath life, improve rinse efficiency, counter-current cleaning Recycle/reuse: Recycle/reuse cleaners and rinses
Pattern Printing and Masking/Acid fumes/ organic vapors; vinyl polymers spent resist removal solution; spent acid solution; waste rinse water	Reduce hazardous nature of process: Aqueous processable resist Screen printing versus photolithography Dry photoresist removal Recycle/reuse: Recycle/reuse photoresist stripper
Electroplating and Electroless Plating/ Plating solutions and rinse wastes	Eliminate process: Mechanical board production Materials substituion: Noncyanide baths Noncyanide stress relievers Extend bath life; reduce drag-in: Proper rack design/maintenance, better precleaning/ rinsing, use of demineralized water as makeup, proper storage methods

Table 6 (continued)

Waste Origin/Type	Pollution Prevention and Recycling Methods
Electroplating and Electroless Plating/ Plating solutions and rinse wastes	Extend bath life; reduce drag-out: Minimize bath chemical concentration, increase bath temperature, use wetting agents, proper positioning on rack, slow withdrawal and sample drainage, computerized/automated systems, recover drag-out, use airstreams or fog to rinse plating solution into the tank, collect drips with drain boards. Extend bath life; maintain bath solution quality: Monitor solution activity Control temperature Mechanical agitation Continuous filtration/carbon treatment Impurity removal Improve rinse efficiency: Closed-circuit rinses Spray rinses Fog nozzles Increased agitation Countercurrent rinsing Proper equipment design/operation Deionized water use. Turn off rinsewater when not in use. Recovery/reuse: Segregate streams Recover metal values.
Etching/ Etching solutions and rinse wastes	Eliminate process: Differential plating Use dry plasma etching. Materials substitution: Nonchelated etchants Nonchrome etchants. Increased efficiency: Use thinner copper cladding Pattern vs. panel plating Additive vs. subtractive method. Reuse/recycle: Reuse/recycle etchants.

Table 7 Checklist for the Coating Industry

Waste Origin/Type	Pollution Prevention and Recycling Methods
Coating Overspray/ Coating material that fails to reach the object being coated	Maintain 50% overlap between spray pattern. Maintain 6- to 8-inch distance between spray gun and the workpiece. Maintain a gun speed of about 250 feet/minute. Hold gun perpendicular to the surface. Trigger gun at the beginning and end of each pass. Properly train operators. Use robots for spraying. Avoid excessive air pressure for coating atomization. Recycle overspray. Use electrostatic spray systems. Use turbine disk or bell or air-assisted airless spray guns in place of air-spray guns. Install on-site paint mixers to control material usage. Inspect parts before coating.
Stripping Wastes/ Coating removal from parts before applying a new coat	Avoid adding excess stripper. Use spent stripper as rough prestrip on next item. Use abrasive media paint stripping. Use plastic media bead-blasting paint stripping. Use cryogenic paint stripping. Use thermal paint stripping. Use wheat starch media blasting paint stripping. Use laser or flashlamp paint stripping.
Solvent Emissions/ Evaporative losses from process equipment and coated parts	Keep solvent soak tanks away from heat sources. Use high-solids coating formulations. Use powder coatings. Use water-based coating formulations. Use UV cured coating formulations.
Equipment Cleanup Wastes/ Process equipment cleaning with solvents	Use light-to-dark batch sequencing. Produce large batches of similarly coated objects instead of small batches of differently coated items. Isolate solvent-based paint spray booths from water-based paint spray booths. Reuse cleaning solution/solvent. Standardize solvent usage. Clean coating equipment after each use.
Source Reduction	Reexamine the need for coating, as well as available alternatives. Use longer lasting plastic coatings instead of paint.

Appendix C: Pollution Prevention Program

Top Management Support

☐ Written policy statement supporting pollution prevention

   ☐ Distribute statement to all employees

Getting Your Program Started

☐ Commit to implementation

☐ Designate a pollution prevention coordinator

☐ Develop a pollution prevention team

☐ Set goals

☐ Increase Employee Awareness

☐ Train employees

☐ Reward pollution prevention successes

Understanding Processes and Wastes

☐ Gathering background information

   ☐ Raw materials   ☐ Production Mechanisms

   ☐ Waste Generated   ☐ Process Interrelationships

☐ Characterize general process

☐ Examine unit processes

☐ Perform materials balance

☐ Define production unit

Identify Pollution Prevention Opportunities

- ☐ Begin assessments
- ☐ Prioritize waste streams
- ☐ Generate reduction options

Cost Considerations

- ☐ Determine full cost of waste
- ☐ Develop economics
- ☐ Establish cost allocation system

Identifying and Implementing Pollution Prevention Projects

- ☐ Determine benefits
- ☐ Conduct technical evaluation
- ☐ Conduct economic evaluation

Program and Project Evaluation

- ☐ Evaluate program
- ☐ Modify program as needed
- ☐ Determine methods to measure waste reduction

Sustain the Pollution Prevention Program

- ☐ Rotate pollution prevention team
- ☐ Train employees as needed
- ☐ Publicize success stories

Appendix C: Continue

Appendix D: U.S. EPA's current voluntary pollution prevention programs

The programs expect that voluntary goals and commitments can achieve real environmental results in a timely and cost-effective way. EPA is building cooperative partnerships with a variety of groups, including small and large business, citizen groups, state and local governments, universities and trade associations. EPA views partnership efforts as key to the future success of environmental protection. EPA's voluntary P2 programs are briefly described below (USEPA, 1996):

30/50 Program

EPA challenged business and industry to reduce toxic emissions, to use whatever methods were appropriate, but to consider and adopt source reduction whenever possible. The program set national priorities for preventing chemical releases to the environment by targeting 1.5 billion lbs. of 17 priority pollutants reported to TRI in 1988 for reduction by 33% in 1992 and 50% in 1995. For more information, call 202-554-1404 or 202-260-6907.

AgSTAR Program

It is a voluntary EPA, U.S. Department of Agriculture, and Department of Energy sponsored program that promotes cost-effective methods for reducing methane emissions through manure management. The main focus of the program is on the swine and dairy industries. Its goal is to reduce methane emissions by 2.25 million metric tons of carbon equivalent by the year 2000. For more information, call 202-233-9569.

Climate Wise Recognition Program

It is a key part of the nation's Climate Change Action Plan and reinforces and supports provisions of the 1993 Energy Policy Act. The program provides technical assistance and puts companies in touch with financial services to "jump start" energy efficiency and pollution prevention actions. Shifting the focus from specific technologies to performance, the program allows companies to pursue common sense approaches to achieving environmental and economic results. It helps companies turn energy efficiency and pollution prevention into a corporate asset. Its goals are to encourage immediate reduction of energy use and greenhouse gas emissions; change the way companies view and manage for environmental performance; and foster innovation. For more information, call 202-586-9260 or 202-260-4407.

Coalbed Methane Outreach Program

The program raises awareness of opportunities for profitable investment. Coal mines are encouraged to recover methane for safety and profit. Local communities and other industries are encouraged to invest coalbed methane development that creates jobs and revenues for the local economy. Its goal is to identify barriers and remove obstacles to profitable methane recovery at coal mines through dissemination of unbiased technical and economic information. For more information, call 202-233-9468.

Common Sense Initiative

EPA has convened representatives from governments, industries, consulting firms, academic and research institutes to examine the full range of environmental requirements impacting six pilot industries: automobile manufacturing, computers and electronics, iron and steel, metal finishing, petroleum refining, and printing. The six groups are looking for opportunities to change complicated and inconsistent environmental regulations into comprehensive strategies for environmental and health protection with an emphasis on pollution prevention. Its goal is a cleaner environment at less cost to taxpayers and industry. For more information, call 202-260-7417.

Design for the Environment (DfE)

EPA is promoting the incorporation of environmental consideration into the design and redesign of products, processes, and technical and management systems. The program aims to encourage pollution prevention and efficient risk reduction in a wide variety of activities. Its goals are to encourage voluntary reduction of the use of specific hazardous chemicals; change the way businesses, governments and other organizations view and manage for environmental protection by demonstrating the benefits; and develop effective voluntary partnerships with businesses, labor organizations, government agencies, and environmental and community groups to implement DfE projects and other pollution prevention activities. For more information, call 202-260-1678.

Energy Star Buildings

The program is a multi-stage strategy that considers system interactions, resulting in additional energy savings, while lowering capital expenditures and preventing pollution. Its implementation focuses on use of proven technologies to prevent pollution through profitable investment opportunities. It also gives commercial and industrial building owners an opportunity to act as responsible, proactive corporate citizens. The program goal is to maximize energy efficiency and profits while reducing atmospheric pollution.

Energy Star Residential Programs

The programs are market-based initiatives to prevent pollution by reducing energy use in the residential sector. EPA is working through a combination of several other energy voluntary programs for implementation. Its goal is to promote residential energy efficiency because household energy se contributes to air pollution, including 20% of all U.S. emissions of carbon dioxide. It accounts for 26% of sulfur dioxide emissions and 15% of nitrogen oxides emissions. By using more energy efficient appliances and heating and cooling equipment, and by constructing more energy efficient homes, we can reduce pollution and save money. For more information, call 202-775-6650.

Energy Star Office Equipment

The program is a voluntary program for computer and office equipment manufacturers. They are asked to develop desktop computers, monitors, printers, fax machines and copiers that can power-down while not in use. It can reduce energy consumption by approximately 50%. EPA urges governments, private and public organizations to commit to a similar Energy Star purchasing policy. For more information, call 202-775-6650.

Energy Star Transformer Program

The program has three distinct membership categories: utility partners, manufacturing partners, and allies. Its goal is to encourage the use of high-efficiency distribution transformers by utilities where they are cost-effective. Utility partners have agreed to analyze their transformer purchases using the industry's highest standards and buy cost-effective transforms. Manufacturing partners agree to produce and market Energy Star transformers. Energy star allies agree to produce transformer components and materials which play a critical role in determining transformer efficiency. For more information, call 202-775-6650.

Environmental Accounting Project

Pollution prevention would not be adopted as the first choice of environmental management by the industry until the environmental costs of non-prevention approaches and the economic benefits of pollution prevention could be seen by managers making business decisions. The project's goals are to encourage and motivate business to understand the full spectrum of their environmental costs, and integrate these costs into decision making. Implementing environmental accounting will make environmental costs more visible to company managers, thus making those costs more manageable and easier to reduce. For more information, call 202-260-0178.

Environmental Leadership Program (ELP)

It is designed to recognize and provide incentives to facilities willing to develop and demonstrate innovative approaches to establishing accountability for compliance with existing laws. It is one of several new environmental initiatives announced as part of the Clinton Administration's reinvention of regulation to achieve environmental results at the least cost. The goals are to

- determine what should be the basic components of compliance programs and environmental management systems;
- identify the verification procedures to ensure that ELP is working;
- establish measures of accountability so the program will be credible to the public; and
- promote community involvement in understanding and supporting innovative approaches to compliance. For more information, call 202-564-0034.

Green Chemistry Program

The program collaborates with other federal agencies, industry, and academia to promote the use of chemistry for pollution prevention through completely voluntary partnerships. The goals are to:

- promote the development products and processes that reduce or eliminate the use or generation of toxic substances associated with the design, manufacture, and use of chemicals;
- recognize and promote fundamental breakthroughs in chemistry that accomplish pollution prevention in a cost effective manner;
- seek to support research in the area of environmentally benign chemistry and promote partnerships with industry in developing green chemistry technologies; and
- work with other federal agencies in building green chemistry principles into their operations. For more information, call 202-260-2257.

Green Lights Program

EPA has developed a series of Ally Programs with the lighting industry and with organizations that help promote the Green Lights idea. The program's goal is to prevent pollution by encouraging U.S. institutions to use energy-efficient lighting technologies. EPA provides technical assistance including: a decision support software package, lighting upgrade workshops and manuals, a financing registry, and ally programs. EPA also encourages participants to take advantage of their own opportunities for public recognition through appropriate use of the Green Lights logo and other materials that can be incorporated into internal communications, public relations, marketing, and advertising.

Indoor Environments Program

It is a non-regulatory program which employs a cooperative partnership model, enlisting national medical, consumer, public interest, and private sector groups to pursue common goals of public health protection and good business practices. Using the best science available, the program develops and disseminates information, guidance and solution-based technologies. The program serves as a catalyst for action by guiding research, using innovative and creative risk communication tools, and building public and/or private partnerships. Its goal is to ensure that the air quality in all indoor environments will protect and promote human health and welfare.

Landfill Methane Outreach Program

Landfills are the largest source of anthropogenic methane emissions in the U.S., constituting almost 40% of these emissions each year. Recovery and use of methane from landfills substantially reduces these emissions while capturing their energy value. The program goal is to spur development of environmentally and economically beneficial landfill gas-to energy projects across the country by overcoming barriers. EPA recognized those program participants through newsletters, articles, medical events, and public service advertisements, increasing public awareness of their efforts to reduce greenhouse gas emissions while developing a renewable energy resource. EPA also developed many handbooks and information tools including landfill gas energy recovery project evaluation software.

Natural Gas Star Program

It is a voluntary program that works closely with the natural gas industry to reduce emissions of methane. The program consists of two initiatives, one focused on the transmission and distribution sectors and the other concentrating on the production and processing sectors. Leaks and emissions from the natural gas industry are large sources of anthropogenic methane emissions in the U.S., constituting almost 11% of these emissions each year. The program's goal is to reduce emissions of natural gas using cost-effective pollution prevention processes and technologies. For more information, call 202-233-9370.

Pesticide Environmental Stewardship Program

It is a public/private partnership program. All organizations with a commitment to pesticide use and/or risk reduction are eligible to join the program. The program goal is to reduce pesticide use and risk in both agriculture and nonagricultural settings. EPA, USDA, and FDA jointly announced a Federal commitment to two major goals: (1) developing specific use and/or risk reduction strategies that include reliance on biological pesticides and other approaches to pest control; and (2) having 75% of US agricultural acreage adopt integrated pest management programs by the year 2000. For more information, call 703-308-7022.

Project XL Program

This program is designed to give regulated sources the flexibility to develop alternative strategies that will replace or modify specific regulatory requirements on the condition that they produce greater environmental benefits. EPA is choosing 50 projects to be part of the XL pilot. Each project should be able to achieve environmental performance that is superior to what would be achieved through compliance with current and reasonably anticipated future regulation. A successful proposal will develop alternative pollution reduction strategies that meet eight criteria: better environmental results; cost savings and paperwork reduction; stakeholder support; test of an innovative strategy; transferability; feasibility; identification of monitoring, reporting, and evaluation methods; and avoidance of shifting risk burden. Hopefully, the Project XL will reduce the regulatory burden and promotes economic growth while achieving better environmental and public health protection. For more information, call 202-260-3761.

The Ruminant Livestock Methane Program

The program is a collaborative effort between EPA and USDA that promotes cost-effective methods for reducing methane emissions from ruminant livestock. Domesticated ruminant livestock are responsible for approximately 21% of U.S. methane emissions. The general strategy for reducing methane emissions is to allow more of the carbon in the feed to be routed toward milk and meat production, and less of it toward methane. This program encourages livestock producers to adopt best management practices for improving the efficiency of beef and dairy production thereby reducing methane emissions. The goal is to reduce 1.6 million metric tons of carbon equivalent of methane emissions by the year 2000. For more information, call 202-233-9043.

State and Local Outreach Program

The program forms partnerships with state and local governments to help them build their capacities for understanding the impacts of climate change and reducing their emissions of greenhouse gases. Its goal is to reduce greenhouse gas emissions by providing them with specialized products and services. Through EPA's efforts, states and localities (1) can increase their understanding of risks and impacts of climate change; (2) assess and develop mitigation and adaptation strategies that are cost-effective, environmentally sound, and equitable; and (3) implement, evaluate, and document program results. For more information, call 202-260-4314.

Transportation Partners Program

The program supports the voluntary efforts of local officials, citizens, and businesses to improve the efficiency of transportation systems and reduce the demand for vehicle travel. Its goal is to reduce toxic emissions (carbon oxides, nitrogen oxides, and particulates) from the transportation sector by reducing vehicle miles traveled. Effective measures such as telecommuting, transit- and pedestrian-oriented community design, and market-based reforms also have significant side benefits of reducing traffic congestion, increasing worker productivity, making neighborhoods safer and more livable, and generating revenues that can reduce the funding for transportation infrastructure without increasing general taxes. For more information, call 202-260-3729.

U.S. Initiative on Joint Implementation (USIJI)

The program is a pilot program to help establish an empirical basis for considering approaches to joint implementation. It provides a flexible, non-regulatory approach to encouraging international partnerships in environmentally sound projects that reduce or sequester greenhouse gas (GHG) emissions and promote sustainable development. USIJI has several objectives including encouraging the development and implementation of cooperative, voluntary projects between U.S. and foreign partners aimed at reducing or sequestering GHG emissions. For more information, call 202-260-6803 or 202-426-8677.

Voluntary Aluminum Industrial Program (VAIP)

It is an innovative environmental stewardship and pollution prevention program, developed jointly by the EPA and the U.S. primary aluminum industries, to make reductions in perfluorocarbon (PFC) gas emissions, potent greenhouse gases. The VAIP Program's goal is to reduce PFC emissions from primary aluminum smelting 45% by 2000 using cost-effective approaches that make economic and environmental sense for the Partners. For more information, call 202-233-9569.

Voluntary Standards Network

The EPA Standards Network is a cross-Agency mechanism to coordinate EPA interests in international voluntary standards. A major focus of the Network is Agency participation in the development and implementation of the ISO 14000 Environmental Management Standards. The Standards Network also coordinates with other Agency initiatives such as Project XL, the Environmental Leadership Program and the Common Sense Initiative. The Network aims to further Agency-wide objectives within the Standards development process. Thus, the Network advocates pollution prevention, continuous environmental improvement, the use of environmental technologies, greater stakeholder participation in the standards development process and community involvement in standards implementation. For more information, call 202-260-0178.

Waste Minimization National Plan

The plan promotes a long-term effort to minimize the generation of hazardous constituents by emphasizing source reduction and environmentally sound recycling over waste management. The plan assists the goals of the current EPA's Reinventing Government Program in providing flexibility to comply with environmental regulations. The Plan's goals for reducing constituents in hazardous waste that are persistent, bioaccumulative, and/or toxic around 25% by the year 2000; 50% by the year 2005. For more information, call 202-308-8433.

Water Alliances for Voluntary Efficiency (WAVE)

The program was designed to be similar to Green Lights (an energy-saving program), but in this case to promote more efficient water use. Hotels and lodging associations were targeted by EPA for this program with the hope that they could help educate the public. Hotels participating in the WAVE program provide information about water efficiency to their customers. The goals are to reduce water and energy consumption, link water-use efficiency to reduce costs, and inform hotel guests and employees about the benefits of water efficiency. EPA hopes to expand the scope of the WAVE program to include schools, hospitals and businesses. For more information, call 202-260-1827.

WasteWi$e Program

EPA's research showed that reducing materials used and solid waste generated could save companies money on purchasing, mailing, disposal, labor, and transportation costs. Through consultation with businesses and others, EPA designed a flexible program which allows companies to set their own waste reduction goals based on their circumstances. Through reducing municipal solid waste, energy and natural resources are conserved, and pollution is prevented. WasteWi$e partners reduce municipal solid waste in three ways: waste prevention, collecting recyclables, and increasing the manufacture or purchase of recycled products. For more information, call 703-308-7273.

Reference

USEPA (1996) Partnerships in Preventing Pollution: A Catalogue of The Agency's Partnership Programs, EPA 100-B-96-001

Appendix E: The USEPA's Toxic Chemical Release Inventory Reporting Form R Part II. Chemical-Specific Information

⬦EPA
United States
Environmental Protection
Agency

Appendix E: **EPA FORM R**

PART II. CHEMICAL-SPECIFIC INFORMATION

TRI FACILITY ID NUMBER

Toxic Chemical, Category, or Generic Name

SECTION 1. TOXIC CHEMICAL IDENTITY   (Important: DO NOT complete this section if you complete Section 2 below.)

1.1 CAS Number (Important: Enter only one number exactly as it appears on the Section 313 list. Enter category code if reporting a chemical category.)

1.2 Toxic Chemical or Chemical Category Name (Important: Enter only one name exactly as it appears on the Section 313 list.)

1.3 Generic Chemical Name (Important: Complete only if Part I, Section 2.1 is checked "yes." Generic Name must be structurally descriptive.)

SECTION 2. MIXTURE COMPONENT IDENTITY   (Important: DO NOT complete this section if you complete Section 1 above.)

2.1 Generic Chemical Name Provided by Supplier (Important: Maximum of 70 characters, including numbers,letters, spaces, and punctuation.)

SECTION 3. ACTIVITIES AND USES OF THE TOXIC CHEMICAL AT THE FACILITY
(Important: Check all that apply.)

3.1 Manufacture the toxic chemical:

a. ☐ Produce
b. ☐ Import

If produce or import:
c. ☐ For on-site use/processing
d. ☐ For sale/distribution
e. ☐ As a byproduct
f. ☐ As an impurity

3.2 Process the toxic chemical:

a. ☐ As a reactant
b. ☐ As a formulation component

c. ☐ As an article component
d. ☐ Repackaging

3.3 Otherwise use the toxic chemical:

a. ☐ As a chemical processing aid
b. ☐ As a manufacturing aid

c. ☐ Ancillary or other use

SECTION 4. MAXIMUM AMOUNT OF THE TOXIC CHEMICAL ON-SITE AT ANY TIME DURING THE CALENDAR YEAR

4.1   (Enter two-digit code from instruction package.)

☣EPA

United States
Environmental Protection
Agency

EPA FORM R

**PART II. CHEMICAL-SPECIFIC
INFORMATION (CONTINUED)**

TRI FACILITY ID NUMBER
Toxic Chemical, Category, or Generic Name

SECTION 5. RELEASES OF THE TOXIC CHEMICAL TO THE ENVIRONMENT ON-SITE

		A. Total Release (pounds/year) (enter range code from instructions or estimate)	B. Basis of Estimate (enter code)	C. % From Stormwater
5.1	Fugitive or non-point air emissions ☐ NA			
5.2	Stack or point air emissions ☐ NA			
5.3	Discharges to receiving streams or water bodies (enter one name per box)			
5.3.1	Stream or Water Body Name			
5.3.2	Stream or Water Body Name			
5.3.3	Stream or Water Body Name			
5.4	Underground injections on-site ☐ NA			
5.5	Releases to land on-site			
5.5.1	Landfill ☐ NA			
5.5.2	Land treatment/ application farming ☐ NA			
5.5.3	Surface impoundment ☐ NA			
5.5.4	Other disposal ☐ NA			

☐   Check here only if additional Section 5.3 Information is provided on page 5 of this form.

EPA Form 9350-1 (Rev. 12/4/92) - Previous editions are obsolete.

Range Codes: A = 1 - 10 pounds;   B = 11
C = 500 - 999 pounds.

♦EPA	EPA FORM R	TRI FACILITY ID NUMBER
United States Environmental Protection Agency	PART II. CHEMICAL-SPECIFIC INFORMATION (CONTINUED)	Toxic Chemical, Category, or Generic Name

SECTION 5.3   ADDITIONAL INFORMATION ON RELEASES OF THE TOXIC CHEMICAL TO THE ENVIRONMENT ON-SITE

5.3	Discharges to receiving streams or water bodies (enter one name per box)	A. Total Release (pounds/year) (enter range code from instructions or estimate)	B. Basis of Estimate (enter code)	C. % From Stormwater
5.3.__	Stream or Water Body Name			
5.3.__	Stream or Water Body Name			
5.3.__	Stream or Water Body Name			

SECTION 6. TRANSFERS OF THE TOXIC CHEMICAL IN WASTES TO OFF-SITE LOCATIONS

6.1 DISCHARGES TO PUBLICLY OWNED TREATMENT WORKS (POTW)

6.1.A Total Quantity Transferred to POTWs and Basis of Estimate

6.1.A.1 Total Transfers (pounds/year) (enter range code or estimate)	6.1.A.2 Basis of Estimate (enter code)

6.1.B POTW Name and Location Information

6.1.B.__	POTW Name	6.1.B.__	POTW Name
Street Address		Street Address	
City	County	City	County
State	Zip Code	State	Zip Code

If additional pages of Part II, Sections 5.3 and/or 6.1 are attached, indicate the total number of pages in this box [ ] and indicate which Part II, Sections 5.3/6.1 page this is, here. [ ]
(example: 1, 2, 3, etc.)

EPA Form 9350-1 (Rev. 12/4/92) - Previous editions are obsolete.   Range Codes: A = 1 - 10 pounds; B = 1 C = 500 - 999 pounds.

♻EPA	EPA FORM R	TRI FACILITY ID NUMBER
United States Environmental Protection Agency	**PART II. CHEMICAL-SPECIFIC INFORMATION (CONTINUED)**	Toxic Chemical, Category, or Generic Name

SECTION 6.2   TRANSFERS TO OTHER OFF-SITE LOCATIONS

6.2 __   Off-site EPA Identification Number (RCRA ID No.)

Off-Site Location Name

Street Address

City		County	

State	Zip Code	Is location under control of reporting facility or parent company?	☐ Yes ☐ No

A. Total Transfers (pounds/year) (enter range code or estimate)	B. Basis of Estimate (enter code)	C. Type of Waste Treatment/Disposal/ Recycling/Energy Recovery (enter code)
1.	1.	1. M
2.	2.	2. M
3.	3.	3. M
4.	4.	4. M

SECTION 6.2   TRANSFERS TO OTHER OFF-SITE LOCATIONS

6.2 __   Off-site EPA Identification Number (RCRA ID No.)

Off-Site Location Name

Street Address

City		County	

State	Zip Code	Is location under control of reporting facility or parent company?	☐ Yes ☐ No

A. Total Transfers (pounds/year) (enter range code or estimate)	B. Basis of Estimate (enter code)	C. Type of Waste Treatment/Disposal/ Recycling/Energy Recovery (enter code)
1.	1.	1. M
2.	2.	2. M
3.	3.	3. M
4.	4.	4. M

If additional pages of Part II, Section 6.2 are attached, indicate the total number of pages in this box ☐ and indicate which Part II, Section 6.2 page this is, here. ☐ (example: 1, 2, 3, etc.)

EPA Form 9350-1 (Rev. 12/4/92) - Previous editions are obsolete.   Range Codes: A = 1 - 10 pounds;   B = 11
C = 500 - 999 pounds.

♻EPA

United States
Environmental Protection
Agency

EPA FORM R

PART II. CHEMICAL-SPECIFIC INFORMATION (CONTINUED)

TRI FACILITY ID NUMBER
Toxic Chemical, Category, or Generic Name

SECTION 7B. ON-SITE ENERGY RECOVERY PROCESSES

☐ Not Applicable (NA) - Check here if <u>no</u> on-site energy recovery is applied to any waste stream containing the toxic chemical or chemical category.

Energy Recovery Methods [enter 3-character code(s)]

1 [_____]   2 [_____]   3 [_____]   4 [_____]

SECTION 7C. ON-SITE RECYCLING PROCESSES

☐ Not Applicable (NA) - Check here if <u>no</u> on-site recycling is applied to any waste stream containing the toxic chemical or chemical category.

Recycling Methods [enter 3-character code(s)]

1 [_____]   2 [_____]   3 [_____]   4 [_____]   5 [_____]

6 [_____]   7 [_____]   8 [_____]   9 [_____]   10 [_____]

EPA Form 9350-1 (Rev. 12/4/92) - Previous editions are obsolete.

♻EPA

United States
Environmental Protection
Agency

EPA FORM R

**PART II. CHEMICAL-SPECIFIC
INFORMATION (CONTINUED)**

TRI FACILITY ID NUMBER

Toxic Chemical, Category, or Generic Name

SECTION 7A. ON-SITE WASTE TREATMENT METHODS AND EFFICIENCY

☐ **Not Applicable (NA) - Check here if <u>no</u> on-site waste treatment is applied to any
waste stream containing the toxic chemical or chemical category.**

a. General Waste Stream (enter code)	b. Waste Treatment Method(s) Sequence [enter 3-character code(s)]		c. Range of Influent Concentration	d. Waste Treatment Efficiency Estimate	e. Based on Operating Data?
7A.1a	**7A.1b** 1 [ ] 2 [ ] 3 [ ] 4 [ ] 5 [ ] 6 [ ] 7 [ ] 8 [ ]		**7A.1c**	**7A.1d** %	**7A.1e** Yes ☐ No ☐
7A.2a	**7A.2b** 1 [ ] 2 [ ] 3 [ ] 4 [ ] 5 [ ] 6 [ ] 7 [ ] 8 [ ]		**7A.2c**	**7A.2d** %	**7A.2e** Yes ☐ No ☐
7A.3a	**7A.3b** 1 [ ] 2 [ ] 3 [ ] 4 [ ] 5 [ ] 6 [ ] 7 [ ] 8 [ ]		**7A.3c**	**7A.3d** %	**7A.3e** Yes ☐ No ☐
7A.4a	**7A.4b** 1 [ ] 2 [ ] 3 [ ] 4 [ ] 5 [ ] 6 [ ] 7 [ ] 8 [ ]		**7A.4c**	**7A.4d** %	**7A.4e** Yes ☐ No ☐
7A.5a	**7A.5b** 1 [ ] 2 [ ] 3 [ ] 4 [ ] 5 [ ] 6 [ ] 7 [ ] 8 [ ]		**7A.5c**	**7A.5d** %	**7A.5e** Yes ☐ No ☐

EPA Form 9350-1 (Rev. 12/4/92) - Previous editions are obsolete.

<table>
<tr><td colspan="2">

✦EPA

United States
Environmental Protection
Agency

</td><td colspan="2">

EPA FORM R

**PART II. CHEMICAL-SPECIFIC
INFORMATION (CONTINUED)**

</td><td>

TRI FACILITY ID NUMBER

Chemical, Category, or Generic Name

</td></tr>
</table>

SECTION 8. SOURCE REDUCTION AND RECYCLING ACTIVITIES

All quantity estimates can be reported using up to two significant figures.	Column A 1991 (pounds/year)	Column B 1992 (pounds/year)	Column C 1993 (pounds/year)	Column D 1994 (pounds/year)
8.1 Quantity released *				
8.2 Quantity used for energy recovery on-site				
8.3 Quantity used for energy recovery off-site				
8.4 Quantity recycled on-site				
8.5 Quantity recycled off-site				
8.6 Quantity treated on-site				
8.7 Quantity treated off-site				
8.8 Quantity released to the environment as a result of remedial actions, catastrophic events, or one-time events not associated with production processes (pounds/year)				
8.9 Production ratio or activity index				

8.10	Did your facility engage in any source reduction activities for this chemical during the reporting year? If not, enter "NA" in Section 8.10.1 and answer Section 8.11.		
	Source Reduction Activities [enter code(s)]	Methods to identify Activity (enter codes)	
8.10.1		a. b.	c.
8.10.2		a. b.	c.
8.10.3		a. b.	c.
8.10.4		a. b.	c.

8.11	Is additional optional information on source reduction, recycling, or pollution control activities included with this report? (Check one box)	YES ☐ NO ☐

* Report releases pursuant to EPCRA Section 329(8) including "any spilling, leaking, pumping, pouring, emitting, emptying, discharging, injecting, escaping, leaching, dumping, or disposing into the environment." Do not include any quantity treated on-site or off-site.

EPA Form 9350 - 1 (Rev. 12/4/92) - Previous editions are obsolete.

Appendix F: The New Jersey Department of Environmental Protection and Energy's Release and Pollution Prevention Report (DEQ-114) Section B

(RELEASE & POLLUTION PREVENTION REPORT FOR 1992)

SECTION B — *Complete one Section B Form for each New Jersey Environmental Hazardous Substance for 1992.*

1.1 Name and Location of Plant	1.2 NJEIN	2. Reporting Year
		1992

3.1 Environmental Hazardous Substance Name	3.2 CAS No.	3.3 RTK Substance No.

4.	ACTIVITIES AND USES OF THE SUBSTANCE AT THE FACILITY (Check all that apply.)

4.1	Manufacture the substance: a. ☐ Produce b. ☐ Import	If produce or import: c. ☐ For on-site use/processing e. ☐ As a byproduct	d. ☐ For sale/distribution f. ☐ As an impurity

4.2	Process the substance:	a. ☐ As a reactant b. ☐ As a formulation component c. ☐ As an article component d. ☐ Repackaging

4.3	Otherwise use the substance:	a. ☐ As a chemical processing aid b. ☐ As a manufacturing aid c. ☐ Ancillary or other use

4.4	Date Substance was First Used on Site (if known)
5.1	Principal Method of Storage:
5.2	Frequency of Transfer from Storage: _____ times per _____
5.3	Methods of Transfer:

		QUANTITY (lbs.)	Basis of Estimate
INVENTORY AND THROUGHPUT	6.1 Maximum Daily Inventory of Substance		
	6.2 Average Daily Inventory within all Process Equipment		
	6.3 Average Daily Inventory (excluding 6.2)		
	7. Starting Inventory of Substance		
	8. Quantity Produced on Site		
	9. Quantity Brought on Site		
	10. Quantity Consumed on Site (chemically reacted; NOT product)		
	11. Quantity Shipped off Siteas (or in) Product		
	12. Ending Inventory of Substance		
	13. Quantity Recycled and Used on Site		
	14. Quantity Destroyed through On-Site Treatment		
Air Emissions	15. Total Stack or Point Source Emissions		
	16. Total Fugitive or Non-Point Source Emissions		
Wastewater Discharges	17. Total Discharge to On-Site Treatment or Pretreatment System		
	18. Total Discharge to Publicly Owned Treatment Works (POTW)		
	19. Total Discharge to Surface Waters		
	20. Total Discharge to Ground Water		

Chemical or Category Name

21. On-Site Land Disposal and/or Transfers of Wastes to Off-Site Locations: ☐ N/A

Table A Physical State	Table B Storage Method	Management Facility Information Name, Address (Street, City, State, Zip) and USEPA ID #	Quantity of Substance in Waste (lbs/yr)	Table C Mgmt. Method
		☐ Check if final disposal is "On site"		
		☐ Check if final disposal is "On site"		
		☐ Check if final disposal is "On site"		

21.1   TOTAL QUANTITY OF SUBSTANCE IN WASTE ☐

SELF VERIFICATION STATEMENT: The sum of starting inventory, quantity produced and quantity brought on site should approximately equal the sum of quantity consumed (chemically altered), quantity shipped off site as (or in) product and as (or in) waste, quantity destroyed through on-site treatment, total air emissions, total wastewater discharges, and ending inventory. (See worksheet in the instructions.)

* Quantity released to the environment as a result of remedial actions, catastrophic events, or one-time events not associated with production processes (pounds/year)	

	QUANTITY (lbs.)	Basis of Estimate
** Total Non-product Output (NPO) Generated (pounds/year)		

	QUANTITY	UNITS	PRODUCT DESCRIPTION
22.a.	1992 Quantity and Units of Production* Associated with the Substance			
22.b.	1992 Quantity and Units of Production* Associated with the Substance			
23.a.	1991 Quantity and Units of Production* Associated with the Substance			
23.b.	1991 Quantity and Units of Production* Associated with the Substance			

*PRODUCTION: Whenever possible, "UNITS" should be mass or surface area units only, such as pounds of material manufactured or square footage of product involved.

Chemical or Category Name

POLLUTION PREVENTION ACTIVITIES

For the purpose of this question, pollution prevention means " any method or technique at or before the point of generation, the application of which reduces or eliminates the use or generation of hazardous substances prior to treatment, storage, out-or-process recycling or disposal". Pollution prevention is NOT any type of treatment, out-of-process recycling, incineration, or transfer of releases to different media.

---

24. Has any pollution prevention method been employed to reduce the quantity of this Substance during 1992 relative to 1991 levels?

☐ Yes ☐ No   If "Yes", fill in the table below:

POLLUTION PREVENTION METHODOLOGY (Complete all appropriate sections)	Quantity of Substance Reduced (lbs.) (1991 to 1992)	Basis of Estimate
24.1 Material-Related Change (changes in the amount of Substance used due to substitution of other substance)		

Name and Quantity of Substitute Substance

	CAS NUMBER	SUBSTANCE	QUANTITY (lbs.)
a)			
b)			
c)			

	Quantity of Substance Reduced (lbs.) (1991 to 1992)	Basis of Estimate
24.2 Reformulation or Redesign of Product (resulting in the reduction of Substance generated)		
24.3 Process or Procedure Modifications (using existing equipment to reduce Substance generated)		
24.4 Equipment or Technology Modifications (using new equipment or technology to reduce Substance generated)		
24.5 Improved Operations (due to housekeeping, training, material handling or inventory control to reduce Substance generated)		
24.6 Discontinuance of Operations, excluding operations transferred to or undertaken by another facility		
24.7 Export of Use		
24.8 Miscellaneous		

25. Does your facility anticipate reducing the generation of the Substance (as a waste) in the future due to pollution prevention? ☐ Yes ☐ No   If "Yes", indicate your projections in the table below:

PROJECTION	YEAR	
	1994	1997
Quantity of Substance Reduced per Year due to Pollution Prevention (lbs.)		

Appendix G: The United States University-Affiliated Pollution Prevention Research and Training Centers

University-Affiliated Pollution Prevention/Research and Training Assistance Centers

This section lists organizations involved in more research or training in source reduction and recycling. Some programs may provide assistance to small and/or medium sized businesses. Although these university centers are often partially funded by EPA or State agencies, they operate as independent entities.

Alabama

University of Albama

Environmental Institute for Waste Management Studies (EIWMS)
Activities include policy research, technology transfer, and basic research. Their Hazardous Material & Management and Resource Recovery (HAMMARR) program provides regulatory information, waste exchange and technical assistance for waste minimization, and workshops for small quantity generators and local businesses. Many of the 1992 workshops will focus specifically on the metal casings industry. The University's College of Continuing Education also offers courses on pollution prevention.

Contact: Dr. Robert Griffin, Director
            Hazardous Materials Management and Resource
            Recovery Program (HAMMARR)
            University of Alabama
            275 Mineral Industries Building
            Box 870203
            Tuscaloosa, Alabama 35487-0203
            205-348-8403

Gulf Coast Hazardous Substance Research Center (GCHSRC)
The University of Alabama is a member of the GCHSRC, which is located at Lamar University in Beaumont, Texas (see the listing under Texas).

California

University of California

Environmental Hazards Management Program
The University of California at Berkeley, Davis, Irvine, Los Angeles, Santa Cruz, Riverside, Santa Barbara, and San Diego offers post-graduate continuing education courses on toxic materials that devote some time to pollution prevention issues. Many of the courses give certificates in hazardous material management and air quality management. Some locations offer environmental auditing and other related topics.

Contact: Jon Kindschy, Statewide Coordinator
            Environmental Hazards Management Program
            University of California Extension
            Riverside, California 92521-0112
            714-787-5804

University of California at Los Angeles
Center for Waste Reduction Technologies
The center conducts industry-supported research into waste reduction technology.

Contact: Dr. David Allen
           University of California, Los Angeles
           Los Angeles, California 90024
           213-206-0300

Colorado

Colorado State University
Waste Minimization Assessment Center (WMAC)
WMAC is managed through the University City Science Center of Philadelphia. The center conducts detailed waste minimization assessments at small- to medium-sized manufacturing companies, training workshops for the Department of Health personnel, and training for EPA Region VIII RCRA inspectors. The center is also performing solvent use reduction audits at two manufacturing plants and will develop technical information on solvent use practices for small- to medium-sized manufacturing plants. In addition, the Center conducts training workshops for Department of Health personnel to develop technical expertise in pollution prevention. Contact Region VIII for information on these workshops.

Contacts: Dr. Harry Edwards, Director
          Waste Minimization Assessment Center
          Mechanical Engineering Department
          Colorado State University
          Fort Collins, Colorado 80523
          303-491-5317

          Marie Zanowich, Project Officer
          U.S. EPA Region VIII
          999 18th Street, Suite 500
          Denver, Colorado 80202-2505
          303-294-1065

Connecticut

University of Connecticut
Pollution Prevention Research and Development Center (PPRDC)
An EPA funded research center, the PPRDC will support state government and industry in reducing toxic emissions by encouraging existing and start up companies to provide services and equipment necessary for pollution prevention technologies, and by creating new jobs to meet the demands of this industry. PPRDC's goal is to work with industry to develop pollution prevention technology and a manufacturing base in the region.

Contact: Dr. George Hoag
          Director, Pollution Prevention
          Research and Development Center
          Environmental Research Institute
          Box U-120, Route 44, Longley Building 146
          University of Connecticut
          Storrs, Connecticut 06269-3210
          203-486-4015 Fax 203-486-2269

Waterbury State Technical College
Industrial Environmental Management (IEM)
Waterbury State Technical College offers a waste minimization course as part of its Industrial Environmental Management certificate level and associate degree level programs. Other courses include environmental regulations, safe handling of hazardous wastes, and environmental control processes.

Contact: Cynthia Donaldson, Chairperson
          Industrial Environmental Management
          Waterbury State Technical College
          750 Chase Parkway
          Waterbury, Connecticut 06708-3089
          203-596-8703/575-8089

District of Columbia

Howard University
The Great Lakes and Mid-Atlantic Hazardous Substance Research Center
The center is funded by EPA and focuses on the unique problems of EPA Regions III and V. Research is conducted on hazardous substances and related environmental problems. Among other projects, the center is developing materials for a hazardous waste workshop and videotapes on waste minimization information and training. The University of Michigan and Michigan State University are also members of the center.

Contact: Dr. James H. Johnson, Jr., Assistant Director
           The Great Lakes and Mid-Atlantic Hazardous
           Substance Research Center
           Department of Civil Engineering
           Howard University
           Washington, D.C. 20059
           202-806-6570

Florida

Florida Institute of Technology
Research Center for Waste Utilization
The center offers classroom training in waste utilization at the undergraduate and graduate levels. In addition, the center is involved in research in the areas of municipal solid waste (MSW), industrial solid waste, and pollution prevention. Specific studies include heavy metal sources in the MSW stream, uses of ash from waste-to-energy plants, biological toxicity of ash residues, and degradable plastics characteristics after disposal.

Contact: Edwin Korzun, Executive Director
           Research Center for Waste Utilization
           Department of Marine and Environmental Sciences
           Florida Institute of Technology
           150 West University Boulevard
           Melbourne, Florida 32901-6988
           305-768-8000

University of Central Florida
Gulf Coast Hazardous Substance Research Center (GCHSRC)
The University of Central Florida is a member of the GCHSRC, which is located at Lamar University in Beaumont, Texas (see the listing under Texas).

University of Florida
Center for Training, Research, and Education for Environmental Occupations
The center's activities include developing a statewide training action plan for business, government, and the public; providing RCRA hazardous waste regulation training; developing a university-level waste reduction curriculum; sponsoring a 2-day symposium; and developing a training program for three specific industries.

Contact: Dr. James O. Bryant, Jr., Director
           Center for Training, Research, and
           Education for Environmental Occupations
           Division of Continuing Education
           University of Florida
           3900 S.W. 63rd Boulevard
           Gainesville, Florida 32608-3848
           904-392-9570

Florida Center for Solid & Hazardous Waste Management
The Center coordinates the State's solid and hazardous waste research efforts, including management practices for waste reduction, reuse, recycling, and improved conventional disposal methods.
Contact: Dr. James O. Bryant, Jr.
           Florida Center for Solid and Hazardous Waste Management
           University of Florida
           3900 S.W. 63rd Boulevard
           Gainesville, Florida 32608-3848
           904-392-9570

Georgia

Georgia Institute of Technology (Georgia Tech)
Environmental Science and Technology Laboratory
The institute provides continuing education workshops on a wide variety of waste reduction and pollution prevention topics, including hazardous waste reduction planning requirements. As part of a U.S. EPA grant with the Georgia Hazardous Waste Management Authority, the institute is offering workshops to help industry write proposals for grants implementing new waste minimization technologies. Within the Hazardous Materials Group of the Laboratory are the Hazardous Waste Technical Assistance Program (HWTAP) and the Pollution Prevention Program. These programs provide technical assistance to Georgia industry to encourage voluntary waste reduction and minimization, as well as compliance with hazardous waste regulations. Activities include onsite assistance, telephone consultations, information dissemination, multi-media information releases, short courses, and annual seminars. The Pollution Prevention Program is funded by EPA grants, while HWTAP is paid for through general funds.

Contact: Carol Foley
          Georgia Tech Research Institute
          Environmental Science and Technology Laboratory
          Atlanta, Georgia 30332
          404-894-3806

Illinois

Illinois Institute of Technology

Industry Waste Elimination Research Center (IWERC)
The center's research priorities include recycling or reusing industrial by-products and developing manufacturing processes that avoid generating wastes or pollutants. In conjunction with the Department of Environmental Engineering, graduate programs are offered with an option in hazardous waste management.

Contact: Dr. Kenneth E. Noll, Director
          Industrial Waste Elimination Research Center
          Pritzker Department of Environmental Engineering
          IIT Center
          Chicago, Illinois 60616
          312-567-3536

University of Illinois

Hazardous Waste Research & Information Center (HWRIC)
The center combines research, education, and technical assistance in a multidisciplinary approach to manage and reduce hazardous waste. HWRIC collects and shares information through its library/clearinghouse and a computerized Waste Reduction Advisory System (see description in Section 7, Pollution Prevention Clearinghouses and Associations).

Contact: Dr. David Thomas, Director
          Hazardous Waste Research and Information Center
          One East Hazelwood Drive
          Champaign, Illinois 61820
          217-333-8940

Indiana

Purdue University
Pollution Prevention Program
The Pollution Prevention Program provides outreach and technical assistance efforts to industry (including onsite assessments conducted by graduate students) on pollution prevention opportunities. Purdue University and the Indiana Department of Environmental Management sponsor both general and specific workshops on pollution prevention and recycling.

Contact: Rick Bossingham, Coordinator
          Pollution Prevention Program
          Purdue University
          2129 Civil Engineering Building
          West Lafayette, Indiana 47907-1284
          317-494-5038

Iowa

University of Northern Iowa
Iowa Waste Reduction Center
This EPA funded center is designed to be a technology transfer center, utilizing research findings from across the globe to benefit existing and potentially new Iowa businesses and industries.

Contact: Dr. John L. Konefes
          Director, Recycling and Reuse Technology
          Transfer Center
          Iowa Waste Reduction Center
          75 BRC
          University of Northern Iowa
          Cedar Falls, Iowa 50614-0185
          319-273-2079 Fax 319-273-6494

Kansas

Kansas State University
Hazardous Substance Research Center (HSRC)
This EPA-funded center provides research and technology transfer services for pollution prevention and other waste management techniques. HSRC programs include outreach to industry, assistance to government, videos, radio programs, written materials, data bases, and workshops on pollution prevention and hazardous waste remediation. One pollution prevention focus of this center is on soils and mining waste.

Contact: Dr. Larry E. Erickson, Director
          Hazardous Substance Research Center
          Durland Hall, Room 105
          Kansas State University
          Manhattan, Kansas 66506-5102
          913-532-5584

University of Kansas
Center for Environmental Education and Training
In cooperation with the Kansas Department of Health and Environment, the center offers Hazardous Waste Regulatory Training Conferences. Conference topics include waste minimization, regulatory compliance, and technology transfer components.

Contact: Lani Heimgardner
          Center for Environmental Education and Training
          Division of Continuing Education
          University of Kansas
          6330 College Boulevard
          Overland Park, Kansas 66211
          913-491-0810

Kentucky

University of Louisville

Kentucky PARTNERS - State Waste Reduction Center
This center conducts general and industry-specific seminars and workshops on environmental regulations and pollution prevention methods. Another service is free, non-regulatory pollution prevention services for all Kentucky industries and business. In addition, Kentucky PARTNERS publishes a newsletter and performs onsite assessments.

Contact: Joyce St. Clair
           Executive Director
           Kentucky PARTNERS - State Waste Reduction Center
           Ernst Hall, Room 312
           University of Louisville
           Louisville, Kentucky 40292
           502-588-7260

Waste Minimization Assessment Center
WMAC is managed through the University City Science Center in Philadelphia. The center conducts quantitative, on-site, waste minimization assessments for small to medium sized generators located within a 150 mile radius of Louisville. In addition, the center incorporates risk reduction and pollution prevention into the undergraduate and graduate engineering curricula. Professionals are encouraged to participate in these courses. Engineering students also conduct waste minimization projects at manufacturing plants.

Contact: Marvin Fleischman, Director
           Waste Minimization Assessment Center
           Department of Chemical Engineering
           University of Louisville
           Louisville, Kentucky 40292
           502-588-6357

Louisiana

Louisiana State University (Shreveport)
Hazardous Waste Research Center (HWRC)
Categories of research conducted by faculty and students include incineration and combustion, alternative methods of treatment and destruction, and transport of leachate and wastes from pits and spills.

Contact: David Constant, Director
          Hazardous Waste Research Center
          3418 CEBA Building
          Louisiana State University
          Baton Rouge, Louisiana 70803
          504-388-6770

Louisiana State University (LSU)
Gulf Coast Hazardous Substance Research Center (GCHSRC)
LSU is a member of the GCHSRC, which is located at Lamar University in Beaumont, Texas (see the listing under Texas).

Southern University at Baton Rouge
Center for Energy and Environmental Studies
The center will support individual pollution prevention, treatment technology and socio-economic policy research projects.

Contact: Dr. Robert L. Ford
          Director, Center for Energy and Environmental Studies
          Southern University at Baton Rouge
          Cottage #8, P.O. Box 9764
          Baton Rouge, Louisiana 70813
          504-771-4723 Fax 504-771-4722

Maine

University of Maine (UM)
Chemicals in the Environment Information Center
The Center provides courses, conferences, presentations and brochures emphasizing pollution prevention. Courses are Issues in Environmental Pollution; Pollution Prevention - Changing Ourselves and Changing Society (Honors students); and Pollution Prevention through Understanding and Managing the Chemicals in Our Lives (teachers). Conferences are for business, e.g. Pollution Prevention in the Home, Workplace and Community. Work is carried out in cooperation with state agencies, Cooperative Extension and Maine Waste and Toxics Use Reduction Committee.

Contact: Marquita K. Hill, PhD, Director
          University of Maine
          5737 Jenness Hall
          Orono, Maine 04469-5737
          207-581-2301

Massachusetts

Massachusetts Institute of Technology (MIT)
Center for Technology, Policy and Industrial Development
Along with the Center, the Technology, Business and the Environment Group conducts research and offers workshops in pollution prevention. Pollution prevention concepts are also included in some undergraduate and graduate courses.

Contact: John Ehrenfeld
          Technology, Business and the Environment Group
          Center for Technology, Policy and Industrial Development
          E40-241
          Massachusetts Institute of Technology
          Cambridge, Massachusetts 02139
          617-253-7753

Tufts University

Tufts Environmental Literacy Institute (TELI)
The Institute is conducting a demonstration project, Tufts CLEAN to analyze the energy and materials flow at the university. Funded by EPA's Office of Pollution Prevention, this project involves students in audit design, data collection and analysis, implementation, and evaluation.

Contact: Dr. Anthony Cortese
           Dean of Environmental Programs
           Tufts University
           Office of Environmental Programs
           474 Boston Avenue, Curtis Hall
           Medford, Massachusetts 02155
           617-627-3452

The Center for Environmental Management
The purpose of this center is to develop a multidisciplinary approach to environmental problems through health effects research, technology research, policy analysis, education and training programs, and information transfer. Pollution prevention is emphasized throughout center programs.

Contact: Dr. Kurt Fischer
           Tufts University
           Center for Environmental Management
           474 Boston Avenue, Curtis Hall
           Medford, Massachusetts 02155
           617-627-3452 FAX 617-627-3084

University of Massachusetts • Lowell

Toxics Use Reduction Institute
The Massachusetts Toxics Use Reduction Institute promotes reduction in the use of toxic chemicals or the generation of toxic by-products in Massachusetts industry. The Institute is a multidisciplinary research, education, training and technical support center located at the University of Massachusetts • Lowell.

Contact: Dr. Jack Luskin
           Associate Director for Education and Training
           Toxics Use Reduction Institute
           University of Massachusetts • Lowell
           Lowel, Massachusetts 01854-2881
           508-934-3275 FAX 508-453-2332

Michigan

Grand Valley State University
Waste Reduction and Management Program (WRMP)
The WRMP is a university-based pollution prevention program that conducts research and provides technical assistance to Michigan industry. "Design for Recycling: Solving Tomorrow's Problems Today," a 1-year waste reduction research and demonstration project, is funded by the Padnos Foundation and the Michigan Department of Natural Resources as part of the Quality of Life Bond Program. The overall objective of the project is to reduce the future generation of solid waste by infusing undergraduate engineering curricula with the concept of design for the entire product life-cycle. This project includes the following activities: identifying and prioritizing 10 products that have the greatest potential for design change to promote recycling; and developing a series of seminars to focus Michigan manufacturers, engineers, and engineering faculty on "cutting edge" design approaches; developing engineering curricular materials to assist faculty in developing student awareness and skill in designing products with end-stage product management in mind.

Contact: Dr. Paul Johnson, Associate Professor
           Grand Valley State University
           School of Engineering
           301 W.Fulton, Room 617
           Grand Rapids, Michigan 49504
           616-771-6750
           1400 Townsend Drive
           Houghton, Michigan 49931
           906-487-2098

Michigan Technological University
Center for Clean Industrial and Treatment Technologies (CCITT)
The emphasis of this center is pollution prevention through identification of alternatives , balanced assessment and targeted research and development. Ultimately, the goal is to develop and advocate methods to fully utilize raw materials and produce products which are largely recyclable and/or exhibit minimal lifetime environmental risk. This is to be accomplished by acting as a sort of "analytical bridge" between industry, government and academia to promote practical means of total quality management and environmental equity.

Contact: Dr. John C. Crittenden, Director
           Environmental Engineering Center
           Michigan Technological University
           1400 Townsend Drive
           Houghton, Michigan 49931
           906-487-3143 FAX 906-487-2061

University of Detroit Mercy
Center of Excellence in Polymer Research and Environmental Study
The center is a partnership of university, industry, and government whose purpose it is to conduct high technology research that addresses environmental problems related to polymer wastes and proposes the development of new environmentally responsible and safe polymer products. The center is also committed to the transfer of pollution prevention and waste management technologies to commercial application in products and processes through their industry partners.

Contact: Dr. Daniel Klemper, Director
                 Center of Excellence in Polymer Research
                 and Enviromental Study
                 University of Detroit Mercy
                 4001 W.McNichols Road
                 Detroit, Michigan 48219-3599
                 313-993-1270 FAX 313-993-1409

University of Michigan
EPA Pollution Prevention Center for Curriculum Development and Dissemination
The purpose of this center is to develop pollution prevention curriculum modules for undergraduate and graduate courses in engineering business and science (see description in Section 6, U.S.EPA's Environmental Education Activities).

Contact: Dr. Gregory A. Keoleian, Manager
                 School of Natural Resources
                 University of Michigan
                 Dana Building
                 430 E. University
                 Ann Arbor, Michigan 48109-1115
                 313-764-1412

The Great Lakes and Mid-Atlantic Hazardous Substance Research Center (GLMA-HSRC)
A cooperative research consortium comprising the University of Michigan, Michigan State University, and Howard University, this center supports hazardous substance training, technology transfer, and research.

Contact: Dr. Walter Weber, Director
                 Hazardous Substance Research Center
                 University of Michigan
                 Sute 181 Engineering 1-A
                 Ann Arbor, Michigan 48109-2125
                 313-763-2274

Minnesota

University of Minnesota
Minnesota Technical Assistance Program
Using EPA's Toxic Release Inventory (TRI), the program provides technical transfer, workshops, and fact sheets encouraging decreased use of TRI chemicals through use of alternatives and waste minimization.

Contact: David Simmons
           Public Relations Representative
           Minnesota Technical Assistance Program
           1315 5th St., S.E., Suite 207
           University of Minnesota
           Minneapolis, Minnesota 55414-4504
           612-627-4646

Mississippi

Mississippi State University
Missisipi Technical Assistance Program and Mississippi Solid Waste Reduction Assistance Program
These programs work cooperatively to provide pollution prevention research, onsite waste assessments, workshops, conferences, employee and student education materials, a waste exchange, technology data bases, and a monthly newsletter.

Contact: Dr. Don Hill, Dr. Caroline Hill,
           or Dr. June Carpenter
           Mississippi Technical Assistance Program and
           Mississippi Solid Waste Reduction Assistance Program
           P.O. Drawer CN
           Mississippi State University, Mississippi 39762
           601-325-8454

Mississippi State University
Gulf Coast Hazardous Substance Research Center (GCHSRC)
MSU is a member of the GCHSRC, which is located at Lamar University in Beaumont, Texas (see the listing under Texas).

Nevada

University of Nevada at Reno
Nevada Small Business Development Center
The Nevada Small Business Development Center, in cooperation with the Nevada Division of Environmental Protection, offers free pollution prevention services to industry and businesses, including seminars, workshops, onsite evaluations, fact sheets, and a newsletter. The center also maintains a Hazardous Waste Information Line, assisting businesses with regulations, alternative product use, and pollution prevention.

Contact: Kevin Dick, Manager
Business Environmental Program
Nevada Small Business Development Center
University of Nevada - Reno
Reno, Nevada 89557-0100
702-784-1717

New Jersey

New Jersey Institute of Technology
Hazardous Substance Management Research Center
Areas of research include incineration, biological/chemical treatment, physical treatment, site assessment remediation, health effects assessment, and public policy/education.

Contact: Dick Magee
Advanced Technology Center Building
323 Martin Luther King Boulevard
University Heights
Newark, New Jersey 07102
201-596-5864

New Mexico

New Mexico State University
Waste-Management Education and Research Consortium (WERC)
WERC is a waste management education and research consortium established by
New Mexico State University (NMSU) under a U.S. Department of Energy grant in
1990. Consortium members include NMSU, the University of New Mexico, the
New Mexico Institute of Mining and Technology, the Navajo Community College,
the Los Alamos National Laboratories, and the Sandia National Laboratories. The
mission of WERC is to expand the Nation's capability to address the issues related
to management of all types of waste (hazardous, solid, and radioactive). WERC activities involve all waste management options, including pollution prevention.
Some of the major programs undertaken by WERC are the following:

- Education and curricula development in waste management by the consortium universities (graduate, undergraduate, and associate degrees with concentrations in environmental management);
- A professional development teleconference series for industry and government;
- Research programs that provide training to faculty and students.

Contact: John S. Townsend, Assistant Director
           WERC
           New Mexico State University
           Box 30001
           Department 3805
           Las Cruces, New Mexico 88003-0001
           505-646-2038

New York

Clarkson University
Hazardous Waste and Toxic Substance Research and Management Center
This center coordinates and mobilizes funding for multidisciplinary research at
Clarkson University. Projects currently being conducted include a wide range of
basic research, applied engineering, and technology development topics. Many
of these projects address waste minimization and pollution prevention either directly or indirectly.

Contact: Thomas L. Theis, Director
           Hazardous Waste and Toxic Substance Research
           and Management Center
           Rowley Laboratories
           Clarkson University
           Potsdam, New York 13699
           315-268-6542

Cornell University
Waste Management Institute
The institute coordinates interdisciplinary research on waste reduction and management options for hazardous, agricultural, solid, industrial, and sludge wastes. Numerous fact sheets and publications are made available on topics ranging from source reduction opportunities for shoppers to waste minimization opportunity assessment for communities and businesses.

Contact: Richard Schuler, Director
           Waste Management Institute
           313 Hollister Hall
           Cornell University
           Ithaca, New York 14853
           607-255-8674

North Carolina

North Carolina State University
EPA Research Center for Waste Minimization and Management
U.S. EPA is sponsoring a major university-based research center that focuses specifically on the challenge to minimize and manage hazardous substances. Located at North Carolina State University, the center involves Texas A&M University and the University of North Carolina at Chapel Hill. The mission of the center is to develop practical means for industry to eliminate the use and generation of hazardous substances, treat those wastes that cannot be eliminated, and provide secure containment for treatment residues. The major research focus at the center will be the elimination or reduction in discharge of hazardous substances to all environmental media. A strong commitment also will be made to technology transfer and training.

Contacts: Dr. Michael Overcash
           Dr. Cliff Kaufman
           Center for Waste Minimization and Management
           North Carolina State University
           Box 7905
           Raleigh, North Carolina 27695-2325
           919-515-2325

University of North Carolina - Chapel Hill
EPA Research Center for Waste Minimization and Management
The University of North Carolina at Chapel Hill is a member of the U.S. EPA Research Center located at North Carolina State University in Raleigh, North Carolina (see the listing under "North Carolina State University").

Contacts: Dr. William H. Glaze
Department of Environmental Science & Engineering
University of North Carolina - Chapel Hill
Chapel Hill, North Carolina 27514
919-966-1024

North Dakota

University of North Dakota
Energy and Environmental Research Center (EERC)
The EERC features an integrated systems approach to energy and environmental research and technology development beginning with fundamental evaluation and characterization of earth resources, followed by research and development of innovative technologies to extract and utilize these resources in an efficient and environmentally acceptable manner, and culminating in the utilization of safe disposal of wastes generated in using natural resources.

Contact: Dr. Gerald Groenwald, Director
Energy and Environmental Research Center
Center of Excellence for Toxic Metal Emissions
University of North Dakota
15 North 23rd Street, Box 8213
University Station
Grand Forks, North Dakota 58202-8213
701-777-5131 FAX 701-777-5181

Ohio

University of Cincinnati
American Institute for Pollution Prevention (AIPP)
The AIPP is located at the University of Cincinnati (see description in Section 7, Pollution Prevention Clearinghouses and Associations).

Contact: Jean Boddocsi, Director
American Institute for Pollution Prevention (AIPP)
Office of the University Dean for Research
University of Cincinnati
Cincinnati, Ohio 45221
513-556-4532

University of Findlay
RCRA Generator Training Program
Workshops introduce U.S. EPA's Pollution Prevention Program for personnel at industries and commercial businesses that generate hazardous waste. Training courses assist generators in developing waste minimization strategies such as source reduction with the goal of eliminating waste generation. Regulation and compliance are also discussed. Workshops consist of 2-3 day sessions.

Contact: George Kleevic
           Workshop Instructor
           RCRA Generator Training Program
           P.O. Box 538
           St. Clairsville Ohio 43950
           614-695-5036

Oklahoma

Oklahoma State University
Center for Resource Conservation and Environmental Research (CRCER)
The goal of this Center is to establish and maintain a "center without walls" to provide Oklahoma, the Southwest region, and the nation with benefits of a co-ordinated, multidisciplinary, multi-institutional research, analysis, and evaluation of the technical, policy and managerial issues related to resource conservation and reduction; reduction/disposal of municipal and industrial wastes; and avoidance/correction of pollution or air, land and water. The Center will accomplish its technical studies and policy analyses primarily through the resources of Oklahoma State University, the University of Oklahoma, and the University of Tulsa.

Contact: Mr. Robert Fulton
           Vice President, Oklahoma Alliance for Public Policy Research
           2630 Northwest Expressway, Suite B
           Oklahoma City, Oklahoma 73112
           405-943-8989 FAX 405-840-0061

Pennsylvania

University of Pittsburgh
Center for Hazardous Materials Research (CHMR)
The center conducts applied research, health and safety training, education, and international technology transfer projects involving hazardous and solid wastes. It also provides technical assistance, onsite assessments, and fact sheets and manuals on pollution prevention for industries in Pennsylvania.

Contact: Dr. Edgar Berkey
           Center for Hazardous Materials Research
           University of Pittsburgh Trust
           Applied Research Center
           320 William Pitt Way
           Pittsburgh, Pennsylvania 15238
           412-826-5320

Rhode Island

University of Rhode Island
Chemical Engineering Department
Advanced students and their professors develop and evaluate pollution prevention engineering solutions for Rhode Island firms. These firms are referred by the Rhode Island Department of Environmental Management's voluntary pollution prevention technical assistance program.

Contact: Prof. Stanley M. Barnett, Chairman
           Chemical Engineering Department
           Crawford Hall
           University of Rhode Island
           Kingston, Rhode Island 02881
           (401) 792-2443

South Carolina

University of South Carolina
Hazardous Waste Management Research Fund
The fund sponsors research and educational programs in the area of hazardous waste reduction. Research priorities include technology transfer, assessment training, site remediation, recycling and reuse strategies, and policy issues. Topics to be covered in the educational programs include vehicle/auto service shops, textiles, metal fabrication and machine shops, painting and coating, solvent use reduction, and developing a site specific waste reduction program. The fund has also established educational programs at Clemson University in Clemson, South Carolina.

Contact: Doug Dobson, Executive Director
           Institute of Public Affairs
           University of South Carolina
           Gambrell Hall, 4th Floor
           Columbia, South Carolina 29208
           803-777-8157

Tennessee

University of Tennessee
Center for Industrial Services (CIS)
The center sponsors an extensive waste reduction assessment training program that includes indepth waste reduction assessment courses. This training program was originally developed to instruct retired industrial engineers and managers, who became a highly skilled waste reduction assessment team. A key program for the center has been waste reduction assessments by full-time field engineers and retired engineers.

Contact: Cam Metcalf
           Center for Industrial Services
           University of Tennessee
           226 Capitol Boulevard Building
           Suite 606
           Nashville, Tennessee 37219
           615-242-2456

Waste Minimization Assessment Center (WMAC)
Managed by the University City Science Center in Philadelphia, WMAC is staffed by engineering students and faculty who have considerable expertise with process operations in manufacturing plants and who also have the skills needed to minimize waste generation. These staff members perform quantitative waste minimization assessments for small to medium sized generators.

Contact: Dr. Richard J. Jendrucko, Director
           Department of Engineering
           Science and Mechanics
           University of Tennessee
           310 Perkins Hall
           Knoxville, Tennesse 37996-2030
           615-974-7682

Texas

Texas A & M University

EPA Research Center for Waste Minimization and Management
Texas A & M University is a member of the U.S. EPA Research Center located at North Carolina State University in Raleigh, North Carolina (see the listing under North Carolina).

Contact: Dr. Kirk Brown
           Department of Soil and Crop Science
           Texas A & M University
           College Station, Texas 77843
           409-845-5251

Gulf Coast Hazardous Substance Research Center (GCHSRC)
Texas A & M is a member of the GCHSRC, which is located at Lamar University in Beaumont, Texas (see the listing under "Lamar University").

Texas Tech University
Center for Environmental Technologies
The center coordinates conferences, short courses, and lectures that address environmental concerns, pollution prevention, pollution controls, and Federal, State, and local regulations. Conferences and short courses are offered for State and municipal audiences, professional and civic groups, and industry. The center is also conducting at least 15 different research projects involving pollution prevention in such areas as storm water discharge, groundwater monitoring, and pesticides.

Contact: Dr. John R. Bradford
           Center for Environmental Technologies
           Texas Tech University
           P.O. Box 43121
           Lubbock, Texas 79409-3121
           806-742-1413

Lamar University
Gulf Coast Hazardous Substance Research Center (GCHSRC)
The GCHSRC is a research consortium of eight universities, with its center located at Lamar University. Its purpose is to conduct research to aid in more effective hazardous substance response and waste management. The center's efforts are concentrated in the areas of waste minimization and alternative technology development. The center receives funding from the U.S. EPA and the State of Texas, with a majority of those funds being pledged to pollution prevention for the petrochemical and microelectronic industries. At this time, the center has some 60 projects in progress in a joint Federal, State, and industry effort at Texas Universities, and at research centers outside the State. The other members of the consortium are Louisiana State University, Mississippi State University, University of Alabama, University of Central Florida, University of Houston, University of Texas - Austin, and Texas A & M.

Contact: Mr. Tom Pinson, Assistant Director
           Gulf Coast Hazardous Substance Research Center
           Lamar University
           P.O. Box 10613
           Beaumont, Texas 77710
           409-880-8707 Fax 409-880-2397

University of Houston
Gulf Coast Hazardous Substance Research Center (GCHSRC)
The University of Houston is a member of the GCHSRC, which is located at Lamar University in Beaumont, Texas (see the listing under Lamar University).

University of Texas - Arlington
Environmental Institute for Technology Transfer (EITT)
EITT was established to facilitate research, technical assistance, and the dissemination of environmental knowledge to assist business and industry in finding cost-effective and environmentally acceptable solutions to compliance problems. In addition to offering training courses that address pollution prevention, the institute provides a forum for industry and regulators to address common concerns through workshops, seminars, and conferences.

Contacts: Dr. Gerald I. Nehman, Director
Dr. Victorio Argento, Associate Director
Environmental Institute for Technology Transfer
University of Texas at Arlington
Box 19050
Arlington, Texas 76019
817-273-2300

University of Texas - Austin
Gulf Coast Hazardous Substance Research Center (GCHSRC)
The University of Texas - Austin is a member of the GCHSRC, which is located at Lamar University in Beaumont, Texas (see the listing under Lamar University).

Utah

Weber State University
Center for Environmental Service
Environmental management training and technical assistance are available with a special emphasis on the needs of Northern Utah's small and medium-sized businesses and manufacturers as well as its cities and towns. Pollution Prevention opportunities are among the topics covered by the Center's services.

Contact: Dianne Siegfreid, Director
Barbara A. Wachocki, Director
Center for Environmental Services
Weber State University
Ogden, Utah 84408-2502
801-626-7559

Wisconsin

University of Wisconsin - Madison
Engineering Professional Development Program
The College of Engineering offers intensive, short courses on waste minimization, environmental compliance, industrial environmental engineering, and pollution prevention from the design aspect.

Contact: Pat Eagan
           Engineering Professional Development Program
           College of Engineering
           University of Wisconsin at Madison
           432 North Lake Street
           Madison, Wisconsin 53706
           608-263-7429

Solid and Hazardous Waste Education Center
In cooperation with the Wisconsin Department of Natural Resources, the Extension Office offers workshops in solid waste reduction, recycling, composting, as well as general and industry-specific (electroplating and metal finishing, auto repair, local government, and schools) workshops on waste minimization and pollution prevention. The center also works directly with industry and government to provide technical assistance.

Contacts: David Liebel
           Wayne Pferdehirt
           Solid and Hazardous Waste Education Center
           University of Wisconsin - Extension
           529 Lowell Hall
           610 Langdon Street
           Madison, Wisconsin 53703
           608-265-2360

Appendix H: Reading suggestions

AIA (1998) American Institute of Architects Environmental Design Charrette's WWW home page: http://www.aia.org/edc/homepage.htm

AIChE (1992) Center for Waste Reduction Technologies, American Institute of Chemical Engineers, Request for Proposal, "Estimates of the True, Current and Future Cost of Waste Emissions," January

Allen A, N Bakashani and K Sinclair (1991) "Pollution Prevention: Homework and Design Problems for Engineering Curricula." Rosselot Department of Chemical Engineering, University of California, Los Angeles

Anderson D (1992) "Economic Growth and the Environment." Policy Research Working Paper 979, World Bank, Washington, D.C.

Byers RL (1991) "Regulatory Barriers to pollution Prevention." J. Air & Waste Management Assoc. 41:418

Cairncross F (1992) "Costing the Earth: The Challenge for Governments, The Opportunities for Business." Harvard Business School Press, Cambridge, Mass.

Carson R (1962) "Silent Spring." Houghton Mifflin, Boston, Mass.

CMA (1992) Turning Commitment to Action: 1992 Responsible Care Progress Report. Chemical Manufacturers Association, Washington, D.C.

Cohrssen J, Covello V (1989) Risk Analysis. White House Council on Environmental Quality, Washington, D.C.

Covello VT (1983) "The Perception of Technological Risks: A Literature Review." Technological Forecasting and Social Change, vol. 23, pp. 285–297

Covello VT and M Merkhofer (1992) Risk Assessment Methods. Plenum Press, New York

Craig J, R Baker and J Warren (1991) "Evaluation of Measures Used to Assess Pollution Prevention Program in the Industrial Sector," Final Report, EPA contract No. 68-W8-0038, EPA Office of Pollution Prevention, January

Cruz MC, CA Mayer, R Repetto and R Woodward (1992) "Population Growth, Poverty and Environmental Stress: Frontier Migration in the Philippines and Costa Rica, 1992." World Resources Institute, Washington, D.C.

CSA (1993) Design for the Environment, 4th Draft. Canadian Standards Association, Toronto

Curran MA (1993) "Broad-Based Environmental Life-Cycle Assessment." Environmental Science and Technology, Vol. 27, No. 3, pp. 430–438

Deland MR (1991) "Anounce of Prevention...After 20 years of cure." Environmental Science and Technology, Vol. 25, No. 4

DIEE (1992) Extended Producer Responsibility as a Strategy to Promote Cleaner Products, Trolleholm Castle, Sweden, 4 May 1992. Dept. of Industrial Environmental Economics, Lund University, Lund, Sweden

DeSimone LD and F Popoff (1997) Eco-efficiency: The Business Link to Sustainable Development. The MIT Press, Cambridge, Mass.

DOE (1998) Center of Excellence for Sustainable Development, Department of Energy, WWW page for basic information including case study links: http://www.sustaina-ble.doe.gov/

Dorfman MH et al. (1992) "Environmental Dividends: Cutting More Chemical Wastes." IN-FORM, New York, N.Y.

Economist (1992) "Pollution and the Poor." The Economist, February 15

ENR/Engineering News-Record (1992) ENR Special Report – The Top 500 Design Firms, April 6

EnvironTech (1998) Environmental Technologies WWW page: http://www.gnet.org/GNET

EnvironLink (1996) A Newsletter for Educators in the Field of Business and the Environment, Spring

Fava JA, R Denison, B Jones, MA Curran, B Vigon, S Selke and J Barnum (1991) A Technical Framework for Life-Cycle Assessments. Society of Environmental Toxicology and Chemistry workshop held in Smuggler's Notch, Vermont, August 18–23

Freeman HM and GE Hunt (1992) "Industrial Pollution Prevention in the U.S. Environmental Protection Bulletin." Institution of Chemical Engineers, Rugby, U.K.

Friedlander SK (1989) "The implementation of environmental issues for engineering R&D and education." Chem. Engineer. Progr., November

FSI (1993) The Product Ecology Report: Environmentally Sound Product Development Based on the EPS System. Federal Swedish Industries, Stockholm, Sweden

Hearne SA, Ancott M (1992) "Source Reduction vs. Release Reduction: Why the TRI cannot Measure Pollution Prevention." in: Pollution Prevention Review, Winter 1991/92

Healey MJ, D Watts et al. (1998) Pollution Prevention Opportunity Assessments. Wiley, New York, ISBN 9-471-2926-5

Higgins T (1989) Hazardous Waste Minimization Handbook. Lewis Publisher, Chelsea, Mich.

Hirschhorn S and KU Oldenburg (1991) "Prosperity Without Pollution: The Prevention Strategy for Industry and Consumers." Van Nostrand Rheinhold, New York

Hunt GE (1991) "Waste Reduction Techniques: An Overview." In: Pollution Prevention Review, Winter 1990/91

International Chamber of Commerce (1989) "ICC: The business approach to sustainable development." Development, Journal of the Society for International Development Vol. 2, No. 3, pp. 37–39

International Chamber of Commerce (1991) "The Business Charter for Sustainable Development." Paris, France

Irwin FH (1989) "Could there be a better law?" EPA Journal, Vol. 15, No. 4, July/August

Kolluru R (ed) (1996) Risk Assessment and Management Handbook for Environmental, Health, and Safety Professionals. McGraw-Hill, New York

Lachman BE (1997) Linking Sustainable Community Activities to Pollution Prevention. Critical Technologies Institute, RAND, http://www.rand.or/publications/MR/MR855, ISBN: 0-8330-2500-7

Levinson A, Sudhir S (1992) "Efficient Environment Regulation: Case Studies of Urban Air Pollution." Policy Research Working Paper 942, World Bank, Washington, D.C.

Licis I (1991) "Industrial Pollution Prevention Opportunities for the 1990's." U.S. EPA, EPA/600/8-91/052, August

Ling JT (1984) "The Impact of Environmental Policy on Industrial Innovation." Keynote Statement at the International Conference on Environment and Economics, Organization of Economic Cooperation and Development (OECD), Paris, June 18–21

Lucas Robert EB (1992) "Toxic Releases by Manufacturing: World Patterns and Trade Policies." Policy Research Working Paper 964, World Bank, Washington, D.C.

Martin P et al. (1992) "The Industrial Pollution Projection System: Concept, Initial Development and Critical Assessment." World Bank, Environment Development, Assessments and Programs Division, Washington, D.C.

McKinsey & Co. (1991) The Corporate Response to the Environmental Challenge. McKinsey & Company, Amsterdam

Meadows DH, DL Meadows and J Randers (1992) "Beyond the Limits: Confronting Global Collapse, Envisioning a Sustainable Future." Chelsea Green, Post Mills, Vt.

Miller GT (1990) "Living in the Environment: An Introduction to Environmental Science." 6th Edition. Wadsworth, Belmont, Cal.

Mink S (1992) "Poverty, Population and the Environment." Discussion Paper 189, World Bank, Washington, D.C.

NATO (1991) "Pollution Prevention Strategies for Sustainable Development Newsletter." NATO/CCMS, NATO, Brussels, Winter

Natural Resources Defense Council et al. (1992) "Environmental Safeguards for the North American Free Trade Agreement." NRDC, New York, June

Navin-Chandra D (1994) "The Recovery Problem in Product Design." Journal of Engineering Design Vol. 5, No. 1, pp. 67–87

Oakely BT (1993) "Total Quality Product Design: How to Integrate Environmental Criteria into Product Realization." Total Quality Environmental Management, Spring, pp. 309–321

Organization for Economic Cooperation and Development (1991) Environment Committee Secretariat, "Trade and Environment: Major Environmental Issues Note." OECD, Paris, March

Organization for Economic Cooperation and Development (1992) "The OECD Environment Industry: Situation, Prospects, Government Policies." OECD, Paris

OTA (1986) "Serious Reduction of Hazardous Waste: For Pollution Prevention and Industrial Efficiency." U.S. Congress Office of Technology Assessment Washington, D.C., September

OTA (1992) "Trade and Environment: Conflicts and Opportunities." U.S. Congress, Office of Technology Assessment, ITA-BP-ITE-94 (Washington, D.C., Government Printing Office), May

OTA (1992) Green Products by Design: Choices for a Cleaner Environment, US Congress Office of Technology Assessment, US Government Printing Office, Washington, D.C.

OTA (1994) Industry, Technology, and the Environment. OTA-ITE586, GPO, January

Pojasek RB (1988) "Implementing a Waste Reduction Program," Proceedings of Hazardous Waste Minimization: Corporate Strategies and Federal/State Initiatives. Government Institutes Inc., Washington, D.C., April

Pojasek RB (1991) "Waste Reduction Audits." in: Environmental Risk Management – A Desk Reference, RTM Communications, Alexandria, Va.

Pojasek RB (1991) "Pollution Prevention Progression." in: Environmental Risk Management – A Desk Reference. RTM Communications, Alexandria, Va.

Porter G and JW Brown (1991) "Global Environment Politics." Westview Press, Boulder, Colo.

Porter ME (1989) "The Competitive Advantage of Nations." The Free Press, New York

Post JE (1990) "The Greening of Management." Issues in Science and Technology, pp. 68–72, Summer

PPIC (1998) Pollution Prevention Information Clearinghouse, U.S. EPA, Office of Pollution Prevention and Toxics, EPA/742/F-98/004, Summer

Purcell AH (1988) "Waste Minimization and the Economic Imperative." Proceedings from Hazardous Waste Minimization: Corporate Strategies and Federal State Initiatives, Government Institutes Inc., Washington, D.C., April

Ragsdale M (1994) "U.S. refiners Choosing Variety of Routes to Produce Clean Fuels." Oil & Gas Journal, March 21

RIVM (1992) The Environment in Europe: A Global Perspective. ISBN 90-8980-031-5

Sarokin DJ, WR Muir, CG Miller and SR Serber (1985) Cutting Chemical Wastes: What 29 Organic Chemical Plants Are Doing to Reduce Hazardous Wastes. INFORM, New York

Sarokin D (1992) Toxic Releases from Multinational Corporations: Does the Public Have a Right to Know. Friends of the Earth, Washington, D.C.

Shafik N and S Bandyopadhyay (1992) "Economic Growth and Environmental Quality: Time Series and Cross-Country Evidence." Policy Research Working Paper 904, World Bank, Washington, D.C.

Shen TT (1990) "Educational Aspects of Multimedia Pollution Prevention." in: Environmental Challenge of the 1990's, Proceedings of the International Conference on Pollution Prevention: Clean Technologies and Clean Products, U.S. EPA/600/9-90/039, September

Shen TT (1996) "Critical Environmental Issues Worldwide and Pollution Prevention Strategies," presented at the Academia Sinica in Taipei, Taiwan, December 10

Shen TT (1997) "Let Us Avoid Pollution Prevention Barriers," presented at the 10th Annual Pollution Prevention /conference in Albany, New York, June 3–4. Proceedings prepared by the Pollution Prevention Office of NYSDEC

Shen TT (1997) "Multimedia Pollution Prevention: The Need for Education," presented at the CAAPS Annual Conference in New York City, September 13–14

Silverstein M (1993) Environmental Economic Revolution. St. Martin's Press, New York

Slovic P (1987) "Perception of Risk." Science, Vol. 236, pp. 280–285

Smart B (ed) (1992) "Beyond Compliance: A New Industry View of the Environment." World Resources Institute, Washington

Sorsa P (1992) "The Environment – A New Challenge to GATT?" Policy Research Working Paper 980, World Bank, Washington, D.C.

Speth JG (1990) "Six Steps toward Environmental Security." Christian Science Monitor, January 22

Tellus Institute (1991) "The Tellus Institute Packaging Study." Tellus Institute, prepared for the Council of State Governments, November

United Nations Center on Transnational Corporations (1992) Transnational Corporations and Sustainable Development: Recommendations of the Executive Director (E/C. 10/1992/2). United Nations, New York

UNCED (1992) United Nations Conference on Environment and Development Agenda for the 21st Century. A/CONF.151/26 (Vol. I–III). United Nations, New York

UNEP (1998) International Declaration on Cleaner Production, Recent Programme Highlights, April 1998. http:www.unepie.org/cp/cp rph.html

UNWCED (1987) United Nations World Commission on Environment and Development. Our Common Future, Oxford University Press, New York

U.S. (1990) "Pollution Prevention Act of 1990", 42U.S. C 13101

U.S. Congress, General Accounting Office (1992) "International Environment: International Agreements Are Not Well Monitored." GAO RCED-92–43. Government Printing Office, Washington, D.C., January

USEPA (1986) "Report to Congress." Minimization of Hazardous Waste. Office of Solid Waste and Emergency Response, Washington, D.C, EPA/530/SW/86/033, October

USEPA (1989) "Pollution Prevention Benefits Manual." (Draft). Vol. 1, U.S. EPA, Report No. WAM-1. PPIC, Falls Church, Va., October

USEPA (1990) Reducing Risk: Setting Priorities and Strategies for Environmental Protection. Science Advisory Board, Washington, D.C.

USEPA (1990) Procurement Guidelines for Government Agencies, EPA 530-SW-91-011. U.S. EPA, Washington, D.C. December

USEPA (1990) "Environment Investments: The Cost of a Clean Environment." Report of the EPA Administrator to U.S. Congress

USEPA (1990) "Environmental Protection Agency Pollution Prevention Directive." U.S. EPA, May 13

USDOE (1991) "Industrial Waste Reduction Program." U.S. DOE/CE-0344, October

USEPA (1991) "Pollution Prevention Research Needs," ORD Report to Congress

USEPA (1991) "Pollution Prevention 1991: Progress on Reducing Industrial Pollutants." U.S. EPA, EPA 21P-3003, October

USEPA (1991) "Achievements in Source Reduction and Recycling for Ten Industries in the United States." U.S. EPA/600/2-91/051, September

USEPA (1992) "Pollution Prevention Research Branch Current Projects Listing." RREL, Cincinnati, Oh., January

USEPA (1992) "Pollution Prevention Resources and Training Opportunities in 1992." Office of Toxic Substances, TS-792A, January

USEPA (1994) Directory of Pollution Prevention in Higher Education Faculty & Programs, EPA//742/B-94/006

USEPA (1994) Pollution Prevention Directory, EPA/742/B-94/005

USEPA (1995) Cleaner Technologies substitutes Assessment: A Methodology and Resources Guide, EPA/744/R-95/002

USEPA (1995) Introduction to Pollution Prevention: Training Manual, EPA/742/B-95/003

USEPA (1995) Environmental Accounting Case Studies: Full Cost Accounting for Decision Making at Ontario Hydro, EPA/742/R-95/004

USEPA (1995) Environmental Cost Accounting for Capital Budgeting: A Benchmark Survey of Management Accountants, EPA/742/R-95/005

USEPA (1995) Profile of the Inorganic Chemical Industry, EPA/310-R-95-004, September

USEPA (1996) Pollution Prevention Success Stories, EPA/742/R96/002

USEPA (1996) Directory of EPA's Environmental Accounting Network, EPA/742/B-96/006

USEPA (1996) EPA's Design for the Environment Program: Partnerships for a Cleaner Future, EPA/744/F-96/018

USEPA (1997) ISO 14000: Resource Directory, EPA/625/R-097/003

USEPA (1997) Partners for the Environment: A Catalogue of the Agency's Partnership Programs, EPA/742/B-97/003

USEPA (1997) A Study of State and Local Government Procurement Practices that Consider Environmental Performance of Goods and Services, EPA/742/R-96/007

USEPA (1997) Pollution Prevention 1997: A National Progress Report. Office of Pollution Prevention and Toxics, EPA 742-R-97-00, June

USEPA (1998) Enviro$ense: Common Sense Solutions to Environmental Problems, EPA/335/F-97/001

USEPA (1998) Environmental Management systems Voluntary Project Evaluation Guidance, EPA/742/B-98/002

USEPA (1998) Pollution Prevention Information Clearinghouse (PPIC) General Information Packet, EPA/742/E-98/003

USEPA (1998) EPA Community-Based Environmental Protection. WWW page: www.epa.gov/ecosystems/idex.html

USPS (1992) "Waste Reduction Guide." U.S. Postal Service, AS 552, February

Van Weenen JC (1990) "Waste Prevention: Theory and Practice." Castricum Publishers, Delft, The Netherlands

Vinyl Institute (1991) "Vinyl Product Life-cycle Assessment," prepared for the Vinyl Institute, Chem Systems, September 17

Vogel D (1986) National Styles of Regulation: Environmental Policy in Great Britain and the United States. Cornell University Press, London

Warren J (1992) "Why Not Zero? 20 Questions About Pollution Prevention." Keynote presentation at Pollution Prevention: Making It Happen. Engineering Foundation Conference, January

Williams ME (1992) "Why and How to Benchmark for Environmental Excellence." Total Quality Environmental Management, Winter, pp. 177–185

Welter TR (1990) "Beyond the Dumpster." Industry Week, Vol. 239, pp. 53–54

Wheeler R (1991) "Design for reliability reshapes designing." Electronic Design, Vol. 39, p. 121

Wheeler D and P Martin (1991) "Prices, Policies, and the International Diffusion of Clean Technology: The Case of Wood Pulp Production." Presented at the World Bank Symposium on International Trade and the Environment, Washington, D.C.

WBCSD (1998) The Sustainable Business Challenge Newsletter, January, http://challenge.bi./sbc/Newsletter.htm

Willis DG (1991) "Pollution Prevention Plan – A practical Approach." In: Pollution Prevention Review, Autumn

World Bank (1992) "World Development Report 1992: Development and the Environment." Oxford University Press, Oxford. May

World Bank (1992) "World Development Report 1992: Development and the Environment." Oxford University Press, New York

World Bank (1992) "The World Bank and the Environment, Fiscal 1992." Washington, D.C.

World Commission on Environment and Development (1987) "Our Common Future." Oxford University Press, New York

Zosel TW (1991) "How 3M makes pollution prevention pay big dividends." In: Pollution Prevention Reviews, Winter 1990/91

Brief Biography

Thomas T. Shen is an independent environmental consultant and President of the Phi Tau Phi Scholastic Honor Society of America, East American Chapter. He received his advanced studies in Northwestern University in Evanston, Illinois with M.Sc. in Sanitary Engineering; and his Ph.D. degree in Environmental Engineering from Rensselaer Polytechnic Institute in Troy, New York. Dr. Shen has served as a senior research scientist with the New York State Department of Health and Department of Environmental Conservation (1966 to 1993). Meanwhile, he has taught graduate courses at Columbia University for 12 years (1981 to 1993) and lectured in more than two dozens of universities in China, Korea, Taiwan, Hong Kong, Malaysia, Singapore, Japan, Australia, Scotland, Finland, and Argentina. Dr. Shen served as a member of USEPA's Science Advisory Board for 3 years (1986 to 1988). He also served as a consultant to the United Nations' Development Program and World Health Organization on environmental projects for 11 years (1982 to 1993). Dr. Shen was one of the Technical Reviewers for President Bush's First and Second Annual President's Environment and Conservation Challenge Awards in 1991 and 1992.

Dr. Shen has involved more than a dozen of professional and service organizations. He has received several Service Awards from organizations such as the Air and Waste Management Association, American Society of Civil Engineers, the Chinese Institute of Engineers, the Overseas Chinese Environmental Engineers and Scientists Association, The Chinese American Academic and Professional Society, Chinese American Alliance in Capital District of New York, Delmar Rotary Club International among others. In 1993, he received the National Award of Industrial Waste Minimization Excellent Performance, jointly presented by the Taiwan's Ministry of Economic Affairs and Environmental Protection Administration in Taipei. In the same year, he also received the Frank Chamber Scientific Award from the International Air & Waste Management Association in Denver, Colorado.

Dr. Shen has authored and co-authored three books, three book chapters, more than one hundred articles, reports, documents, and training manuals for short-courses. His recent book, INDUSTRIAL POLLUTION PREVENTION, was published by the Springer Verlag International Technical Publisher, Heidelberg, Germany in early 1995. It has been translated into Chinese by the Foundation of Taiwan Industry Service in late 1995.

Subject Index

Environmental Engineering

The protection of our environment is one of the most important challenges facing today's society. Sustainable development and pollution control are the key factors in the development of strategies for the solution of these problems.
Based on the most recent research, **Environmental Engineering** aims to present the technologies developed for: Air quality measurement and control; Waste water treatment, disposal and reuse; Treatment, disposal and recycling of solid and hazardous wastes; Contamination localization and decontamination of soils; Pollution prevention; and Industrial safety engineering.
The series addresses all those responsible for environmental issues in industry, consulting services and administration, as well as environmental agencies and associations. The books are written in a concise manner and in simple language, and range from introductory texts to practical guides.

Please order from
Springer-Verlag
P.O. Box 14 02 01
D – 14302 Berlin, Germany
Fax: +49 / 30 / 872 87 301
e-mail: orders@springer.de
our through your bookseller

Springer